Liposomes

Liposomes

edited by

Marc J. Ostro

The Liposome Company
Princeton, New Jersey

MARCEL DEKKER, INC. New York and Basel

Library of Congress Cataloging in Publication Data

Main entry under title:

Liposomes.

Includes index.
I. Liposomes. I. Ostro, Marc J.
QH601.L567   1983   574.87'34   83-1959
ISBN 0-8247-1717-1

MARCEL DEKKER, INC.
270 Madison Avenue, New York, New York 10016

Current printing (last digit):
10 9 8 7 6 5 4 3 2 1

PRINTED IN THE UNITED STATES OF AMERICA

to

Annette, Henry, and Evelyn

Preface

I have been told that one usually includes prefaces in scientific works in order to
point out to the reader the abundance of knowledge assembled within the text.
This seems somewhat unnecessary considering the estimable credentials of the
contributors, so I will dispense with this task by simply thanking all the authors
for their excellent contributions and move on to a brief mention of an observation I made at a recent conference on liposomes held in San Francisco. At that
conference I was struck by what would have been a rather incongruous observation had it been made 5 years ago—30% of all those attending the meeting were
affiliated with commercial laboratories. Whether corporate interest in liposomes
can be taken as an indication that practical applications of liposome technology
are perceived by industry to be forthcoming or is a reflection of the current tight
money policies in Washington is unclear. What is clear is that recent advances in
the understanding of membrane structure-function relationships have made the
engineering of liposomes to fit specific, well-defined roles possible.

With the numerous methods available for producing liposomes (Chapter 1),
one is able to regulate the size, charge, and the number of lipid bilayers comprising the vesicle. One can manipulate the membrane lipids and thereby effectively regulate permeability (Chapter 7) to both large and small molecules.
Several complex macromolecules can be effectively anchored in liposome membranes—glycolipids (Chapter 6), lipoproteins (Chapter 6), and antibodies (Chapters 3, 6, and 7), enabling one to develop targeting strategies as well as specific
vaccines. Liposomes can be reconstituted so that cell membrane proteins are
functionally expressed on their surface, thereby providing a model system for
studying receptor-membrane interactions (Chapter 4); they can be used as
models to study the effect of drugs on cell membranes (Chapter 2); and
can be valuable tools for delivering RNA, proteins, and, most importantly, DNA
into cells (Chapter 5).

What we have attempted to do in this book is to assemble reviews of various
areas of liposome research in such a way that the total manuscript represents a
thorough compilation of the major work done in the field since 1965. We have
included titled reference lists which we hope will, in themselves, be useful to
investigators well versed in "liposomology" as well as those who wish to enter
the field.

Finally, we would like to pay a special tribute to Dr. A. D. Bangham for his introductory chapter. Dr. Bangham was given a rather poorly defined commission to write an opening to this book that would be at once informative and amusing, while at the same time giving the reader a sense of historical perspective. Dr. Bangham has accomplished this task with his typical style and humor, for which we are grateful.

Marc J. Ostro

Contents

Contributors

Carl R. Alving, M.D., Department of Membrane Biochemistry, Walter Reed Army Institute of Research, Washington, District of Columbia

A. D. Bangham, M.D., F.RS., Biophysics Unit, A.R.C. Institute of Animal Physiology, Babraham, Cambridge, United Kingdom

David W. Deamer, Ph.D., Department of Zoology, University of California at Davis, Davis, California

D. Giacomoni, Ph.D., Department of Microbiology and Immunology, College of Medicine, University of Illinois, Chicago, Illinois

Leaf Huang, Ph.D., Department of Biochemistry, University of Tennessee, Knoxville, Tennessee

R. L. Juliano, Ph.D., Department of Pharmacology, The University of Texas Medical School at Houston, Houston, Texas

P. Malathi, Ph.D., Department of Physiology and Biophysics, College of Medicine and Dentistry of New Jersey, Rutgers Medical School, Piscataway, New Jersey

Eric Mayhew, Ph.D., Department of Experimental Pathology, Roswell Park Memorial Institute, Buffalo, New York

Marc J. Ostro, Ph.D., The Liposome Company, Princeton, New Jersey

Demetrios Papahadjopoulos, Ph.D., Cancer Research Institute and Department of Pharmacology, University of California, Medical School, San Francisco, California

Roberta L. Richards, Ph.D., Department of Membrane Biochemistry, Walter Reed Army Institute of Research, Washington, District of Columbia

Paul S. Uster, Ph.D., Department of Zoology, University of California at Davis, Davis, California

Liposomes

Liposomes: An Historical Perspective

A. D. Bangham / A.R.C. Institute of Animal Physiology, Babraham, Cambridge, United Kingdom

Biophysics Unit
A.R.C. Institute of Animal Physiology
Babraham, Cambridge CB2 4AT

1 March 1981

Dear Dr. Ostro,

Almost exactly a year has passed since you telephoned and asked me whether I would be willing to write an introductory chapter to your book entitled *Liposomes.* "How extraordinary" was my immediate reply and I went on to explain that I had that very day completed a chapter for another book (1). I rang off without commitment and forwarded you a gleaming Xerox copy of the finished article with the hope that you would appreciate my dilemma. I felt quite sure that you would let me off the hook and get someone like Demetri Papahadjopoulos or Gerry Weissmann to write it; after all, they were casting around in my laboratory before liposomes were even called liposomes! Alas, my manuscript obviously did not have the hoped-for effect and you were on the phone again explaining that you really had in mind that I should fill in with some anecdotal stuff and leave the heavy material to the others. "That sounds better," I thought to myself. Confirmation of this somewhat unusual open-ended commission arrived in the form of a letter enclosing abstracts of some seven chapters and a casual comment to the effect that "the introduction by Dr. Bangham required no outline since overlap with other chapters will probably be minimal." You are dead right! Here, then, is my contribution.

You see, there really is a problem concerning the writing of reviews and introductory chapters on the same subject year by year. But since you asked me to write specifically about the historical perspective of liposomes I shall exercise my discretion to the limit, for accuracy's sake, and include a quotation of 249 words from our own original *J. Mol. Biol.* 13:238 (1965) paper (2). Our paper, according to the Journal, was first received on 16 March 1965 and in a much improved form on 6 May 1965. I can very well remember Malcolm Standish and myself spending a whole afternoon in the Department of Physiology, Cambridge with one of the referees, now a most distinguished professor, writing and rewriting, clarifying and polishing the manuscript. What a pity I cannot identify this referee-cum-coauthor! The section I would like to quote verbatim is the part that attempts to describe how membrane molecules, and phospholipids in particular, interact with water—the nub of all liposome phenomena. We wrote (2):

The liquid crystal (smectic mesophase) is a preferred phase structure of many biological lipids in the presence of water or salt solutions. [D. G. Dervichian, 1964]. It may be ascribed to the nature and heterogeneity of the hydrocarbon moieties and to the possession by the lipids of either polar, ionogenic or both types of head group. The precise geometry of the structures depends upon the relative concentration of the two principal components (lipid and water), the temperature, the composition of the lipid and the salt concentration of the aqueous phase [A. D. Bangham, 1963; D. A. Haydon and J. Taylor, 1963; A. S. C. Lawrence, 1961; V. Luzzati and F. Husson, 1962]. Over a wide range of such variables, however, the commonest phase structure appears to be that of a layer lattice giving rise to spherulites and myelins, both composite structures consisting of many concentric bimolecular layers of lipid each separated by an aqueous compartment. For thermodynamic reasons [D. A. Haydon and J. Taylor, 1963], it is probable that at equilibrium each and every lipid bilayer forms an unbroken membrane—there being no exposed hydrocarbon/water interfaces—from which it follows that every aqueous compartment would be discreet and isolated from its neighbour, including a complete separation of the outermost aqueous compartment of the whole structure from the continuous aqueous phase in which it is suspended. Since these liquid crystalline structures form spontaneously when dry lipids are allowed to swell in aqueous salt solutions, it seemed reasonable to test for the integrity of unbroken membranes by measuring the amount of electrolyte . . .

and at this point I must paraphrase, "which have become entrapped within the liposomes." Goodness knows how many variations on that bit of descriptive writing there have been over the past 15 years!

While we are on the subject of the *J. Mol. Biol.* 13:238 (1965) paper (2)

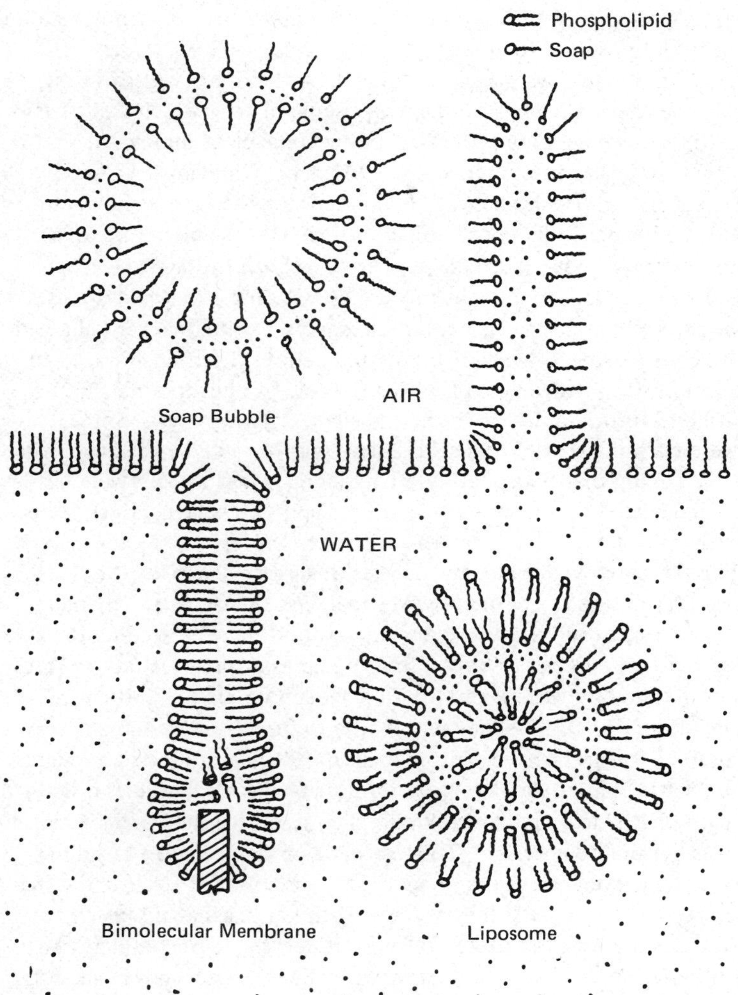

Figure 1 Preferred orientation of soap and phospholipid molecules at air-air, air-water, and water-water interfacial regions. (Adapted from K. J. Mysels, 1968. J. Gen. Physiol. 52:113.

entitled "Diffusion of univalent ions across the lamellae of swollen phospholipids," it would be appropriate to acknowledge a crucial technical tip which Demetri Papahadjopoulos offered us in his capacity as a recently installed visitor to our laboratory. His suggestion involved the use of Sephadex G-50 as a rapid

dialysis system. Without it we would not have been able to say very much about
the permeability of liposomes to anions.

But what, you may ask, was Demetri doing in my laboratory at this time,
anyway? Well, the answer lies in a bit of preliposome history which I think is
probably worth recounting. Perhaps I should now come clean and admit that I
really only made my first liposome in about 1961, having previously relied exclusively on their preparation by Norma Hemington, long-serving technician to
Rex (Dawson), with whom I had been collaborating. Our problem then concerned certain phospholipases and their sensitivity to both the sign and magnitude of the surface charge of their substrates (4-6). Having developed the idea
of varying the surface charge of egg lecithin liposomes (normally a zero-charged
zwitterion) by codispersing long-chain anions and cations, I felt encouraged to
enter a contemporary controversy concerning the role of phospholipids in the
formation of blood thromboplastin, in the so-called "stypven" test. Some
authorities were claiming that phosphatidylserine was the active substance, whereas others were claiming that phosphatidylethanolamine was more active. All
parties concerned agreed that a pure egg lecithin dispersion was inactive. It occurred to me that it might, as had been established for at least two phospholipases (5,6) be the sign, magnitude, and species of surface charge of the phospholipid dispersions (structure as yet unknown) which was the operative characteristic. Since this became my very own problem, I found myself bleeding Rosalind,
my wife, daily before breakfast to obtain lipid-free plasma, and testing for thrombo-
plastic activity initiated by various mixtures of phospholipids as "liposomes" prepared by Nigel Miller or myself. They worked, but not before we had a narrow es-
cape from failure. You see, before writing it all up I decided to send a preparation of the synthetic "blood platelets" to Professor Gwyn MacFarlane of Oxford,
the authority on blood clotting, to check out. To my dismay he reported that the
preparation was "virtually inactive." To Rosalind's dismay I had to bleed her
again daily before she even cooked my bacon and eggs so that I could show that
the liposome preparations were still very active with her plasma and in our test.
Again I opted to send a preparation to Oxford but decided to accompany the
"liposome" preparation and see how Professor MacFarlane carried out his tests.
I must have taken one of the last through trains from Cambridge to Oxford—the
track is now occupied by an aerial array belonging to the University of Cambridge, Mullard Radio Astronomy Department. Once again, my preparation
proved to be inactive in MacFarlane's test and I was about to take the slow train
back when I noticed that MacFarlane was adding Ca^{2+} at twice the concentration
I had been using! "Phew," as we would say, or "wow," as you might—an adjust-
ment in the concentration of long-chain anions and all was well: the test worked
and the paper appeared in *Nature* (7). Now the point of this story in relation to
the historical perspective of liposomes is that Demetri, then a postdoctoral fellow
in Professor Hanhahan's department in Seattle and working on the same theme

for other very personal reasons, read my paper and decided to visit me. Easier done than said in those days! By the time he had negotiated his American Heart Association Research Fellowship and turned up in my laboratory (1963-1964) we were well into the discovery of liposomes and had lost interest in the thromboplastin story. Poor Demetri was immediately recruited to measure birefringence of various phospholipid mixtures in salt solutions and has never, to my knowledge, had the opportunity of resuming his studies in the role of phospholipids in blood clotting! But, as I have already mentioned, he was soon offering us some very good advice and helping us prepare and purify our own phospholipids.

Looking back through my records, I am reminded of the fact that I was partly responsible for recruiting Robert Horne to Babraham at about this time. Robert was appointed as electron microscopist and brought with him the (then) very latest techniques of negative staining. We soon put them to good use by looking at the phospholipid dispersions Rex Dawson had taught me to prepare, directly in the diluted negative stain. We had a wonderful time; not least was the moment, sometime in 1962, when we reproduced the marvelous Dourmashkin artifact (8) but using lecithin, cholesterol, and saponin only (9). I actually took a wet plate over to the Strangeways laboratory—some 4 miles away—to show Jack Lucy and Audrey Glauert, only to be shown an identical one! (11) With hindsight, this was a most bizarre but persuasive piece of evidence that liposomes, if prepared with lecithin and cholesterol, were valid models of real, cholesterol-containing membranes. Ultimately, these observations encouraged van Zutphen, Laurens van Deenen, and Steve Kinsky to pinpoint the site of action of such antibiotics as amphotericin, nystatin, and some filipins (10).

I also recollect Dave Robertson at a meeting we were both attending looking somewhat reassured to see his "triple line" revealed in a different guise! He was under attack for a long time and, believe it or not, still is (12).

Apart from Demetri's prolonged visit, there were three other visitors to Babraham who, during this period, played a crucial role in the development of the liposome closed-membrane concept. Chronologically, they were Brian Chappell, Gerald Weissmann, and Dikran Dervichian.

I should remind you that at about this time (1964-1965), one of the most promising lines of biochemical research being pursued concerned the effect of certain cyclic polypeptides, such as the valinomycins, nonactins, enniatins, and gramicidins, on membranes. The initiative for testing these compounds with artificial (nonprotein) membranes derived from the controversy concerning their mode of action on complete biological systems such as mitochondria and erythrocytes (13-17). The facts were not controversial—but the mechanisms were. Gramicidin A made both mitochondria and erythrocytes permeable to Na^+ and K^+, whereas valinomycin caused them to be selectively permeable to K^+. And so, looking for a pure lipid model system, Chappell turned up in my laboratory with

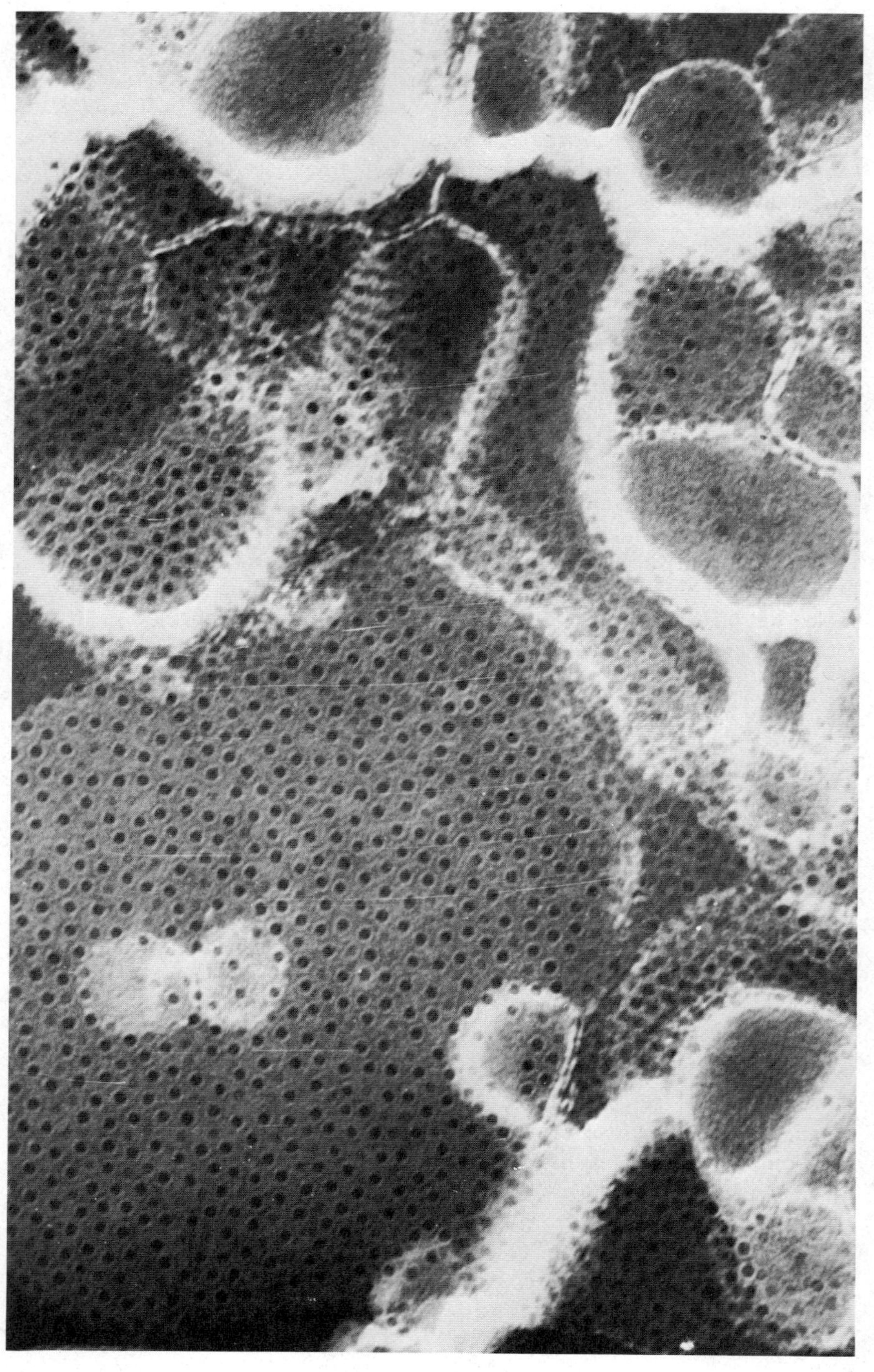

a large cardboard box containing two space-filling models and two small glass ampules containing a few milligrams of whitish powders. Chappell had no idea that we had by then worked out the liposome system, and he was easily persuaded to part with a few milligrams of his precious valinomycin and gramicidin. Within a day or so we were able to establish that both ionophores worked as expected and, importantly, on a pure lipid system, the liposome. Although these and later experiments were published in 1966 by Chappell and Crofts (16), they enabled me to substantiate the proposition that liposomes were indeed closed membrane structures, and we were able to describe their properties with some conviction to a symposium arranged by K. R. Porter in Frascati, Rome, in June 1965 (18). There was a nightmare moment on our way to that meeting because the particular Alpine pass (Petite St. Bernard) Rosalind and I had elected to cross into Italy was "officially closed" when we reached the foot of it. Miraculously, they opened it an hour or so later and we became the "first" over the top for 1965! Had it not opened we would have been too late for the meeting and the opportunity of proclaiming the virtues of liposomes for the first time would have passed.

Chappell, who moved to Bristol, lost Haarhoff, his first Ph.D., almost as soon as he had learned how to prepare and manipulate liposomes. Fortunately, he soon recruited both McGivan and Henderson. McGivan was entrusted with the liposome model and produced an outstanding Ph.D. thesis. Indeed, his work, published in a paper together with Henderson and Chappell (19) entitled "The action of certain antibiotics on mitochondria, erythrocytes and artificial membranes," did more to enhance the validity of liposomes during that year than did any other. Incidentally, this paper provided great comfort to Peter Mitchell, who, by this time, had privately published his chemiosmotic coupling theory of oxidative phosphorylation (20,21) and who badly wanted tangible evidence that a simple phospholipid bilayer was, indeed, capable of containing protons while being selectively permeable to other ions.

Meanwhile, of course, the black lipid membrane (BLM) model, first described in 1962 by Rudin, Mueller, Ti Tien, and Westcott (22), was rapidly gathering adherents all over the world and providing a wealth of transient and steady-state electrical data highly relevant to the known properties of biomembranes. As exponents of the liposome model we greatly admired the elegance of the BLM and were not entirely convinced that we had its match in versatility. Have there been any more books about the BLM? Tien published one some years ago (23).

Figure 2 Micrograph shows a dispersion of 2 mol of lecithin and 1 mol of cholesterol in water, treated with an equal volume of 0.2% (w/v) saponin, and mixed with an equal volume of 2% potassium phosphotungstate. Magnification 100,000X.

But I digress. Gerry Weissmann was the second visitor to our laboratory that summer (1964). Strictly speaking, he was visiting Honor Fell's laboratory on the outskirts of Cambridge to continue his collaborative work with her on the properties of lysosomes which had started a year earlier. Weissmann also arrived carrying a cardboard box containing an extraordinary collection of compounds, most of which I had never heard of! Cortisone, yes, I knew what that was because I had worked with it in 1940, but "allopregnanolone," "etiocholanolone," "pregnanolone," and "diethylstilboestrol" were all tongue-twisters to me! As with Chappell, Weissmann's problem was whether these compounds manifested their lytic properties for cells and cell organelles within and/or upon the membrane phase or elsewhere. I remember phrasing the introduction to our paper very carefully, and I here quote again without paraphrase (24):

> The interpretation of experiments devised to investigate the lytic mechanisms of biological cells and subcellular organelles has been bedevilled by the uncertainty surrounding the precise structure and composition of their membranes. Few workers in this field have been ready to discount the presence or role of proteins as an integral part of membrane structure, and inevitably, conclusions are shrouded in reservations concerning the behaviour of lipo-proteins or lipoprotein complexes.

Unlike Chappell, who was in the process of setting up a biochemistry department in Bristol, Gerry Weissmann worked his way through his box of compounds before returning to New York for the fall term and we were able to write up the results while still putting the finishing touches to our original paper; both papers were published simultaneously (2,24)! Gerry Weissmann wasted no time in getting liposome technology off the ground and within the next two years had published some half dozen highly enterprising papers (25-29). Furthermore, he and Sessa also wrote an encouraging review promoting the new model and irrevocably renaming the smectic mesophase of a phospholipid, by referring to "liposomes" in their title (30). Colloquially, they had become known as Bangosomes, a patronymic, coined, I think, by Al Lehninger (31) after reading our original paper. Personally, I think all these names are inappropriate to describe structures having such exquisite and precise macro and molecular architecture. Perhaps Gerry Weissmann will get around to writing one of his monthly essays in *Hospital Practice* on this subject! Which reminds me of one of Weissmann's most charming claims, namely to have prepared boy and girl liposomes (29)! I wonder what they are doing now; they must surely have grown up by now!

I would like to think that our own paper entitled "Osmotic properties and water permeability of phospholipid liquid crystals," published in the first volume of *Chemistry and Physics of Lipids* (32), was a further landmark in establishing the validity of the model system and for a number of reasons. First, and within the constraints of having to use a very primitive experimental method, it was able to assign a permeability coefficient for water in actual units; second, it established that liposomes behave as perfect osmometers when alkali metal salts, glucose, sucrose, and manitol are used as solutes; and third, it showed that other small nonelectrolytes exhibited a graded permeability. Finally, it raised the problem as to whether or not the membranes were permeable to protons. For example, neither ammonium chloride nor potassium acetate could be shown to be permeable, whereas ammonium acetate was highly so. Potassium acetate was later shown to be permeable but only when valinomycin and an uncoupling agent were present (33). All very difficult and confusing, particularly when one remembers that fluxes are often coupled and of course concentration dependent. My coauthors on this paper were both remarkable men. Guy Greville, then head of the Biochemistry Department at Babraham, was, at the time, writing the first interpretive text of the chemiosmotic coupling theory (34) and must have been in touch with Peter Mitchell as our results came out. Greville was obsessive; he could only use glassware which he had himself washed endlessly and meticulously, and we could only use solutions which he had himself made up. He was mesmerized by the accuracy which he could achieve in a particular measurement, ar d it was always difficult to move him on to the next concentration point. Not so Hans de Gier, who was over from Holland for six months to learn all about these newfangled liposomes. De Gier was from the Department of Biochemistry Utrecht, famous by this time for its work on phospholipids. De Gier was practical, objective, and anxious to establish the properties of the experimental material to solve the pressing problems of real membranes, like erythrocytes. Ours was one of those papers that was written before the measurements were made!

Before concluding this letter I must tell you about Dikran Dervichian's visit in, I think, 1967, and of the crucial question he asked me. "If, as you claim, liposomes are so impermeable to ions," he asked me, "how do they get in to the structures in the first place?" This was a stunner and I had to do some hard thinking before I offered an explanation in a review I wrote about a year later (81). I should have expected something like this from Dervichian because, of course, he knew a great deal about phospholipids in water (3) and had the reputation of an "enfant terrible" at meetings of the Faraday Society. What I did not know was that he had once been a pupil of my father.

Enough for the moment, I have become engrossed by an enchanting book entitled *The Harvest of a Quiet Eye, A Selection of Scientific Quotations* by Alan L. Mackay from which I certainly would wish to quote far in excess of the alloted space. Allow me just one little poem.

> Twinkle, twinkle, little star
> I don't wonder what you are,
> For by spectroscopic ken
> I know that you are hydrogen.

Yours sincerely,

A. D. Bangham

17 High Green
Great Shelford
Cambridge

14 March 1981

Dear Ostro,

Anxiety is one of the most potent inhibitors of writing that I know of! I haven't written a word since I signed off my first letter to you a fortnight ago and the cause of anxiety has been a failure to repeat experiments I did about two years ago. Actually, Nigel Miller, my assistant for more than 20 years, won me a reprieve just before the weekend and I've dared to take up my pen again. The experiments are pretty basic ones as far as liposomes are concerned but essential to a novel theory we have contrived concerning the mechanism whereby general anaesthetics render you unconscious! The crucial question is whether liposomes are exceptionally permeable to protons and hydroxyls, compared to other univalent ions and whether anaesthetics stimulate this permeating mechanism or not. My despair was not entirely relieved when I came across the following quatrains in a delightful book all scientists should have on their writing desks. *A Random Walk in Science* is an anthology compiled by R. L. Weber and edited by E. Mendoza and published by the Institute of Physics (1973).

> Grant, oh God, Thy benedictions
> On my theory's predictions
> Lest the facts, when verified,
> Show Thy servant to have lied.

> May they make me B.Sc.,
> A Ph.D. and then
> A D.Sc., and F.R.S.,
> A *Times* Obit. Amen.

Oh, Lord, I pray, forgive me please,
My unsuccessful syntheses,
Thou know'st, of course—in Thy position—
I'm up against such competition.

Let not the hardened Editor,
With referee to quote,
Cut all my explanation out
And print it as a Note.

Entitled "The Researcher's Prayer," it was published first in the Proceedings of the Chemical Society, January 1963, as part of a competition.

Our anxiety slowly changed to one of elation as we gradually regained confidence in the observation. Indeed, by using reverse-phase evaporation (REV) liposome preparations and an autotitrator we were able to quantify rather precisely the accelerated collapse of a pH gradient by the presence of general anaesthetics. Don't tell Dave Deamer, but we find that the REV liposomes are absolutely great! Their trapping capacity and ion integrity have even permitted us to demonstrate that liposomes fully conform with the Donnan theory and I'm in the process of telling both McGivan (39) and Peter Nicholls (40) that they *will* generate a pH gradient in response to a K^+ gradient. Perhaps you were unaware that liposomes had, hitherto, failed this acid test. I am very anxious to get on with these observations just as soon as I have finished writing these letters and have cultivated my vegetable garden for the forthcoming season. Despite it being the end of March, the ground has been so wet that no growing time has been lost.

On the other hand, we have had no difficulty in confirming earlier results which showed that H^+ efflux is tightly coupled to cation influx, e.g., K^+, caused by valinomycin, a result which would appear to be thoroughly at odds with the most recent paper on the subject by Wylie Nichols and Dave Deamer (35).

The irony of this situation is, of course, that Dave Deamer visited my laboratory twixt 1974 and 1975, during which time he developed the ether injection method of making jumbo liposomes (36,37) for the primary purpose of using them to study the decay rates of pH gradients. Together we published a paper (38) in which we claimed that liposomes were indeed highly permeable to protons and hydroxyls if and when one took into consideration their true concentrations. However, I would say that there was no evidence in that paper to suggest that proton fluxes were not cation-coupled. In our laboratory we would say that liposomes sustain pH gradients just so long as they remain tight to cations and the corollary may be true for hydroxyls and anions.

The anesthetic/membrane story being told in our laboratory, not unnaturally, has developed from our ongoing research of the permeating mechanisms exhibited by liposomes as a function of permeating species, the concentration of anesthetics, the pressure, temperature, and liposome composition (chronologically referenced as 41-49). While the liposome model seemed to fulfill most of the passive properties required of a biological membrane, it was clearly inadequate to explain the overall behavior of animals—subjected to assaults by anesthetics, warmth, cold, pressure, and combinations of such parameters, widely described by working anesthetists and noted by our own limited experience (50).

We call the theory "The proton pump/leak theory of anesthesia" and the gist of it is outlined below (51). Instead of paraphrasing the experimental evidence concerning each step in the argument, I shall merely reference it.

1. Assumptions: Alertness, awareness, awakefulness, and probably "life" itself depend upon the normal functioning of energy-consuming enzyme systems (pumps) which generate concentration gradients of ions and other solutes across membranes and of which protons might be considered quite important. It is now regarded as highly likely that the packaging and storage of specific neurotransmitters such as adrenaline, noradrenaline, dopamine ("weak bases"), or, for example, γ-aminobutyric acid as a "weak acid" depend upon a steady-state, intravesicular pH which is different from that of the sap of the host cell normally accepted as being close to neutrality.

It is further assumed that if, for one reason or another, the intravesicular concentration of neurotransmitter falls below some critical value, the synaptic transmission will fail and a *deviant pattern of neuroactivity will emerge,* e.g., "anesthesia" when anesthetics are being administered.

2. The Theory: General anesthetics such as chloroform, ether, halothane, nitrous oxide, and the n-alkyl alcohols, by dissolving in bimolecular lipid membranes, promote the extension or initiate the existence of proton and/or cation transmembrane pathways (35,52). In turn, this leads to a rapid partial or complete collapse of any steady-state pH within a vesicle toward that of the cell sap. Since *uncharged* solutes such as undissociated acids (γ-aminobutyric) or bases (adrenaline, noradrenaline, dopamine, etc.) can and do diffuse freely *across* any region of the molecular lipid membrane (48), they would automatically and rapidly redistribute in accordance with their relative abundance determined by their respective pKs and of the prevailing pH inside and outside the vesicle.

Thus, *any* stress which might upset the normal, steady-state equilibrium between pump and leak across the synaptic vesicle membrane would tend to deplete transmitter substances and attenuate synaptic transmission (50,52). Note that the theory envisages circumstances where depletion of transmitter substance would ensue without necessarily administering anesthetics. For example, since enzymes characteristically have positive temperature and (probably negative pressure coefficient, excessive cooling, or exposure to high pressure of the *whole* animal would reduce the activity of the proton pumps. In other words, it is necessary to take into account the combined effects of temperature, pressure, and presence of anesthetics on *both the aqueous and lipid phases* (51).

The theory is epitomized by a consideration of how the brandy carried by the St. Bernard dogs might be lifesaving (53):

You are a climber caught in an avalanche, and you are about to die because your deep body temperature is so low that your proton pumps are unable to sustain adequate concentrations. Your synaptic pathways fail because the respective transmitter concentrations are inadequate. Along comes the St. Bernard dog with its own particular brand of 11 M ethanol. Ethanol is a bad anesthetic because its partition coefficient, membrane/water, is near to unity. (Good anesthetics have much higher coefficients, which means that for a given concentration in the membrane, the aqueous concentration is lower.) You drink the brandy and it very rapidly diffuses into membrane and aqueous space indiscriminately, thereby changing the Gibbs free energy of *both* aqueous and membrane phase. The effect is equivalent to adding antifreeze to a radiator of water at $0°C$! Your pumps are released into activity by the best thermodynamic equivalent of "instant heat" and you retire to the safety of the monastery for further warming!

In writing to you about this exploitation of liposome modeling in such detail, I was chuffed (as is said in East Anglia) to reread the last paragraph of our very first paper on the subject in 1965 (41): "Inherent in these investigations is the belief that narcosis represents a transient, reversible increase in membrane permeability to cation; higher concentrations [of anesthetic] resulting in a degree of permeability which energy-driven pumps cannot keep pace with . . ."! We were nearer a truth then than we might have imagined.

I would remind you that liposomes were also shown to be a very nice preparation for measuring partition coefficients (46) and certainly more realistic than octanol or olive oil! All in all, the liposomes have told us a great deal about the way pressure (54), temperature (43), and anesthetics (49) interact with membranes and the total understanding is now considerable, so much so that I feel that in the future they could and should be used as a primary screening system for all new (general) anesthetic concoctions. The repugnant, but ultimately necessary, LD_{50} test would then only be used for the short-listed compounds.

It was Sheena Johnson, of course, joining me as a postdoctoral fellow, who did most of the gritty groundwork on the effects of temperature, pressure, and anesthetics on liposome permeability (43-45). How much easier it would all have been with REVs.

Subsequently, Martyn Hill joined me as a postdoctoral fellow and was responsible for generalizing all the data available and formulating the Gibbs free energy theory of anesthetic action as it might affect membranes (47). As I have said before, it was the ultimate inadequacy of this theory, when tested out on *whole* animals, which led us to propose the pump/leak hypothesis. Martyn, I

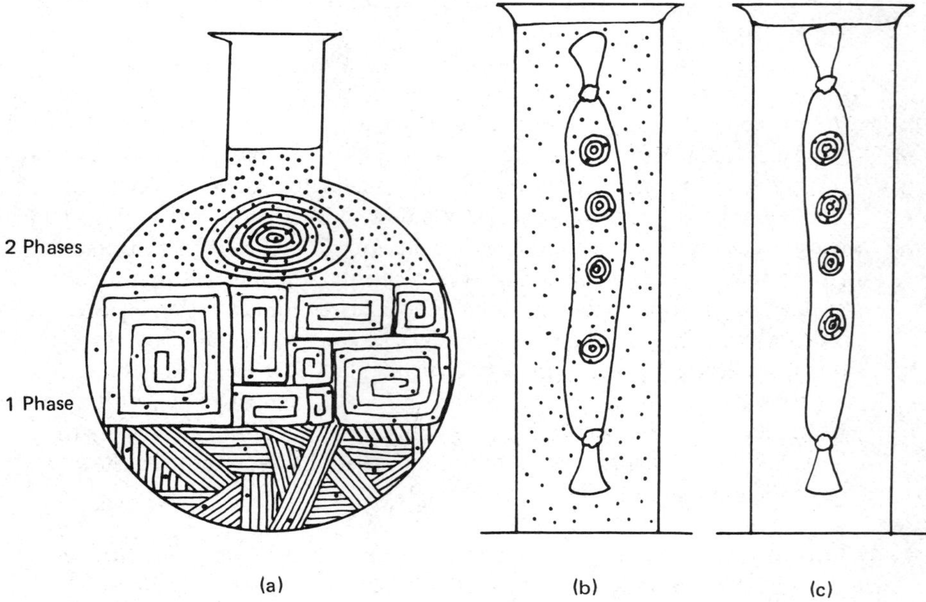

Figure 3 The stages in the preparation of the liposome containing a permeability marker or drug. (a) Sequential rearrangements of the phospholipid molecules as more water and solute are added. (b) The dialysis assay of the untrapped marker or drug. (c) The preparation ready for permeability measures or for delivery of drugs. (From Ref. 53.)

should explain, did his Ph.D. thesis on the subject of liquid helium and has been approaching biological temperatures ever since!

Another area of interest in which our experience of liposome structure, function, and behavior enabled us to offer alternative ideas concerned the properties of lung surfactant. It happened like this: In 1969 money became available for Jeof Watkins to return from Australia to Babraham but on condition that he work on the problem of lung surfactant and not liposomes. He came, and published the first, and to my mind, most convincing evidence that proteins were probably not implicated in the surface film activity of lung alveolae (55). His work did, however, emphasize the dilemma of how rapid spreading and surface replenishment could be achieved from a reservoir of fully hydrated phospholipids whose critical micelle concentration we all knew to be vanishingly low (56). This last fact is easily verified by trying to generate a froth in a tube of liposomes! How, then, are phospholipids made abundantly available to the rapidly extending air/liquid interface when one takes one's first breath? The answer would appear to be that they are stored in vesicles (lamellar bodies) in a limited hydrated state and protected from access to free water; i.e., they have not formed closed membranes and therefore the sheets of phospholipid are immediately free to spread over the extending surface (57,58). Watkins surreptitiously renewed his interest with liposomes and carried out an extended study with Brian Heap on their interactions with estradiol and progesterone (59,60).

And so I bring you more or less up to date. Whether or not liposomes will make anyone a fortune, they have generated a lot of friends and quantities of egg white!

Yours sincerely,

A. D. Bangham

Biophysics Unit
A.R.C. Institute of Animal Physiology
Babraham, Cambridge

31 March 1981

Dear Ostro,

Reading through the abstracts of the succeeding chapters I am grateful to you for your choice of authors and subjects. I am confident that everything worthwhile reporting about liposomes will be reported, which leaves me to fill in with some of my own personal anecdotes. For example, I have a letter from our National Research Development Corporation dated 19 May 1969, which thanks me for suggesting the idea of putting hemoglobin in liposomes with a view of providing a universally compatible blood! The snag with that suggestion, as we all should have known anyway, is that liposomes of whatever composition are instantly recognized as non-you and whipped out of the circulation! As a one-time experimental pathologist, it has always intrigued me to know how to prepare a surface which resembles a group O erythrocyte. Until one has been able to prepare such a thing I do not think it possible to begin to address a liposome to this or that part of the body, or can Eric Mayhew persuade me otherwise in Chapter 7? It saddens me a little to think of the vast amount of effort which has gone into the misdirected use of liposomes as Trojan horses; what an appropriate term to use, by the way, and can it be related to our trio of Greek expatriots who have

worked so hard in this field? I use the work "misdirected" deliberately because
I don't think people have truly realized how effective body defenses are to invasion by particles or "things," as opposed to free molecules, meaning that when
given intravenously, liposomes will invariable end up in the reticuloendothelial
system, or by mouth, will undoubtedly be dismantled into molecular fragments.
Perhaps I could phrase my dismay another way by saying that I am surprised that
liposomes are not by now being used to deliver antimony compounds for the
treatment of human leishmaniasis. Or are they? As leishmaniasis is one of hu-
manity's numerically most common diseases, liposomes would have been worth
discovering for the purpose of helping to cure this disease alone. [*Ed. note:* See
Refs. 242-246 in Chap. 7.]

Equally, I'm surprised that liposomes are not yet being used for the delivery
and slow release of drugs by subcutaneous and/or intramuscular injection. [*Ed.
note:* See Chap. 7.] Surely this state of affairs cannot be related to the existence
of Patent No. 417,715 accepted 1 October 1935 on behalf of I.G. Farben Industry for the purpose of placing drugs in a depot for slow release, and which read
as follows: "An emulsion is prepared from 25 parts of lecithin; 20 parts of water;
1.5 parts of cholestrin; 0.03 parts of strophantin and 0.5 parts of Nipasol (p-
hydroxybenzoic acid normal propyl ester)." I was asked by patent lawyers many
years ago whether this was a reasonable description for making liposomes and I
had to agree. I doubt very much whether the patentees had any real idea as to
the structure of these preparations because their other formulations in the patent
simply consisted of triglycerides stabilized to form classic emulsions. For all I
know, Saunder's original application of lecithin as a carrier for cholesterol may
form the basis of all sorts of cosmetic creams, drug carrier systems, etc. Someone is going to make money out of liposomes, sometime, somewhere, somehow!

Surely the most useful function of liposomes has been to set limits on what
biological membranes might be made of, how fluid, how liquid, how permeable
and to what, how easy to fuse, how miscible with other lipids and proteins. As
Peter Nicholls put it in the Festshrift to James Danielli (61), liposomes have been
widely appreciated as a means of "escaping physiological complexity." Of course
there are dangers in straying too far from the real thing, but with liposomes I
think the model is strong enough to be pushed still harder.

No one over the years has pushed them harder than the physical chemists,
such as Wayne Hubbel and Hardern McConnell (62-66), Dennis Chapman (67,68),
and Ching-Hsien Huang (69). They must have blessed us for encouraging them to
use nice, pure, or simple binary mixtures of pure compounds. Incidentally, I
challenge McConnell when he states (69) that no one really thought that membranes were "liquid." From my experience of phospholipid monolayers (5,6),
I was surprised to learn that they might be solid! The biological significance of
having two phases (64) might represent some form of thermal energy buffer system.

I remain a little anxious that Anne Haywwod's marvelous pictures of Sendai virus interactiong with liposomes are not to be overlooked in this volume (70-73). For me they are far and away the most convincing pictures of liposomes which have clearly fused with something else. For one thing, you can easily identify the exploded DNA, or is it RNA, inside the liposome with the viral spikes stranded on the outer membrane. (I'm enclosing a copy for you to look at and admire!) Incidentally, if you should ever want to know anything about English railways, ask her brother! Then there was Gordon Kepner, who blasted liposomes with megarads of radioactivity and found it had no effect and did not report it. A personal communication by him tactfully passed down the line did, however, cause someone to retract a published claim that radioactivity caused artificial membranes to become more conducting! Kepner also helped me to disentangle the superwater artifact and had me opposing a motion to the effect that the stuff existed, in open debate at the Biophysics Congress, Boston, in 1969. Again, Roger Klein fed a group of chickens on cod liver oil until their eggs were quite inedible, another group being fed with tallow. Their eggs were collected and liposomes prepared from their very contrasting lecithins. In collaboration with Malcolm Moore and Michael Smith he then studied the diffusion of various neutral amino acids across these membranes (74)—and a very interesting study it turned out to be. And Roger Lester paid us a visit from Boston University School of Medicine during 1970-1971 and together with Martyn Hill worked out a rough-and-ready phase diagram of lecithin, a variety of gangliosides and water (75,76), showing that lamellar structures were not compatible with high ratios of the two lipids. In their own way, these experiments provided some explanation for the principal histological findings in advanced cases of Tay-Sachs disease, but offered no succor to the sufferers.

Other short-term visitors included Mike Singer from Queens University, Kingston, Ontario, who showed that when liposomes are made simultaneously permeable to *both* components of a salt, either as independent cations and anions, e.g., K^+ carried by valinomycin and I^- carried by I_2, or as ion pairs, a progressive increase in volume of the liposomes is observed (33). In other words, in multilamellar vesicles the electrostatic repulsions between adjacent lamellae is balanced by an osmotic pressure difference due to the lamellae being permeable to water but not to salt. When the membrane is made permeable, the osmotic pressure difference is dissipated and the electrostatic repulsion of the fixed ions leads to an increase in liposomal volume. Note that "fixed" ions, e.g., dicetylphosphoric acid, also behave as impermeable ions in the sense that they would set up a Donnan equilibrium. Singer continued to work with liposomes (77,78) but I have not been in touch with him recently and he never sends me his reprints!

Not unrelated to Singer's observations were those of a Ph.D. student by the name Tim Kaethner. Kaethner was a plant biochemist and was absolutely deter-

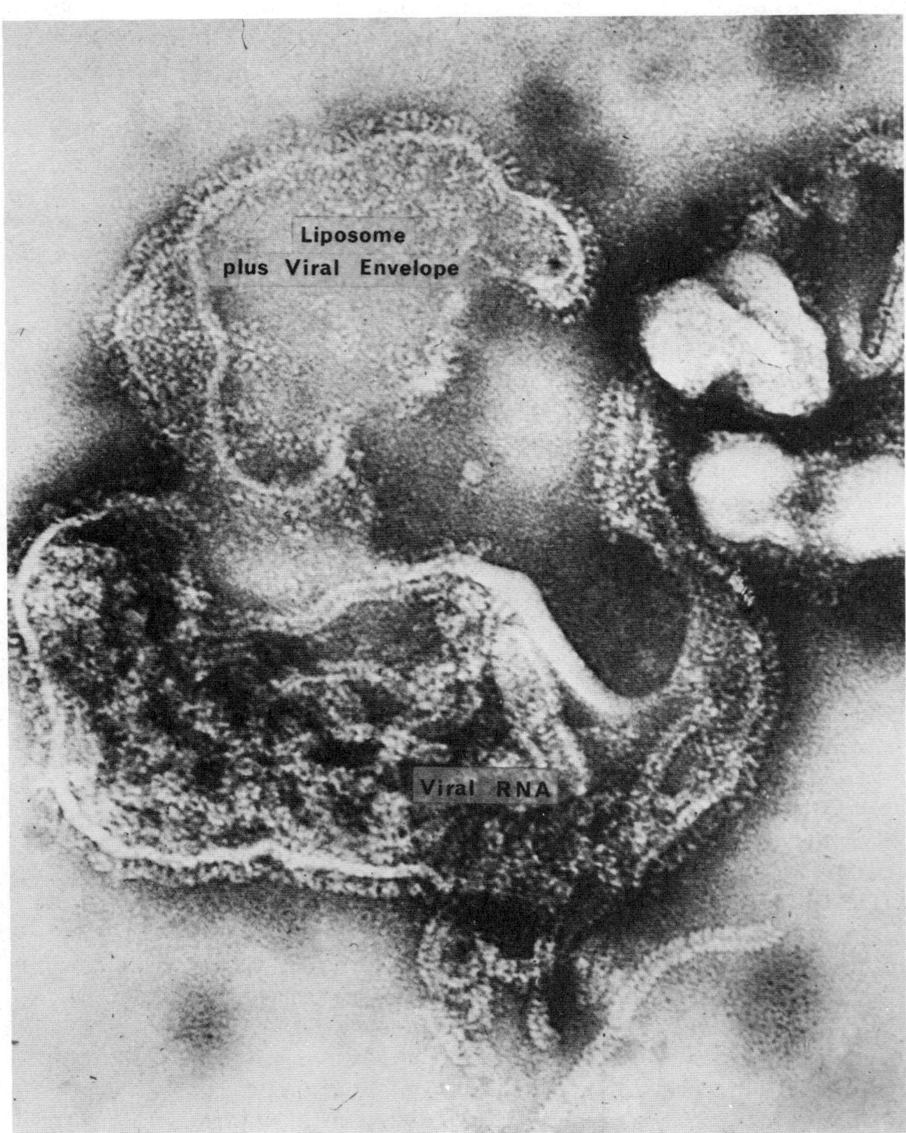

Figure 4 Liposomes were prepared by the method described previously by Haywood (71) from 0.61 μmol of cholesterol, 0.22 μmol of sphingomyelin, 0.44 μmol of phosphatidylcholine, 0.05 μmol of phosphatidylserine, 0.10 μmol of phosphatidylethanolamine, and 0.32 mg of gangliosides containing 0.11 μmol of N-acetylneuraminic acid. Phosphatidylcholine and phosphatidylethanolamine were pre-

mined to solve the problem of how plant auxins worked. Although he failed to achieve his ambition in my laboratory (but wrote an outstanding thesis on the subject), he did make the liposome do some interesting and revealing tricks. Among them was to entrap carbonic anhydrase inside liposomes and then to use CO_2 as a permeant source of dehydrated anions, enabling the system to imbibe salt and water. A K^+ gradient, valinomycin, and an uncoupling agent were also necessary but it should not be forgotten that both CO_2 and carbonic anhydrase are ubiquitous reagents in biological systems (79).

In 1974 Enrique Ochoa showed that 5-hydroxytryptamine but not acetylcholine, carbamylcholine, or L,D-noradrenaline bound to liposomes containing gangliosides; the finding was reversible and could be specifically blocked by 7-methyltryptamine (80).

Anyone reading this letter might get the impression that basic membrane work with liposomes was only going on here at Babraham during the late sixties and early seventies. This was not my intention and I can only suppose that I have omitted reminding you of the very significant and extensive contributions made by people like Johans de Gier and Laurens van Deenen, Steve Kinsky and Carl Alving, Brenda Ryman and Gerry Weissmann, Helmut Hauser, and Yechezkel Barenholz because of the old problem of the paraphrase failing to do justice to seminal reporting. Their work has, indeed, been reported, discussed, and paraphrased in a number of reviews by myself (1,53,81-86) and others (30,61, 87,93) and of course edited into whole books by Papahadjopoulos (94), Gregoriadis and Allison (95), Knight (1) and now your good self, and I am sure, much of their work will again be referred to in ensuing chapters.

Finally, it gives me very great pleasure to place on record the unique contribution which my assistant, Nigel Miller, has made over the past 25 years and without whose skill and unstinted help liposomes might never have been dis-

---

Figure 4   (continued)
pared from egg yolks by the method of Bangham et al. (85) and the method of Papahadjopoulos and Miller (96), respectively. Cholesterol was obtained from Sigma Chemical Co., and gangliosides were obtained from Koch-Light Laboratories, Ltd. Sphingomyelin and phosphatidylserine were obtained from Lipid Products. The phosphatidylserine was further purified by silicic acid chromatography; 358 hemagglutinating units of purified Sendai virus, strain ESWS, was adsorbed to liposomes in phosphate-buffered saline at 0-4°C for 1 hr and then incubated at 37°C for 2 hr.

For negative staining the samples were diluted fivefold into 0.145 M ammonium acetate and 10^{-4}% bovine serum albumin. A drop of the diluted sample was stained for 30 sec with a drop of 2% ammonium molybdate. Magnification 65,000X. [From Ref. 72, with permission from Journal of Molecular Biology 1974. 87:625. Copyright by Academic Press Inc. (London) Ltd.]

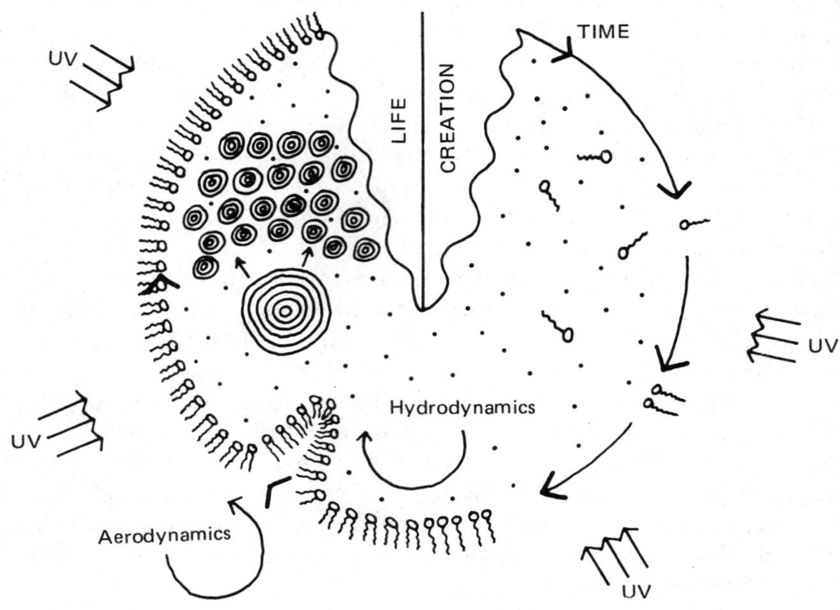

Figure 5 Cartoon illustrating a possible evolution of closed compartment systems in a prebiotic world. (From Ref. 53.)

covered and developed. As many visitors to Babraham would be happy to testify, Nigel is indefatigable. An ever-friendly helper, shrewd observer, careful experimenter, and loyal colleague, he has held the laboratory together through thick and thin times. He alone knows all the preparations, methods, techniques, and, I suspect, souls of the two remaining members of the Biophysics Unit. His early paper with Demetri Papahadjopoulos must be one of the most frequently quoted papers in liposome lore (96).

You asked for a history of liposomes to be placed in perspective—one's eyes are sometimes short—and at other times long-sighted—I hope these letters will convey both perspectives. But let me end with a couple more quotations—like poetry they crystallize ideas and concepts and save space. The first is attributed to Ludwig Buechner (1827-1899), who was supposed to have said: "Ohne Phosphor, kein Gedanke"—"Without phosphorus there would be no thoughts"; and J. L. W. Thudichum, who wrote (97):

Phosphatides are the centre, life, and chemical soul of all bioplasm whatsoever, that of plants as well as animals. Their chemical stability is greatly due to the fact that their fundamental radicle is a mineral acid of strong and manifold dynamicities. Their varied functions are the result of the collusion of radicles of strongly contrasting properties. Their physical properties are, viewed from a teleological point of standing, eminently adapted to their functions. Amongst these properties none are more deserving of further inquiry than those which may be described as their power of colloidation. Without this power no brain as an organ would be possible, as indeed the existence of all bioplasm is dependent on the colloid state.

Incidentally, I'm indebted to Roger Klein for this quotation—he used it in his thesis, which weighed 2.463 kg!

Yours sincerely,

A. D. Bangham

ACKNOWLEDGMENT

I thank Jean Weick for translating my handwritten hieroglyphics into legible text.

REFERENCES

1.   Bangham, A. D. 1981. Introduction. In *Liposomes from Physical Structure to Therapeutic Application,* G. Knight (Ed.). Elsevier North-Holland, New York.

2.   Bangham, A. D., M. M. Standish, and J. C. Watkins. 1965. Diffusion of univalent ions across the lamellae of swollen phospholipids. J. Mol. Biol. 13:238.

3.   Dervichian, D. G. 1964. The physical chemistry of phospholipids. Prog. Biophys. Mol. Biol. 14:265.

4.   Bangham, A. D. and R. M. C. Dawson. 1958. Control of lecithinase activity by the electrophoretic charge on its substrate surface. Nature 182:1292.

5.   Bangham, A. D. and R. M. C. Dawson. 1951. The relation between the activity of a lecithinase and the electrophoretic charge of the substrate. Biochem. J. 72:486.

6.   Bangham, A. D. and R. M. C. Dawson. 1962. Electrokinetic requirements for the reaction between *Cl. perfringens* α-toxin (phospholipase C) and phospholipid substrates. Biochim. Biophys. Acta 59:103.

7.   Bangham, A. D. 1961. A correlation between surface charge and coagulant action of phospholipids. Nature 192:1197.

8.   Dourmashkin, R. R., R. M. Dougherty, and R. J. C. Harris. 1963. The

effect of saponin and digitonin on Rous sarcoma virus and cell membranes; an approach to the cytochemistry of cellular components by the use of negative staining. J. R. Microsc. Soc. 81:215.

9. Bangham, A. D. and R. W. Horne. 1962. Action of saponin on biological cell membranes. Nature 196:952.

10. Van Zutphen, H., L. L. M. van Deenen, and S. C. Kinsky. 1966. The action of polyene antibiotics on bilayer lipid membranes. Biochem. Biophys. Res. Commun. 22:393.

11. Glauert, A. M., J. T. Dingle, and J. A. Lucy. 1962. Action of saponin on biological cell membranes. Nature 196:953.

12. Hillman, H. and P. Sartory. 1980. *The Living Cell*. Packard, Chichester, England.

13. Moore, C. and B. C. Pressman. 1964. Mechanism of action of valinomycin on mitochondria. Biochem. Biophys. Res. Commun. 15:562.

14. Pressman, B. C. 1965. Induced active transport of ions in mitochondria. Proc. Natl. Acad. Sci. USA 53:1076.

15. Pressman, B. C., E. J. Harris, W. S. Jagger, and J. H. Johnson. 1967. Antibiotic mediated transport of alkali ions across lipid barriers. Valinomycin, nigericin. Proc. Natl. Acad. Sci. USA 58:1949.

16. Chappell, J. B. and A. R. Crofts. 1966. Ion transport and reversible volume changes of isolated mitochondria. In *Regulation of Metabolic Processes in Mitochondria*, Vol. 7, BBA Library. J. M. Tager, S. Papa, E. Quagliariello, and E. C. Slater (Eds.). Elsevier, Amsterdam, pp. 293-316.

17. Chappell, J. B. and K. N. Haarhoff. 1967. The penetration of the mitochondrial membrane by anions and cations. In *Biochemistry of Mitochondria*, E. C. Slater, Z. Kaniuga, and L. Wojtczak (Eds.). Academic Press, New York, pp. 75-91.

18. Bangham, A. D., M. M. Standish, J. C. Watkins, and G. Weissmann. 1967. The diffusion of ions from a phospholipid model membrane system. Protoplasma 63:183.

19. Henderson, P. J. F., J. D. McGivan, and J. B. Chappell. 1969. The action of certain antibiotics on mitochondrial, erythrocyte and artificial phospholipid membranes. Biochem. J. 111:521.

20. Mitchell, P. 1961. Coupling of phosphorylation to electron and hydrogen transfer by a chemi-osmotic type of mechanism. Nature 191:144.

21. Mitchell, P. 1966. Chemi-osmotic coupling in oxidative and photosynthetic phosphorylation. Glynn Research Laboratories, Bodmin, Cornwall.

22. Mueller, P., D. O. Rudin, H. T. Tien, and W. C. Westcott. 1962. Reconstitution of cell membrane structure in vitro and its transformation into an excitable system. Nature 194:979.

23. Tien, H. T. 1974. *Bilayer Lipid Membranes—Theory and Practice*. Marcel Dekker, New York.

24. Bangham, A. D., M. M. Standish, and G. Weissmann. 1965. The action of steroids and streptolysin S on the permeability of phospholipid structures to cations. J. Mol. Biol. 13:253.

25. Weissmann, G., G. Sessa, and S. Weissmann. 1966. The action of steroids

and Triton X-100 upon phospholipid/cholesterol structures. Biochem. Pharmacol. 15:1537.

26. Weissmann, G., G. Sessa, and A. W. Bernheimer. 1966. Staphylococcal α toxin: effects on artificial lipid spherules. Science 154:772.

27. Sessa, G. and G. Weissmann. 1967. Effect of polyene antibiotics on phospholipid spherules containing varying amounts of charged components. Biochim. Biophys. Acta 135:416.

28. Sessa, G. and G. Weissmann. 1968. Effect of four components of the polyene antibiotic, filipin, on phospholipid spherules. J. Biol. Chem. 243:4364.

29. Sessa, G. and G. Weissmann. 1968. Differential effects of etiocholanolone on phospholipid/cholesterol structures containing either testosterone or estradiol. Biochim. Biophys. Acta 150:173.

30. Sessa, G. and G. Weissmann. 1968. Phospholipid spherules (liposomes) as a model for biological membranes. J. Lipid Res. 9:310.

31. Lehninger, A. L., E. Carafoli, and C. S. Nossi. 1967. Energy-linked ion movements in mitochondrial systems. Adv. Enzymol. 29:259.

32. Bangham, A. D., J. de Gier, and G. D. Greville. 1967. Osmotic properties and water permeability of phospholipid liquid crystals. Chem. Phys. Lipids 1:225.

33. Singer, M. A. and A. D. Bangham. 1971. The consequences of inducing salt permeability in liposomes. Biochim. Biophys. Acta 241:687.

34. Greville, G. D. 1969. A scrutiny of Mitchell's chemiosmotic hypothesis of respiratory chain and photosynthetic phosphorylation. Curr. Top. Bioenerg. 3:1.

35. Nichols, J. W. and D. W. Deamer. 1980. Net proton-hydroxyl permeability of large unilamellar liposomes measured by an acid-base titration technique. Proc. Natl. Acad. Sci. USA 77:2038.

36. Deamer, D. and A. D. Bangham. 1976. Large volume liposomes by an ether vaporization method. Biochim. Biophys. Acta 443:629.

37. Deamer, D. W. 1978. Preparation and properties of ether-injection liposomes. Ann. N.Y. Acad. Sci. 308:250.

38. Nichols, J. W., M. W. Hill, A. D. Bangham, and D. W. Deamer. 1980. Measurement of net proton-hydroxyl permeability of large unilamellar liposomes with the fluorescent pH probe, 9-aminoacridine. Biochim. Biophys. Acta 596:393.

39. McGivan, J. D. 1968. The movement of ions across artificial phospholipid membranes. Ph.D. dissertation, Univ. of Bristol.

40. Nicholls, P. and N. Miller. 1974. Chloride diffusion from liposomes. Biochim. Biophys. Acta 356:184.

41. Bangham, A. D., M. M. Standish, and N. Miller. 1965. Cation permeability of phospholipid model membranes: effect of narcotics. Nature 208:1295.

42. Bennett, P. B., D. Papahadjopoulos, and A. D. Bangham. 1967. The effect of raised pressure of inert gases on phospholipid membranes. Life Sci. 6:2527.

43. Johnson, S. M. and A. D. Bangham. 1969. The action of anaesthetics on phospholipid membranes. Biochim. Biophys. Acta 193:92.

44. Johnson, S. M., K. W. Miller, and A. D. Bangham. 1973. The opposing effects of pressure and general anaesthetics on the cation permeability of liposomes of varying lipid composition. Biochim. Biophys. Acta 307:42.

45. Johnson, S. M. and K. W. Miller. 1975. Effect of pressure and the volume of activation on the monovalent cation and glucose permeabilities of liposomes of varying composition. Biochim. Biophys. Acta 375:286.

46. Hill, M. W. 1974. The effect of anaesthetic-like molecules on the phase transition in smectic mesophases of dipalmitoyl lecithin. Biochim. Biophys. Acta 356:117.

47. Hill, M. W. 1974. The Gibbs free energy hypothesis of general anaesthesia. In *Molecular Mechanisms of General Anaesthesia,* M. J. Halsey, J. A. Sutton, and R. A. Miller (Eds.). Churchill Livingstone, New York, pp. 132-144.

48. Cohen, B. E. 1975. The permeability of liposomes to nonelectrolytes. I. Activation energies for permeation. II. The effect of nystatin and gramicidin A. J. Membr. Biol. 20:205.

49. Hill, M. W. 1978. Interaction of lipid vesicles with anaesthetics. Ann. N.Y. Acad. Sci. 308:101.

50. Hill, M. W., J. Hoyland, and A. D. Bangham. 1980. Effects of temperature and n-butanol (a model anaesthetic) on a behavioural function of goldfish. J. Comp. Physiol. A 135:327.

51. Bangham, A. D., M. W. Hill, and W. T. Mason. 1980. A biochemical and biophysical view of anesthesia. In *Molecular Mechanisms of Anesthesia. Progress in Anesthesia,* Vol. 2, R. Fink (Ed.). Raven Press, New York, pp. 69-77.

52. Bangham, A. D. and W. T. Mason. 1980. Anesthetics may act by collapsing pH gradients. Anesthesiology 53:135.

53. Bangham, A. D. 1979. Model membranes from membrane molecules. Speaking of Science: Proc. R. Inst. Gr. Br. 51:69.

54. MacNaughton, W. and A. G. MacDonald. 1980. Effects of gaseous anaesthetics and inert gases on the phase transition in smectic mesophases of dipalmitoyl phosphatidylcholine. Biochim. Biophys. Acta 597:193.

55. Watkins, J. C. 1968. The surface properties of pure phospholipid in relation to those of lung extracts. Biochim. Biophys. Acta 152:293.

56. Smith, R. and C. Tanford. 1972. The critical micelle concentration of L-α-dipalmitoyl phosphatidylcholine in water and water/methanol solutions. J. Mol. Biol. 67:75.

57. Morley, C. J., A. D. Bangham, P. Johnson, G. D. Thorburn, and G. Jenkin. 1978. Physical and physiological properties of dry lung surfactant. Nature 271:162.

58. Bangham, A. D., C. J. Morley, and M. C. Phillips. 1979. The physical properties of an effective lung surfactant. Biochim. Biophys. Acta 573:552.

59. Heap, R. B., A. M. Symons, and J. C. Watkins. 1970. Steroids and their interactions with phospholipids. Biochim. Biophys. Acta 218:482.

60. Heap, R. B., A. M. Symons, and J. C. Watkins. 1971. An interaction be-

tween oestradiol and progesterone in aqueous solutions and in a model membrane system. Biochim. Biophys. Acta 233:307.

61. Nicholls, P. 1981. Liposomes—as artificial organelles topochemical matrices and therapeutic carrier systems. Chapter 9, Festshrift to James Danielli. Int. Rev. Cytol. Suppl. 12:327.

62. Hubbell, W. L. and H. M. McConnell. 1969. Motion of steroid spin labels in membranes. *Homarus americanus* nerve fiber. Proc. Natl. Acad. Sci. USA 63:16.

63. Shimshick, E. J. and H. M. McConnell. 1973. Lateral phase separation in phospholipid membranes. Biochemistry 12:2351.

64. Hong-Wei, S. and H. M. McConnell. 1975. Phase separations in phospholipid membranes. Biochemistry 14:847.

65. Kornberg, R. D. and H. M. McConnell. 1971. Inside-outside transitions of phospholipids in vesicle membranes. Biochemistry 10:1111.

66. Kornberg, R. D., M. G. McNamee, and H. M. McConnell. 1972. Measurement of transmembrane potential in phospholipid vesicles. Proc. Natl. Acad. Sci. USA 69:1508.

67. Chapman, D. and S. A. Penkett. 1966. Nuclear magnetic resonance spectroscopic studies of the interaction of phospholipids with cholesterol. Nature 211:1304.

68. Chapman, D., R. M. Williams, and B. D. Ladbrooke. 1967. Physical studies of phospholipids. Part 6. Thermotropic and lyotropic mesomorphism of some 1,2-diacyl-phosphatidylcholines (lecithins). Chem. Phys. Lipids 1:445.

69. Huang, C. 1969. Studies on phosphatidylcholine vesicles. Formation and physical characteristics. Biochemistry 8:344.

70. McConnell, H. M. 1978. Bags of Life. BBC Horizon television program.

71. Haywood, A. M. 1974. Characteristics of Sendai virus receptors in a model membrane. J. Mol. Biol. 83:427.

72. Haywood, A. M. 1974. Fusion of Sendai viruses with model membranes. J. Mol. Biol. 87:625.

73. Haywood, A. M. 1975. "Phagocytosis" of Sendai virus by model membranes. J. Gen. Virol. 29:63.

74. Klein, R. A., M. J. Moore, and M. W. Smith. 1971. Selective diffusion of neutral amino acids across lipid bilayers. Biochim. Biophys. Acta 233:420.

75. Lester, R., M. W. Hill, and A. D. Bangham. 1972. Molecular mechanism of Tay-Sachs disease. Nature 236:32.

76. Hill, M. W. and R. Lester. 1972. Mixtures of gangliosides and phosphatidylcholine in aqueous dispersions. Biochim. Biophys. Acta 282:18.

77. Singer, M. A. 1973. Effect of different anions on sodium-22 permeability of liposomes. Can. J. Physiol. Pharmacol. 51:779.

78. Singer, M. 1979. Interaction of small molecules with phospholipid bilayer membranes: permeability studies. Chem. Phys. Lipids 25:15.

79. Kaethner, T. M. and A. D. Bangham. 1977. Selective compartmentation of hydration products of carbon dioxide in liposomes and its role in regulating water movement. Biochim. Biophys. Acta 468:157.

80. Ochoa, E. L. M. and A. D. Bangham. 1976. N-Acetylneuraminic acid

molecules as possible serotonin binding sites. J. Neurochem. 26:1193.
81.   Bangham, A. D. 1968. Membrane models with phospholipids. Prog. Biophys. Mol. Biol. 18:29.
82.   Bangham, A. D. 1970. The liposome as a membrane model. In *Permeability and Function of Biological Membranes,* L. Bolis, A. Katchalsky, R. D. Keynes, W. R. Lowenstein, and B. A. Pethica (Eds.). North-Holland, Amsterdam, pp. 195-206.
83.   Bangham, A. D. 1972. Lipid bilayers and biomembranes. Annu. Rev. Biochem. 41:735.
84.   Bangham, A. D. 1972. Model membranes. Chem. Phys. Lipids 8:386.
85.   Bangham, A. D., M. W. Hill, and N. G. A. Miller. 1974. Preparation and use of liposomes as models of biological membranes. Methods Membr. Biol. 1:1.
86.   Bangham, A. D. 1980. Development of the liposome concept. In *Liposomes in Biological Systems,* G. Gregoriadis and A. C. Allison (Eds.). Wiley, Chichester, England, pp. 1-24.
87.   Thompson, T. E. and F. A. Henn. 1970. Experimental phospholipid membranes. In *Membranes of Mitochondria and Chloroplasts,* E. Racker (Ed.). Van Nostrand Reinhold, New York, pp. 1-52.
88.   Tyrrell, D. A., T. D. Heath, C. M. Colley, and B. E. Ryman. 1976. New aspects of liposomes. Biochim. Biophys. Acta 467:259.
89.   Kimbelberg, H. K. 1977. The influence of membrane fluidity on the activity of membrane bound enzymes. In *Dynamic Aspects of Cell Surface Organizations: Cell Surface Review,* Vol. 4, G. Poste and G. L. Nicolson (Eds.). Elsevier/North-Holland Biomedical Press, Amsterdam, pp. 205-293.
90.   Pagano, R. E. and J. N. Weinstein. 1978. Interactions of liposomes with mammalian cells. Annu. Rev. Biophys. Bioeng. 1:435.
91.   Ryman, B. E. and D. A. Tyrrell. 1980. Liposomes—bags of potential. Essays Biochem. 16:49.
92.   Szoka, F. and D. Papahadjopoulos. 1980. Comparative properties and methods of preparation of lipid vesicles (liposomes). Annu. Rev. Biophys. Bioeng. 9:467.
93.   Rahman, Y.-E. 1979. Potential of the liposomal approach to metal chelation therapy. In *Lysosomes in Biology and Pathology,* J. T. Dingle, P. J. Jaques, and I. H. Shaw (Eds.). North-Holland, Amsterdam, pp. 625-647.
94.   Papahadjopoulos, D. (Ed.). 1978. *Liposomes and Their Uses in Biology and Medicine.* Ann. N.Y. Acad. Sci., Vol. 308. New York Academy of Sciences, New York.
95.   Gregoriadis, G. and A. C. Allison (Ed.). 1980. *Liposomes in Biological Systems.* Wiley, Chichester, England.
96.   Papahadjopoulos, D. and N. Miller. 1967. Phospholipid model membranes. I. Structural characteristics of hydrated liquid crystals. Biochim. Biophys. Acta 135:624.
97.   Thudichum, J. L. W. 1884. *The Chemical Constitution of the Brain.* Baillière, Tindall, and Cox, London, pp. xii, 1, 10-21.

Liposome Preparation: Methods and Mechanisms

David W. Deamer and Paul S. Uster / University of California at Davis, Davis, California

INTRODUCTION

Liposome preparations are now established as a useful model membrane system. Their demonstrated potential for delivering materials to the intracellular compartment has attracted investigators in experimental cell biology and in medicine as well. As a result, increasing interest has focused on preparation methods which permit liposome properties to be tailored for specific applications. The purpose of this chapter is to introduce the reader to a variety of methods [see also Szoka and Papahadjopoulos (1,2)].

An ideal preparation method would allow the investigator to produce liposomes from a variety of lipid components over a broad range of lipid concentrations. The vesicle size would be under experimental control, and size distribution would be relatively homogeneous. Finally, the method would require minimal time input and would not damage or contaminate the lipid.

No single preparation method encompasses all of these objectives, but there are several general methods now available, one of which will often provide a satisfactory approximation of a desired property. In this chapter we first discuss

the terminology that has evolved to describe liposome preparations and outline
some characteristic and useful parameters. We then review the protocols of the
more commonly used methods, compare certain properties of the resulting liposomes, and provide examples of recent applications.

TERMINOLOGY

Some specialized terminology has come into general use to describe specific
parameters of liposome preparations and it is worthwhile to define such terms
here since they will often be used later in the chapter. The first set of terms describes properties of individual vesicles and the second describes properties of
liposome suspensions.

A *liposome* is defined as any structure composed of lipid bilayers that en-
close a volume. The lipid is not necessarily phospholipid but this is the most commonly used component. Liposomes are characterized by their lipid composition.
In mixed lipid vesicles, composition has been described in terms of mole fraction,
mole ratio, or the mass ratio of one lipid to another. However, expressing the
vesicle lipid composition in terms of stoichiometric proportions is more useful
than mass ratios when comparing the lipid composition. Furthermore, lipid composition is more conveniently expressed as the mole fraction of total lipid rather
than as the mole ratio. For example, assume that one is comparing some property
of liposomes as a function of lipid composition in which a liposome preparation
(3 mM total lipid) composed of 1.5 mM phosphatidic acid and 1.5 mM phosphatidylcholine is contrasted with one of 2 mM phosphatidic acid and 1 mM phosphatidylcholine. The mole ratios of phosphatidic acid:phosphatidylcholine are
1:1 and 2:1, respectively, and the mole fractions are 0.5:0.5 and 0.67:0.33. Mole
fractions are more descriptive of the difference in individual bilayer composition
and the total amount of each lipid in suspension. The contrasts between liposome
preparations of different composition are also conceptualized more readily. This
becomes particularly apparent if one is dealing with lipid mixtures of three or
more components.

When phospholipids are dispersed in an aqueous phase, a heterogeneous mixture of vesicular structures is usually formed, most of which contain multiple
lipid bilayers forming concentric spherical shells. These were the liposomes first
prepared by Bangham and co-workers (3) and now called *multilamellar vesicles*
(MLVs). If a lipid dispersion is sonicated, the MLVs are reduced to much smaller
structures in the size range 25-50 nm diameter. These are termed *small unilamellar vesicles* (SUVs) since they contain only a single bilayer. Both MLVs and SUVs
have certain limitations as model systems and more recently several laboratories
have been able to produce vesicles in the size range 100-500 nm diameter. These
are called *large unilamellar vesicles* (LUVs).

There are two important parameters of liposome preparations that are functions of vesicle size and concentration. The first we will term *captured volume*, defined as the volume enclosed by a given amount of lipid and with units of liters entrapped per mole of total lipid (1 mol^{-1}). The captured volume naturally depends on the radius of the liposomes produced by a given technique and it is important to be aware that the vesicle radius and hence the captured volume are affected by the lipid composition of each vesicle and the ionic composition of the medium. Phosphatidic acid vesicles, for instance, are typically smaller than phosphatidic acid:phosphatidylcholine mixes. These effects are relatively small and for any given lipid composition the captured volume is independent of lipid concentration over a wide range. The second parameter is *encapsulation efficiency* defined as the fraction of the aqueous compartment sequestered by bilayers. The encapsulation efficiency is directly proportional to the lipid concentration; when more lipid is present, more solute can be sequestered within liposomes. The latter two parameters are important to liposome preparation methods for several reasons. For example, if liposomes are used for intracellular delivery of some entrapped compound, an LUV preparation with high captured volume will deliver a much greater volume per lipid mass than a SUV preparation. Furthermore, if the substance is costly, it is clearly desirable to maximize the amount trapped in the liposome volume.

Calculated values for captured volume and encapsulation efficiency are given in Figures 1 and 2. Note that a typical LUV preparation has a captured volume approximately 10 times that of SUVs, and that a 60 mM LUV preparation will encapsulate approximately half of the original volume. (It is interesting to note that the maximum captured volume possible for a liposome preparation is in a single large vesicle of 1 mol of lipid that is 260 m in diameter and encapsulating 6 billion liters.)

There are several methods for determining these parameters, either by direct measurement of an encapsulated compound or electron microscopic observation of individual vesicles. These are discussed later in the section "Ancillary Methods."

ARTIFACTS

Several kinds of artifacts can be introduced into liposome preparations by the methods used to produce and store them. The most common artifacts result from contamination by organic solvents and detergents, lipid peroxidation, and hydrolysis products (fatty acids and lysophosphatides), and trace amounts of polyvalent metals. These have not been investigated in detail but some accumulated practical knowledge can be summarized here.

Solvent Artifacts

Solvent contamination in liposomes can affect experimental results and it is usually worthwhile to reduce trace organic solvent concentrations to minimal levels.

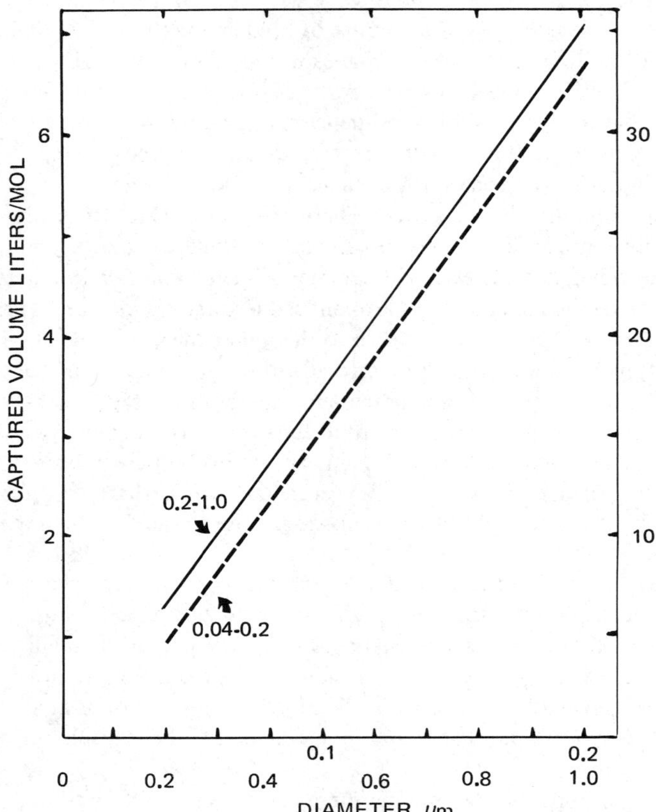

Figure 1 Captured volume of theoretical ideal unilamellar liposome preparations. The lower dashed line shows calculated captured volumes of homogeneous liposomes ranging from 0.04 to 0.2 μm in diameter (left ordinate), and the upper solid line for liposomes between 0.2 and 1.0 μm in diameter (right ordinate).

Trace chloroform is best reduced by an overnight vacuum evaporation (<10 mbar) of the dried lipid before it is used for liposome preparation. It is important to flush the evacuating apparatus with some inert gas such as nitrogen or argon to remove residual molecular oxygen that might readily peroxidize the dried lipids.

The concentrations of other solvents such as ethers and alcohols are reduced by gel filtration or dialysis. However, there are few data on how much can be removed and some investigators have been surprised by the amount of labeled ethanol that remains with liposomes following filtration. In one study (4) diethyl

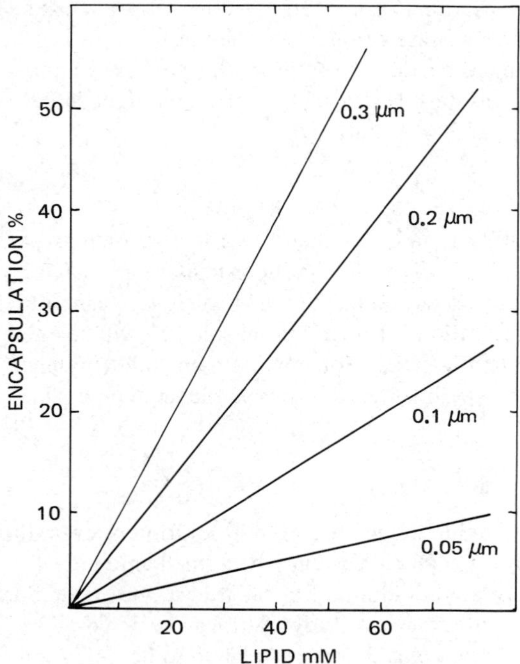

Figure 2   Calculated encapsulation efficiencies for varying concentrations of idealized liposomes. Four homogeneous preparations are shown, with diameters of 0.05, 0.1, 0.2, and 0.3 μm.

ether was shown to be reduced by gel filtration from saturating concentrations (about 0.5 M) to undetectable levels, but the error was such that concentrations up to 1 mM would not have been detected.

In practice, one cannot be certain that all traces of a solvent have been removed. However, it is possible to add small amounts of the potential contaminent to determine whether it affects the parameter being measured. If there is no effect, one may tentatively conclude that traces of that contaminant probably will not produce artifactual data.

Detergents

Cholate and deoxycholate have been used in several preparation methods (5,6). These detergents are typically removed by dilution, dialysis, BioBead adsorption, or gel filtration and have been shown to be reduced to as little as 0.1 mol%. Triton X-100 has also been used in liposome and membrane reconstitution methods and can be removed by BioBead filtration (7,8). A 10 mM concentration can be reduced to about 0.1 mM in a single passage through the column.

More recently octyl glucoside has been found to produce liposomes in a useful size range (9). This detergent has the advantage that it is readily reduced to as little as 0.25 mol% after dialysis and gel filtration.

Divalent and Polyvalent Cations

Surprising amounts of trace metal ions can be introduced into a liposome preparation in which buffers and salts are used. Suppose that liposomes are prepared in 0.1 M NaCl. Reagent-grade NaCl contains about 0.005% of calcium, magnesium, iron, and lead, but the final concentrations of these ions in solution will be a few micromolar, significant in some investigations. Some workers include ethylenediaminetetracetic acid (EDTA) in experimental solutions to reduce the possible effects of trace cations.

Peroxidation Damage and Hydrolysis Products

Unsaturated lipids can undergo peroxidation while stored in solution or even during their isolation and purification. There will certainly be a small amount of peroxidation damage in any unsaturated lipid mixture and at best one may reduce the peroxidation to levels that do not affect the study. This can be achieved by using freshly isolated lipid or synthetic lipids that are not labile to peroxidation. Storage should be in chloroform under nitrogen at $-20°C$ and liposome preparation should be done in an inert atmosphere. Diethyl ether readily accumulates peroxides and if it is used in a preparation method it should be freshly distilled over sodium bisulfite or otherwise purified. A simple test for lipid peroxidation is to monitor the ultraviolet (UV) absorbance ratio at 215 nm/230 nm, which measures diene conjugation produced by oxidative damage (10).

A second potential problem in liposome preparation is hydrolysis. Again, small quantities of hydrolysis products are inevitably present in typical lipid mixtures and one must therefore attempt to minimize possible artifacts. If it can be shown that trace amounts of fatty acids or lysophosphatides affect the liposome property that one is studying, the lipids should be purified by chromatography just before use.

At least one class of lipids does not undergo oxidation or hydrolysis. Branched-chain dialkyl ethers are found in primitive photosynthetic cells such as the halobacteria (11), and should be explored for their potential in liposome preparations. Such lipids serve as a useful control to determine the extent to which peroxidation and hydrolysis contribute to liposome properties.

METHODS

The major methods to be discussed are those that produce MLV, SUV, and LUV preparations. We will also discuss several other preparations that include liposomes composed of single-chain amphiphiles and model phospholipids. Each technique will be considered in terms of the number of lamellae the liposomes have, sources of contamination, ease of preparation, encapsulation efficiency, and captured volume.

Small Unilamellar Vesicles

The first SUV lipid dispersions were prepared by sonication about the same time that MLVs were introduced (12,13). They have since been developed into a standard system for investigating biophysical aspects of lipid bilayers (14-16). In a typical preparation, phospholipid in chloroform is dried under nitrogen and rotary-evaporated to remove trace solvent. The desired aqueous phase is added to produce a lipid concentration in the millimolar range, after which the suspension is sonicated to clarity at 45°C in a probe or bath sonication unit. This might require only a few minutes with the probe but as much as 2 hr with bath sonication. Although it takes considerably longer, the bath sonication has the advantage that it can be carried out in a closed container under nitrogen or argon, does not contaminate the lipid with metal from a probe tip, or produce significant heating. When using a bath sonicator it is important that the closed container be located precisely in the nexus of the ultrasonic waves. At this focal point of input power the lipid dispersion inside will be vigorously and visibly agitated.

When the resulting dispersion is examined by negative staining electron microscopy it is found to contain vesicles ranging from 25 to 50 nm in diameter. Huang (17) showed that these can be brought to near homogeneity by gel filtration. When this is performed, a highly uniform preparation of vesicles in the size range 25 nm is produced. This remains the most homogeneous liposome preparation. A summary of captured volume and encapsulation efficiencies is given in Table 1.

A second method for producing SUVs was developed by Batzri and Korn (18). In this procedure an ethanol solution of lipid is injected quickly into the desired aqueous phase to a final lipid concentration of 3-30 mM. The resulting vesicles range from 30 to 110 nm in diameter, depending on the lipid concentration. A potential drawback of this method is that high concentrations of ethanol must be removed by a subsequent purification step.

A third method for producing SUVs is to pass MLVs through a French press at 20,000 psi four times (19). Once again, the resulting product is in the range 30-50 nm in diameter. This preparation method has not been extensively studied but is well worth pursuing since it sidesteps some of the problems associated with other SUV methods.

Table 1 Summary of Characteristics of Typical Liposome Preparations

Lipid composition[a]	Diameter (μm)	Captured volume (liters/mol)	Encapsulation efficiency (% original volume)	Reference
Small unilamellar vesicles				
PC:PA 90:10 (1 mM)	—	0.8	—	23
PC:DCP:Chol 70:20:10	—	0.7	—	24
PS 20 mM	0.03	0.5	1	32
PC 20 mM	0.03	0.23	—	6
Large unilamellar vesicles				
PC:PA 90:10 (1 mM, diethyl ether injection)	0.05-0.25	17 ± 4	1.7	23
PC:DCP:Chol 7:2:1 (15 mM, ether injection)	0.1-0.4	23-31	38-46	24
PG:PC:Chol 1:4:5 (66 mM, REV method)	0.2-1.0	11.7	35-65	25
PS 20 mM (PS-cochleate method)	0.2-1.0	7	15	32

PC 50 mM (DOC + dialysis)	0.1	2.4	12	6
PC 16 mM (octyl glucoside + dialysis)	0.2-0.3	7-7.4	22	9
Soybean phospholipid 20 mg/ml (FTS method)	–	8	–	34
Single-chain amphiphile (dodecyl sulfate-dodecanol 50 mM)	1-10	–	42	36
Multilamellar vesicles				
PG:PC:Chol 1:4:5 (66mM)	0.4-3.5	3.5	5-15	25
PC:DCP:Chol 7:2:1 (15 mM)	–	6.9	10.4	24
PS 20 mM	–	2.5	5	32
PC:PA 90:10 (1 mM)	–	1.8	–	23

[a]PC, phosphatidylcholine; PA, phosphatidic acid; DCP, dicetyl phosphate; PS, phosphatidylserine; PG, phosphatidylglycerol; DOC, deoxycholate; Chol, cholesterol.

Fendler (10) has shown that dialkyl surfactant compounds, dioctadecyldi-
methylammonium chloride (DODAC) and dihexyldecylphosphate (DHP), are
capable of forming vesicular structures and has termed the resulting vesicles
"membrane mimetic agents." In a typical preparation the vesicles are formed by
simple sonication. They have most of the properties of SUV liposomes produced
from naturally occurring lipids and offer a useful alternative model system. In
particular, the DODAC vesicles carry a positive charge which is not available
when natural lipids are used.

The homogeneity of SUV preparations makes them a preferred system for
biophysical studies of lipid bilayers. Because of their small diameter and low
captured volume they are of limited value as mediators of intracellular solute delivery. Also, due to their high radius of curvature, SUVs may have special properties not shared by more planar membranes, and in some applications this could
limit their usefulness as a model membrane system.

Multilamellar Vesicles

Bangham and co-workers (3) first recognized that lipid vesicles were models of
cell membranes, particularly with regard to permeability studies. Their liposome
system, now described as multilamellar vesicles (MLVs), is still used in certain
applications. In a typical preparation, 10 mg of egg phosphatidylcholine containing 5 mol% of egg phosphatidic acid is evaporated from chloroform to produce a thin film on the wall of a 100-ml round-bottom flask. (The phosphatidic
acid adds a negative charge to the lipid which aids hydration and inhibits aggregation.) The flask is placed under vacuum overnight to remove any remaining
chloroform. The desired aqueous solution is added (5 ml) and the lipid is hydrated for several hours. The milky suspension is then swirled or vortexed to
disperse the lipid.

In a variation of this method, Reeves and Dowben (21) hydrated the dried
lipid with wet nitrogen gas that had passed through a water phase, followed by
addition of water and gentle swirling. The slower hydration was found to produce very large vesicles up to several hundred micrometers in diameter, but only
in the absence of ions or protein.

(The process of lipid hydration is an interesting classroom exercise and an
esthetic pleasure. Lipid hydration can be observed under a phase or polarizing
microscope. A drop of lipid in chloroform is dried on a slide, a coverslip is
placed over it, and a drop of buffer solution is added from the side. Over a
period of 30-60 min, a beautiful array of myelin figures grows out of the lipid
during hydration, often forming remarkably complex geometric structures.)

Dehydration Formation of Multilamellar Vesicles

The following MLV preparation method is unpublished and utilizes de-
hydration to cause vesicle fusion. The resulting liposomes are oligolamellar

and multilamellar and can efficiently encapsulate large macromolecular structures.

One to ten micromoles of phospholipid is dispersed by sonication in 1 ml of water. The material to be encapsulated is added and the water is evaporated under a dry nitrogen stream. As the liposomes are concentrated by the drying process, they fuse to form multilamellar planes and thereby sandwich the added material between the resulting lamellae. After drying is complete, the lipid is rehydrated by placing a moist cotton plug in the tube or by flushing the lipid with a stream of hydrated nitrogen for a few hours. Water is then added and the mixture is vortexed to disperse the lipid vesicles. The resulting liposomes can be sized by passing them through polycarbonate filters. This method has the advantage that it can be used with most phospholipids and does not require any special apparatus other than a sonicator. The vesicles have an encapsulation efficiency of nearly 50% for DNA (1 mg of DNA to 10 mg of lipid) but are not well suited for trapping smaller molecules.

The major advantage of MLV preparation is the simplicity of the procedure and the fact that it is applicable to a wide variety of different lipid mixtures. Certain studies may even require multilamellar structure. For instance, it is likely that the particulate structures seen in some liposomes by freeze-fracture represent points at which neighboring lamellae touch; hence, studies examining this phenomenon require MLVs (22).

The captured volume of MLVs is variable due to changes in lipid composition and hydration techniques. As seen in Table 1, values range from 1.8 (23) to 6.9 liters/mol (24). However, the multilamellar character and heterogeneity of MLVs is a distinct disadvantage in many applications. The MLV character is particularly confusing when working with permeability studies since it is not certain how many of the lamellae are involved in permeation of the fluxing species. Problems may also arise in liposome fusion studies with cells since large amounts of lipid are delivered along with the liposomal content.

Large Unilamellar Vesicles

It became apparent in the 1970s that larger liposomes than SUVs would be useful and therefore several investigators began to explore various methods for preparing LUVs. The three most commonly used principles are infusion procedures, reverse-phase evaporation, and detergent dilution. In the infusion method, a lipid solution in a nonpolar solvent is infused into an aqueous solution under conditions that cause the solvent to caporize, resulting in the formation of liposomes. In the reverse-phase evaporation (REV) method, a dispersion of inverted micelles of lipid is produced in a system containing a mixed organic and aqueous phase. As the solvent is evaporated away, the micelles coalesce to form liposomes. Finally, a variety of methods use different means to dilute detergent originally codispersed with the lipid. This principle has been applied to the first

commercially available liposome preparation apparatus. These three general
methodologies are described in the following section.

Infusion Methods

An infusion method utilizing diethyl ether was published by Deamer and Bangham (23) and has been developed further over the past few years (24). This
method depends on the fact that a solution of lipid in a nonpolar solvent can be
infused into an aqueous phase under conditions that cause the solvent to vapor-
ize. The lipid is left as a thin film at the interface of the vapor bubble and the
aqueous phase. It can be calculated that the film is composed of only a few bilayers of lipid. As the bubble rises through the aqueous phase, the lamellae are
dispersed to form uni- and oligolamellar liposomes. These are then sized by passage through polycarbonate filters.

In a typical run, 4 ml of lipid solution in pentane or diethyl ether (1-2 mM
lipid) is infused slowly (0.2 ml/min) into 4 ml of aqueous phase warmed to 60°C.
This phase is contained in a converted Leibig condenser that has warm water circulating through the outer jacket. The aqueous phase is flushed with a gentle
stream of nitrogen bubbles, which promotes solution mixing, provides an inert
atmosphere at the air-solution interface, and increases the size of the vapor bubble, thereby favoring formation of unilamellar vesicles. When infusion is complete, the liposomes are filtered through a polycarbonate filter to size them and
remove aggregates. This step is important to most of the LUV methods and it is
likely that larger oligolamellar vesicles are sized down as they are extruded through
the pores of the filter. Following this step, the liposomes are put through a gel
filtration step (i.e., Sephadex G-25 or G-50, Biogel P-10) to remove any remaining solvent and to exchange the unencapsulated aqueous phase if desired.

The advantage of the infusion procedure is that it is applicable to a wide
variety of lipids and can be completed in less than an hour. The main limitation
is that the resulting preparations are relatively dilute (1-15 mM lipid) so that the
encapsulation efficiency is low. (However, Schieren et al. (24) reported that they
could obtain 35% encapsulation efficiency using pet ether infusion and 30 mM
final concentration.) If the temperature of infusion might damage heat-sensitive
materials, it is possible to place the column under vacuum so that lower temperatures can be used.

Reverse-Phase Evaporation

Szoka and Papahadjopoulos (25) have designed an alternative LUV methodology
in which lipid in mixed aqueous-nonpolar solvent forms inverted micelles, such
that the lipid tails are inserted into the nonpolar phase and the head groups sur-
round water droplets. As the nonpolar solvent is evaporated away under vacuum
the micelles coalesce to form large uni- and ologolamellar vesicles. In a typical
preparation, 66 μmol (50 mg) of lipid is dispersed by bath sonication in a 1 ml:3

ml mixture of aqueous solution-diethyl ether. The ether is then removed by
rotary evaporation under vacuum followed by polycarbonate filter sizing and
gel filtration to remove trace solvent.

The REV method has the distinct advantages of being applicable to a range
of volumes (as little as 0.2 ml) and in its ability to encapsulate a large fraction of
the original volume (up to 65% in low-ionic-strength media). Its disadvantages
are that components are exposed to organic solvents and sonication which may
be damaging and that the lipid requirement (10 mg per 0.2 ml of solution) is relatively large.

Detergent Dilution

Detergents such as the bile salts (cholate and deoxycholate) and Triton X-100
have been applied to the solubilization and reconstitution of biological membranes with considerable success. In general, the detergent is added until the
membrane suspension, due to the formation of mixed micelles of detergent,
lipid, and protein, has clarified. The detergent concentration is then lowered
and the original lipid and protein form vesicular membrane structures, often with
reconstituted function of the protein.

Detergents have been used less often as a primary liposome preparation
method. However, these procedures do have certain advantages. First, they are
relatively gentle in their action and would not be expected to hydrolyze or peroxidize liposome components, and second, the detergent/lipid ratio is readily
varied, which permits considerable experimental control over the size of the resulting vesicles. In a typical detergent-LUV method (6), egg phosphatidylcholine
is mixed with deoxycholate in 1 ml of aqueous solution (20 μmol of lipid per 10
μmol of detergent). After a few minutes of sonication in a bath sonicator, the
originally turbid suspension becomes opalescent due to the presence of mixed
micelles. The deoxycholate is then removed by gel filtration. These vesicles are
about 100 nm in diameter and would be considered small LUVs. They have a
captured volume of 2-3 liters/mol and successfully encapsulate cytochrome c and
glucose. They do not trap radiolabeled sodium chloride unless it is present during the removal of deoxycholate.

Octyl glucoside has been employed by Mimms et al. (9). In this procedure,
egg phosphatidylcholine is dried as a film in a glass container and 0.4-0.6 ml of
octyl glucoside solution is added. In typical experiments, the amount of lipid
can be varied from 2 to 10 mg, and the optimal detergent/lipid ratio is in the
range 10:1 to 15:1. The solution is briefly sonicated and the clear mixture is
dialyzed twice for 12 hr against 1 liter of buffer which includes the compound
to be encapsulated. During dialysis, the solution becomes turbid and examination shows that vesicles ranging around 0.2 μm in diameter are formed. The captured volume ranges from 6 to 7.5 liters/mol and cytochrome c trapping demon-
strates a 22% encapsulation efficiency. After two gel filtrations, the detergent

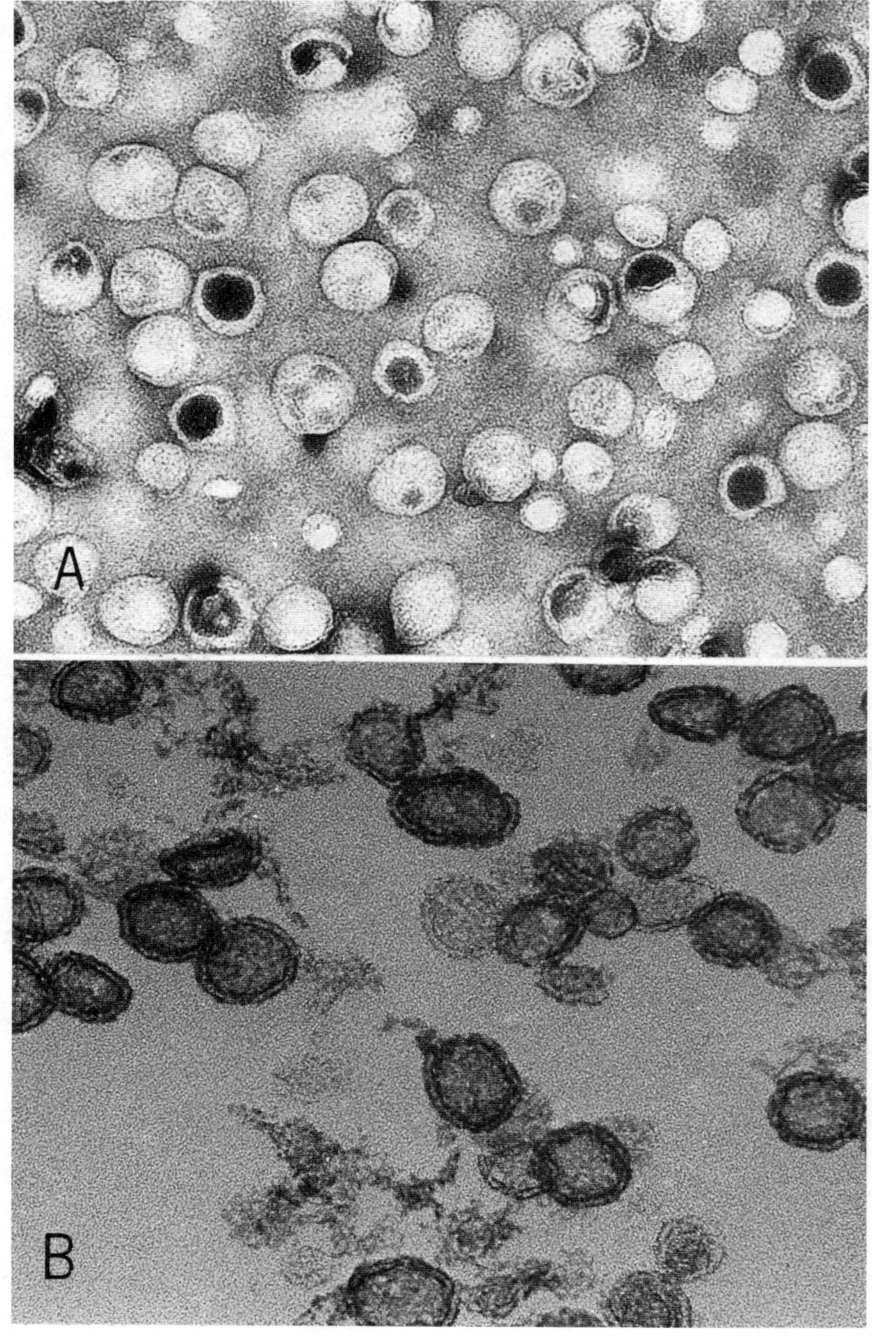

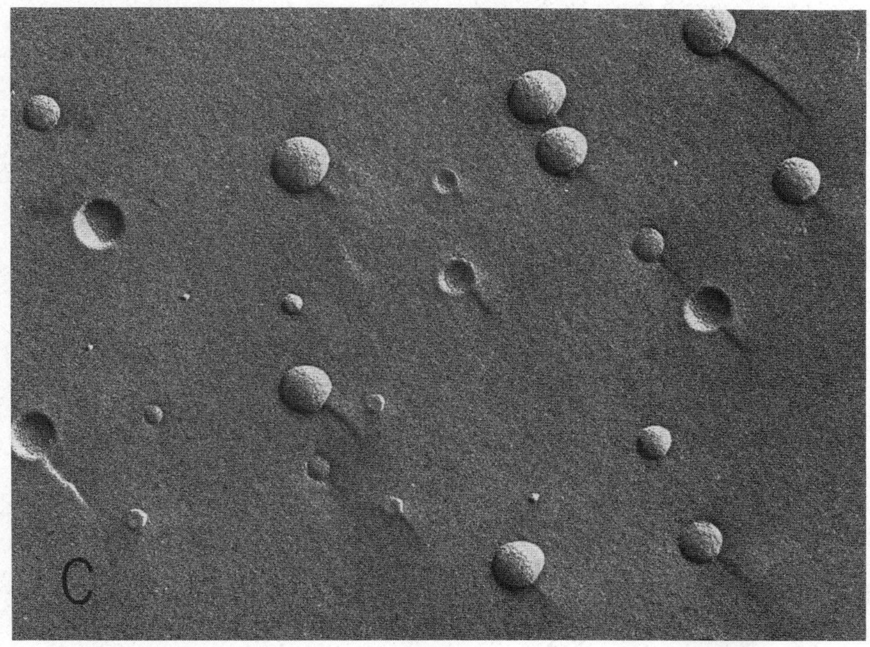

Figure 3 Electron micrographs of LUV preparations: (A) Negative staining; (B) Thin sections; (C) Freeze-fracture. Prepared by the Lipoprep apparatus. (From Ref. 27.)

is reduced to about 1 octyl glucoside molecule per 400 phospholipids molecules. The measured permeability to monovalent cations and anions is in the range observed for SUVs, suggesting that this amount of detergent does not markedly affect the permeability barrier.

Detergent removal has been applied in the only commercial liposome apparatus on the market, called Lipoprep. This device is based on the work of Milsmann et al. (26) and Zumbuehl and Weder (27). In the Lipoprep method, phospholipid is dispersed with detergent in ratios ranging from 1:1 to 2:1. The detergent is removed by passing the mixture through a rapid-flow dialysis cell. The size of the vesicles prepared with cholate is 50-80 nm in diameter (Fig. 3) and the captured volume ranges from 1.8 to 2.4 liters/mol. If octyl glucoside is used rather than cholate, the diameter increases to 180 nm (Fig. 4). The vesicle diameter can be varied from 80 to 200 nm in a predictable and homogeneous fashion by changing the ratio of octyl to heptyl glucoside (28). The homogeneity and the ability to manipulate the mean diameter of the vesicle preparation is a significant advantage of the Lipoprep method, since a number of investigations involve studies of liposome properties that are related to vesicle diameter.

The detergent dilution methods are widely applicable in liposome preparations, particularly when proteins are to be reconstituted into the resulting membrane. The methodological limitations are that preparations using dialysis require hours or even days to be completed, and if the solute to be entrapped is dialysis membrane permeant, the encapsulation efficiency is very low.

Fusion Methods

Phosphatidylserine Cochleate Vesicles If SUVs can be caused to fuse, the fusion process should be able to encapsulate some fraction of the material dissolved in the aqueous phase. Papahadjopoulos et al. (29) mixed SUVs composed of negatively charged phospholipids with calcium followed by EDTA to chelate the calcium. When calcium ion interacts with phosphatidylserine SUVs, the vesicles first aggregate to form a dense precipitate and then fuse to produce multilamellar arrays that often take the form of cochleate cylinders. When the calcium is removed by EDTA, the structures swell and form large uni- and oligolamellar vesicles.

In a typical experiment, 1-10 mM unsaturated phosphatidylserine is used to prepare SUV. Sufficient calcium chloride is added to produce a final calcium/ lipid ratio of 1:2 and precipitate the SUVs. To disperse the precipitate, EDTA is added in slight excess with stirring until the precipitate forms an opalescent suspension of LUVs. The resulting vesicles are able to encapsulate structures as large as viruses (30), nucleic acids (31), and ferritin (32) under relatively mild conditions. The captured volume is in the range of 7 liters/mol, and the encapsulation efficiency is 10-15% in a 20 mM lipid dispersion.

Freeze-Thaw-Sonication A second fusion method is the freeze-thaw-sonicate (FTS) technique. This method was first explored by Kasahara and Hinkle (33) for reconstituting functional membranes and Pick (34) has shown that it is applicable to liposome preparation. In a typical procedure, 30 mg of lipid is dispersed by sonication in 1 ml of buffer, then mixed with the material that is to be encapsulated. The mixture is frozen in liquid nitrogen, thawed and bath-sonicated for 30 sec. (The sonication serves primarily to disperse aggregated material.) The resulting vesicles are reasonably impermeable to ions and have captured volumes in the range of 8 liters/mol. The FTS method has the advantage of being a relatively

---

Figure 4 Diameters of liposome preparations. Bar shows 0.2 μm. (From Ref. 28.)

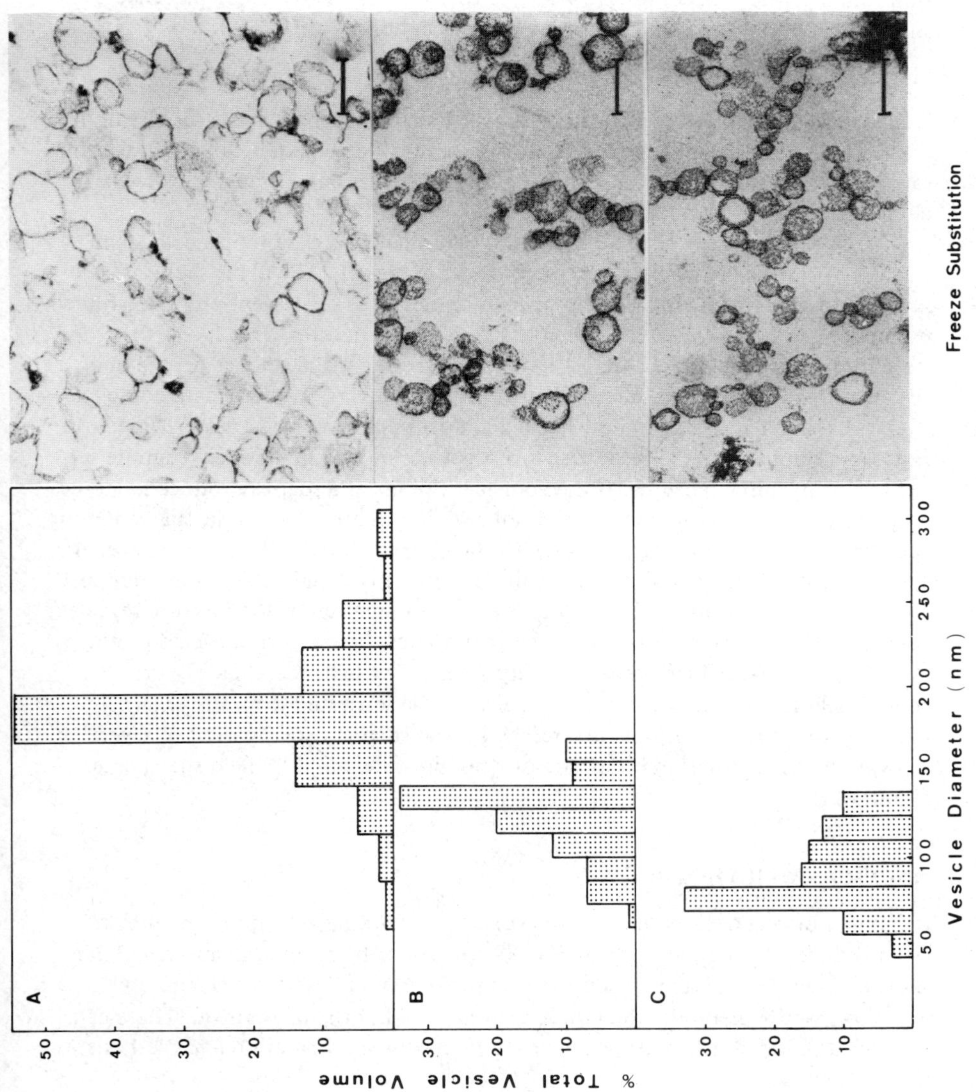
A
B
C
50
40
30
20
10
30
20
10
30
20
10
% Total Vesicle Volume
50
100
150
200
250
300
Vesicle Diameter (nm)
Freeze Substitution

simple and gentle procedure. However, it does not work well in the presence of sugars, high concentrations of ions, or divalent cations.

Single-Chain Amphiphiles

It has recently been found that many single-chain amphiphiles form vesicles under certain conditions. This was first demonstrated by Gebicki and Hicks (35) for un-saturated fatty acids, and later extended by Hargreaves and Deamer (36) to include other single-chain amphiphiles.

To produce oleic acid liposomes, a 50 mM micellar solution of oleate is pre-pated by titrating oleic acid to pH 10 with 0.1 M NaOH. The pH is then carefully lowered to pH 8.5. At this pH the suspension becomes opalescent and if examined by phase microscopy is found to contain vesicular structures. It is likely that the stability of the vesicles depends on the formation of an acid soap at pH 8.5 and that hydrogen bonding stabilizes the structure through oleate-oleic acid complexes.

A striking demonstration of single-chain amphiphile vesicles was provided by Hargreaves and Deamer (36). These investigators showed that dodecyl sulfate is capable of forming liposomes if the equivalent of an acid soap is produced. How-ever, instead of titrating, dodecyl alcohol is added to introduce an uncharged group capable of forming hydrogen bonds with the dodecyl sulfate. In a typical prepara-tion, a 2:1 dodecyl alcohol:dodecyl sulfate mixture (50 mM total amphiphile in aqueous buffer) is sonicated briefly at 55°C. Examination of the mixture reveals vesicles in the size range 0.1-1.0 μm, and after a few hours, fusion events produce even larger vesicles which are remarkably stable. For instance, if carboxyfluores-cein is included in the mixture, the vesicles can be separated from the external phase by gel filtration and the dye remains encapsulated over a period of several weeks. The encapsulation efficiency of a 50 mM total amphiphile dispersion is about 40%.

ANCILLARY METHODS

A number of practical methods have evolved for working with liposome prepara-tions and monitoring their properties. These cannot be covered in any detail in a short chapter, but it seems appropriate to provide brief descriptions and refer-ences to specific methods which have proven useful in the laboratory. These can roughly be divided into methods involved in liposome preparation and methods used in analysis.

Preparation Methods

Gel Filtration

Most investigators are familiar with standard gel filtration columns that can be used to size liposomes and to exchange the external volume. One disadvantage of such methods is that they are often slow and always produce considerable dilution of the liposomes. Fry and co-workers (37) have published a method which does not dilute the sample and is much more rapid, especially if multiple samples must be run. In this, the desired gel is prepared in a plastic syringe tube. The tube is then placed in a conveniently sized centrifuge tube and centrifuged once at a low *g* force to remove all void volume solution. The liposome preparation is placed on top of the gel, followed by a second low-speed centrifugation. The undiluted liposomes appear in the bottom of the tube, leaving the original external phase entrapped in the internal gel volume at the top. It is important to determine empirically the time and *g* force of the low-speed centrifugation for the amount and type of gel used. The retention of the void volume is very dependent on the individual gel characteristics.

Sizing Liposomes

As mentioned earlier, it is convenient to use polycarbonate filters to size a heterogeneous liposome preparation. The liposome dispersion is placed in a glass syringe and forced through the filter under gentle pressure. The filter can range from 0.1 to 1.0 μm, depending on the final size desired. Simple centrifugation can be used to separate SUVs from MLVs, and as described by Huang (17), gel filtration can then size the SUVs to near homogeneity.

Concentrating Liposomes

Occasionally, it is necessary to concentrate liposomes. Sometimes vesicles can be pelleted by high-speed centrifugation but this will not always work with SUV and LUV preparations. An improved way to carry this out is to add a small amount of protamine or polylysine until an obvious flocculant precipitate has formed (38). This can be centrifuged to form a pellet. The pellet can be dispersed by the addition of heparin, which competes with the protamine for binding sites and releases the liposomes. This method is useful only if the liposomes have at least some small amount of net negatively charged lipid (such as 5% phosphatidic acid) and if the presence of small amounts of protamine and heparin do not affect desired vesicle properties.

Monitoring Liposome Properties

Light Microscopy

Sometimes liposomes are large enough to be viewed by phase microscopy (1-10 μm in diameter). However, depending on conditions, the liposomes may not have sufficient phase contrast to be seen readily. We have found that the addition of sucrose or dextran to such dispersions produces a phase contrast so that the vesicles appear phase bright against a dark background (36). An alternative procedure is to include a small amount (<0.5 mol%) of a fluorescent lipid analog and view the liposome preparation by fluorescence microscopy.

Electron Microscopy

Negative staining and freeze-fracture methods are most commonly applied to liposome preparations and the techniques are well known. We have found it useful in negative staining of unsaturated LUVs to prefix with OsO_4 before staining. This prevents the distortion often caused in larger vesicles by the staining process. We have also found that etching of freeze-fracture specimens can provide information about the fracture of vesicles that are unilamellar in a given preparation. In standard freeze-fracture preparations, the vesicles appear only if the fracture plane follows the membranes and cross fractures are not obvious. However, after 5 min of etching, cross-fractured vesicles are clearly seen and the number of lamellae can readily be determined (Fig. 5).

Captured Volume and Encapsulation Efficiency

For simple estimates of captured volume in high-ionic-strength media, liposomes can be prepared in 0.1 M sodium or potassium chromate. Following gel filtration to remove the external chromate, the trapped chromate can be determined by solubilizing an aliquot of the liposomes in 10 mM Triton X-100, followed by measuring A_{380} for chromate. A simple calculation then provides an estimate of captured volume as liters per mole of lipid. The ratio of chromate concentration in the aliquot (Triton added) to the original chromate concentration is divided by the actual lipid concentration of the aliquot. Thus, it is also necessary to determine the final lipid concentration in order to determine the captured volume. The same data can be used to calculate the encapsulation efficiency. The captured volume multiplied by the actual lipid concentration ($\times 100$) gives the encapsulation efficiency. For low-ionic-strength media, 6-carboxyfluorescein can be used as a marker (0.1 mM) and its fluorescence measured to determine the amount entrapped.

Liposome-Liposome Interactions

Since liposomes are dynamic structures, it is worthwhile considering some interactions between vesicles in suspension which might affect a liposome property under study. We conclude this chapter by considering briefly unimolecular exchange of lipid, liposome aggregation, and liposome-liposome fusion.

Unimolecular disassociation from one bilayer, diffusion, and reassociation

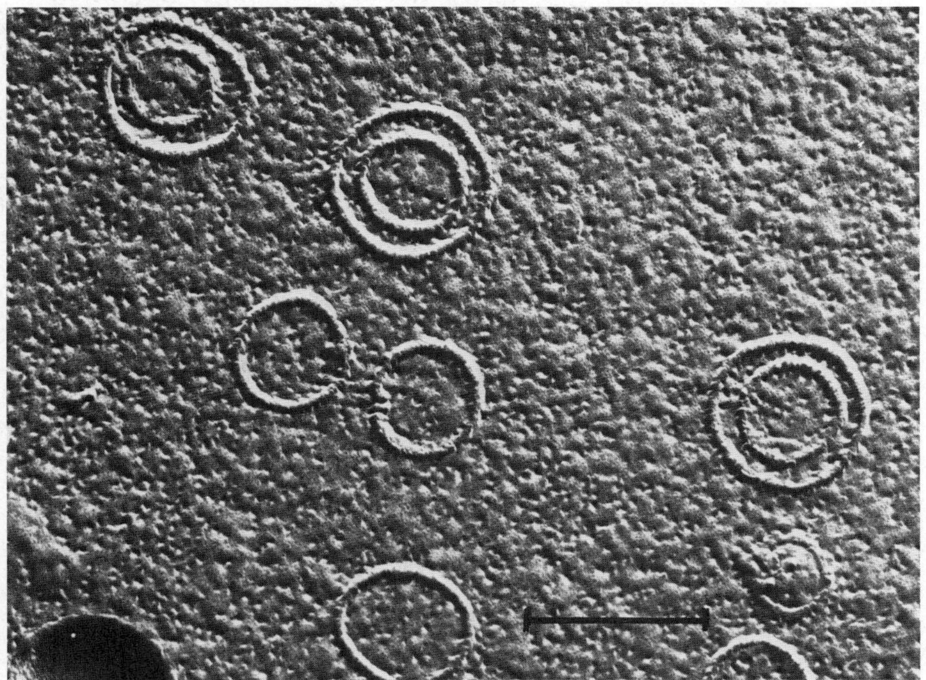

Figure 5 Freeze-etch image of cross-fractured liposomes. Phosphatidylserine liposomes were prepared by diethyl ether infusion method in 10 mM phosphate buffer, pH 7.5. Bar shows 0.2 μm.

can result in changes in vesicle composition. For instance, it has been shown that monomolecular phosphatidylcholine is in equilibrium with vesicular lipid (39,40). There is a vectorial addition of lipid from smaller to larger vesicles, or from short acyl chain vesicles to longer acyl chain pure phosphatidylcholine vesicles. It has also been shown that monomolecular lipid exchange occurs between fluorescently labeled phospholipids and the plasma membrane of cells in culture (41).

The detection of unimolecular exchange is not a trivial procedure. The investigators cited above used differential scanning calorimetry and fluorescence spectroscopy, respectively. With the advent of a variety of synthetic phospholipids that are commercially available, differential scanning calorimetry has become more practical. Other techniques using radiolabeled or fluorescent derivatives of lipids have the inherent problem of separating the labeled fraction from the remainder of the lipid. If it is not possible to use changes in the monomer/ excimer ratio of a labeled analog (40), our experience suggests that monitoring lipid exchange between SUVs or LUVs and MLVs is currently the strongest approach, since it is relatively easy to separate MLVs from the others by centrifugation or filtration.

Aggregation is usually prevented by imparting a net charge on the bilayer and including chelating agents in the buffer to remove trace polyvalent ion effects. For some experimental protocols it is not possible to include these safeguards and it may therefore be necessary to determine if significant liposome-liposome aggregation is occurring during the time scale of the experiment. Aggregation can be monitored by following light-scattering changes at A_{500}. Light-scattering increases also include those due to vesicle size increases from monomolecular diffusion and liposome fusion. The latter possibilities can be excluded by testing for exchange diffusion and vesicle fusion with assays described in this section. Alternatively, a fluorescence resonance energy transfer assay has been suggested which follows vesicle aggregation (42).

The third possible liposome-liposome interaction that we will consider is that of fusion. Two methods are now available which follow the mixing of the entrapped aqueous compartments or mixing of the lipid components. The first assay was developed by Wilschut and Papahadjopoulos (43) and monitors the formation of a complex between the lanthanide ion, terbium, and dipicolinate. This complex is several orders of magnitude greater in fluorescence than the intrinsic fluorescence of terbium. Terbium citrate is trapped in one set of liposomes and sodium dipicolinate is trapped in a second population. The liposomes are then mixed and the rate and extent of fusion is followed by the increase in fluorescence.

An alternative method is to monitor quantitatively the mixing of two fluorescent lipid analogs originally in separate bilayers. This recently developed assay for vesicle fusion (44) is based on the transfer of fluorescence resonance energy from donor probe to acceptor probes in the same bilayer. If a fusion event occurs between a donor-labeled liposome and acceptor-labeled liposome, the proximity of both probes in the same bilayer causes a change in the fluorescence emission spectrum. The extent of probe intermixing can be determined quantitatively by comparing the experimental liposomes against a "mock-fused" standard. Both the aqueous compartment assay and the lipid compartment assay are adaptable to a wide range of experimental conditions and both assays are relatively insensitive to aggregation-induced artifacts.

REFERENCES

1. Szoka, F. and D. Papahadjopoulos. 1980. Comparative properties and methods of preparation of lipid vesicles (liposomes). Annu. Rev. Biophys. Bioeng. 9:467.

2. Szoka, F. and D. Papahadjopoulos. 1981. Liposomes: preparation and chatacteristics. In *Liposomes: From Physical Structure to Therapeutic Applications,* G. Knight (Ed.). Elsevier/North Holland Biomedical Press, Amsterdam, pp. 51–82.

3. Bangham, A. D., M. M. Standish, and J. C. Watkins. 1965. Diffusion of univalent ions across the lamellae of swollen phospholipids. J. Mol. Biol. 13:238.

4. Deamer, D. W. 1978. Preparation and properties of ether-injection liposomes. Ann. N.Y. Acad. Sci. 308:250.

5. Kagawa, Y. and E. Racker. 1971. Partial resolution of the enzymes catalysing oxidative phosphorylation. XXV. Reconstitution of vesicles catalysing $^{32}P_i$-adenosine triphosphate exchange. J. Biol. Chem. 246:5477.

6. Enoch, H. G. and P. Strittmatter. 1979. Formation and properties of 1000-A-diameter, single-bilayer phospholipid vesicles. Proc. Natl. Acad. Sci. USA 76:145.

7. Gerritsen, W. J., A. J. Verkleij, R. F. Zwaal, and L. L. Van Deenen. 1978. Freeze-fracture appearance and disposition of band 3 protein from the human erythrocyte membrane in lipid vesicles. Eur. J. Biochem. 85:255.

8. Peterson, S. W., S. Hanna, and D. W. Deamer. 1978. Comparative studies of detergent effects on the calcium adenosine triphosphatase of sarcoplasmic reticulum. Arch. Biochem. Biophys. 191:224.

9. Mimms, L. T., G. Zampighi, Y. Nozaki, C. Tanford, and J. A. Reynolds. 1981. Phospholipid vesicle formation and transmembrane protein incorporation using octyl glucoside. Biochemistry 20:833.

10. Klein, R. A. 1970. The detection of oxidation in liposomes preparations. Biochim. Biophys. Acta 210:486.

11. Kates, M., B. Palameta, C. N. Joo, D. J. Kushner, and N. E. Gibbons. 1966. Aliphatic diether analogs of glyceride-derived lipids. IV. The occurrence of di-O-dihydrophytylglycerol ether containing lipids in an extreme halophilic bacteria. Biochemistry 5:4092.

12. Saunders, L., J. Perrin, and D. B. Gammack. 1962. Ultrasonic irradiation of phospholipid sols. J. Pharm. Pharmacol. 14:567.

13. Abramson, M. B., R. Katzmann, and H. P. Gregor. 1964. Aqueous dispersions of phosphatidylserine. J. Biol. Chem. 239:70.

14. Papahadjopoulos, D. and N. Miller. 1967. Phospholipid model membranes. I. Structural characteristics of hydrated liquid crystals. Biochim. Biophys. Acta 135:624.

15.   Papahadjopoulos, D. and J. C. Watkins. 1967. Phospholipid model membranes: permeability properties of hydrated liquid crystals. Biochim. Biophys. Acta 135:539.

16.   Johnson, S. M., A. D. Bangham, M. W. Hill, and E. D. Korn. 1971. Single bilayer liposomes. Biochim. Biophys. Acta 233:820.

17.   Huang, C. H. 1969. Studies on phosphatidylcholine vesicles: formation and physical characteristics. Biochemistry 8:344.

18.   Batzri, S. and E. D. Korn. 1973. Single bilayer liposomes prepared without sonication. Biochim. Biophys. Acta 298:1015.

19.   Barenholz, Y., S. Amselem, and D. Lichtenberg. 1979. A new method for preparation of phospholipid vesicles. FEBS Lett. 99:210.

20.   Fendler, J. H. 1980. Surfactant vesicles as membrane mimetic agents: characterization and utilization. Acc. Chem. Res. 13:7.

21.   Reeves, J. P. and R. M. Dowben. 1969. Formation and properties of thin-walled phospholipid vesicles. J. Cell. Physiol. 73:49.

22.   Miller, R. C. 1980. Do "lipidic particles" represent intermembrane attachment sites? Nature 287:166.

23.   Deamer, D. W. and A. D. Bangham. 1976. Large volume liposomes by an ether vaporization method. Biochim. Biophys. Acat 443:629.

24.   Schieren, H., S. Rudolph, M. Finkelstein, P. Colemen, and G. Weissman. 1978. Comparison of large unilamellar vesicles prepared by a petroleum ether vaporization method and multilamellar vesicles. Biochim. Biophys. Acta 542:137.

25.   Szoka, F. and D. Papahadjopoulos. 1978. A new procedure for preparation of liposomes with large internal aqueous space and high capture by reverse phase evaporation. Proc. Natl. Acad. Sci. USA 75:4194.

26.   Milsmann, M. H. W., R. A. Schwendener, and H. G. Weder. 1978. The preparation of large single bilayer liposomes by a fast and controlled dialysis. Biochim. Biophys. Acta 512:147.

27.   Zumbuehl, O. and H. G. Weder. 1981. Liposomes of controllable size in the range of 40 to 180 nm by defined dialysis of lipid/detergent mixed micelles. Biochim. Biophys. Acta 640:252.

28.   Schwendener, R., M. Asanger, and H. G. Weder. 1981. n-Alkyl-glucosides as detergents for the preparation of highly homogeneous bilayer liposomes of variable sizes (60–240 nm) applying defined rates of detergent removal by dialysis. Biochem. Biophys. Res. Commun. 100:1055.

29.   Papahadjopoulos, D., W. J. Vail, K. Jacobsen, and G. Poste. 1975. Cocleate lipid cylinders: formation by fusion of unilamellar lipid vesicles. Biochim. Biophys. Acta 394:483.

30.   Taber, R., T. Wilson, and D. Papahadjopoulos. 1978. The encapsulation of pico-RNA viruses by lipid vesicles: physical and biological properties. Ann. N.Y. Acad. Sci. 308:268.

31.   Mannino, R. J., E. S. Allebach, and W. A. Strehl. 1979. Encapsulation of high molecular weight DNA in large, unilamellar phospholipid vesicles. FEBS Lett. 101:229.

32. Papahadjopoulos, D. and W. J. Vail. 1978. Incorporation of macromole-
     cules within large unilamellar vesicles. Ann. N.Y. Acad. Sci. 308:259.
33. Kasahara, M. and P. C. Hinkle. 1977. Reconstitution and purification of
     the D-glucose transporter from human erythrocytes. J. Biol. Chem. 252:
     7384.
34. Pick, U. 1981. Liposomes with a large trapping capacity prepared by
     freezing and thawing of sonicated phospholipid mixtures. Arch. Biochem.
     Biophys. (in press).
35. Gebicki, J. M. and M. Hicks. 1973. Ufasomes are stable particles sur-
     rounded by unsaturated fatty acid membranes. Nature 243:232.
36. Hargreaves, W. R. and D. W. Deamer. 1978. Liposomes from ionic, single
     chain amphiphiles. Biochemistry 17:3759.
37. Fry, D. W., C. White, and D. J. Goldman. 1978. Rapid separation of low
     molecular weight solutes from liposomes without dilution. Anal. Biochem.
     90:809.
38. Gunter, K. K., T. E. Gunter, A. Jarkowski, and R. N. Rosier. 1981. A
     method for resuspending small vesicles separated from suspension by pro-
     tamine aggregation. Biophys. J. 33:187a.
39. Martin, F. J. and R. C. MacDonald. 1976. Phospholipid exchange between
     bilayer membrane vesicles. Biochemistry 15:321.
40. Roseman, M. A. and T. E. Thompson. 1980. Mechanism of the spontane-
     ous transfer of phospholipids between bilayers. Biochemistry 19:439.
41. Struck, D. K. and R. E. Pagano. 1980. Insertion of fluorescent phospho-
     lipids into the plasma membrane of a mammalian cell. J. Biol. Chem. 255:
     5404.
42. Gibson, G. A. and L. M. Loew. 1979. Phospholipid vesicle fusion moni-
     tored by fluorescence energy transfer. Biochim. Biophys. Res. Commun.
     281:141.
43. Wilschut, J. and D. Papahadjopoulos. 1979. Ca^{82+9-} Induced fusion of
     phospholipid vesicles monitored by mixing of aqueous contents. Nature
     281:690.
44. Uster, P. S. and D. W. Deamer. 1981. Fusion competence of phospha-
     tidylserine-containing liposomes quantitatively measured by a fluorescence
     resonance energy transfer assay. Arch. Biochem. Biophys. 209:385.

Interactions of Proteins and Drugs with Liposomes

R. L. Juliano / The University of Texas Medical School at Houston,
Houston, Texas

INTRODUCTION

Liposomes have attracted a great deal of scientific interest in two quite distinct
roles. First, liposomes can provide an excellent model for cell membranes, allowing exploration of questions concerning membrane lipid chemistry, lipid-protein
interactions, transport phenomena, and ligand binding to membranes (1). Second,
liposomes are being developed as a controlled delivery system for therapeutic
agents (2). In both roles questions concerning interactions of lipids with proteins
and with drugs or other small molecules are of great concern. This review surveys
recent studies (since 1977) on drug and protein interactions with liposomes. It
deals briefly with general concepts of lipid-protein interactions and then examines
the interactions of liposomes with particular groups of proteins, including serum
lipoproteins, blood coagulation components, other serum components, toxins,
and lectins. The review does not deal with other aspects of lipid-protein interaction such as phospholipid exchange proteins (3), actions of immunoglobulins
or complement on liposomes (reviewed by Alving and Richards, Chap. 6), or reconstitution of membrane enzymes and transport systems into liposomes [re-
viewed by Malathi, Chap. 4, and by Racker (4)]. This review also surveys the
interaction of drugs with liposomes, primarily from the point of view of develop-

ing successful approaches for the stable incorporation of various drugs. However, some attention will be given to the use of liposomes as a model for understanding drug-membrane interactions. This review deals with clinically active drugs and will not consider substances such as the polypeptide ionophores which are used primarily for research purposes [reviewed by Shamoo (5)].

PROTEIN-LIPOSOME INTERACTIONS

Basic Aspects

Proteins in Biological Membranes

As a prelude to examination of the details of protein-liposome interactions it would be valuable to review some of the basic concepts of how proteins are organized in natural membranes. The basic paradigm for membrane organization remains the fluid mosaic model (6). In this model the lipid bilayer serves as a viscous matrix for membrane proteins, some of which span the bilayer. Secondary interactions among proteins embedded in the lipid, the intrinsic proteins, and between the intrinsic proteins and nonembedded or peripheral proteins, control the dynamic aspects of membrane behavior (7,8). The lipid matrix serves to modulate the lateral mobility, conformation, and function of the intrinsic membrane proteins, while the proteins, in turn, can affect the basic biophysical parameters of the lipid. Modern concepts of lipid-protein interactions in membranes have been reviewed in detail by Lenaz (9) and Quinn and Chapman (10).

There has recently been a lively controversy concerning the mechanisms for protein assembly into membranes; this controversy bears on some very basic questions of lipid-protein interaction and is worth commenting on here. One model of membrane protein assembly, the "signal hypothesis" (7,11), suggests that intrinsic proteins are assembled into the membrane during protein translation and that distinct "signal" sequences of amino acids are responsible, first for the initial binding of the nascent polypeptide to the membrane (this sequence is later cleaved off) and second for locking the completed polypeptide chain into the membrane environment. The competing model, the "trigger hypothesis" (12), suggests that membrane proteins are synthesized initially in soluble form, undergo a conformational change on encountering a membrane, and then insert themselves into the membrane environment.

While the arrangement of some membrane proteins, such as erythrocyte glycophorin (13) and certain viral spike glycoproteins (14), are quite consistent with the signal hypothesis, the transmembrane organization of other proteins is difficult to explain in this way. For example, certain membrane transport proteins, such as erythrocyte band III (15) and bacterial rhodopsin (12), have polypeptide chains which pass through the membrane several times, and in addition, band III has its amino terminal end on the cytoplasmic side of the membrane; these facts are difficult to explain in terms of a simple signal hypothesis model. As we will see, it is clear that many proteins (not only intrinsic ones) have the capacity to bind to lipid bilayers, alter conformation, and penetrate the lipid matrix, thus conforming to the model suggested by the "membrane trigger" hypothesis.

Perhaps the major point of this discussion is that there is currently an on-going dialogue between workers who study membrane protein synthesis in organisms and those who study lipid-protein interaction in purified, artificial membrane systems, with the resulting exchange of ideas cross-fertilizing both fields.

Biophysical Aspects of Bilayers That Affect Protein-Lipid Interaction

In this section we briefly consider some aspects of the biophysics of lipids which are pertinent to protein-liposome interaction. These topics have been reviewed in far greater detail by Quinn and Chapman (10) and Lenaz (9).

Phase Transitions   Lipids dispersed in water can form a variety of structures; however, at low lipid/water ratios the liposome-type structure, that is, a closed bimolecular lipid membrane, is predominant. As the temperature of the system is increased, bilayer membranes composed of pure phospholipids undergo a discrete order-disorder transition involving primarily an increase in the rotational freedom of the fatty acid side chains and an increase in the area per lipid molecule in the bilayer. The temperature at which the lipid shifts from a condensed gellike state to a more fluid, expanded state is called the critical temperature. This temperature is a function of the chemical properties of the lipid and of the nature of the surrounding environment. For example, the critical temperature for a homologous series of phosphatidylcholines increases as the fatty acid chain length increases; also the critical temperature is higher for saturated lipids than for homologous unsaturated lipids. The nature of the polar head group is also important; thus, the critical temperature for phosphatidylethanolamines is about 20°C higher than for the homologous phosphatidylcholines.

Cholesterol has an important modulatory effect on the phospholipid phase changes. The sterol interacts strongly with the phospholipids and the result of this interaction is to keep the phospholipid in an "intermediate fluid" condition. Thus, at points where the lipid would normally be above its critical temperature, the presence of cholesterol tends to increase the packing and rigidity of the bilayer, while at points where the lipid would be below its critical temperature, cholesterol expands and fluidizes the bilayer (10,16).

Phase Separation, Domains   When phospholipids of different transition temperatures are mixed, there are several possible consequences in terms of phase-transition behavior. Homogenous saturated phospholipids, different by only two carbons in fatty acid chain length, will display a broad transition which can be described in terms of a classic phase diagram. On the other hand, when two phospholipids of very different properties are mixed, discrete phase transitions representing the individual components are observed (7,9,10,17). This implies that the two phospholipid species must sort out or undergo a "phase separation" in the plane of the membrane. A similar event transpires under isothermal conditions when mixtures of a neutral lipid such as a phosphatidylcholine and an anionic lipid such as phosphatidylserine are exposed to Ca^{2+}; this results in a phase separation giving rise to "domains" of gel-state phosphatidylserine–Ca^{2+} complexes intermixed with fluid domains of phosphatidylcholine (18). The possible existence of fluid and solid domains in bilayer membranes has a number of important consequences for protein-liposome interaction, for drug permeation through liposome membranes, and for the general stability of lipid vesicles.

General Aspects of Protein-Liposome Interaction

Modes of Interaction   Interactions of lipid and protein in membranes have been broadly reviewed elsewhere (9,10), while Kimelberg has reviewed the specific topic of protein-liposome interaction (19). The work of Papahadjopoulos, Kimelberg, and their colleagues has suggested that there are three major modes of protein-liposome interaction (19-22). The first type of interaction, exemplified by the soluble protein ribonuclease or by poly-L-lysine, seems to entail simple surface binding due to attraction of unlike charges. This type of interaction is abrogated by raising the ionic strength and has little impact on the phase-transition temperature of the phospholipid bilayer. The second type of interaction is exemplified by the peripheral proteins cytochrome *c* and myelin basic protein. Once again charge interactions are involved in the initial binding but, in addition, penetration of the protein into the bilayer occurs, causing expansion, decrease of the critical temperature of the phase transition, and alteration in bilayer permeability properties. This type of interaction can cause major changes in the properties of the entire bilayer. A third mode of interaction is exemplified by myelin proteolipid, an intrinsic membrane protein. In this case, the interaction is governed by nonionic forces and increases in salt concentration do not prevent it. Insertion of the proteolipid into the bilayer occurs, but this does not produce major changes in the thermotropic phase transition although liposome permeability is increased. It was suggested that the myelin proteolipid, although penetrating the bilayer, interacts only with a "halo" of lipid in its immediate vicinity and does not disrupt the bilayer. The concept of halo or "boundary" lipid surrounding intrinsic membrane proteins is reviewed elsewhere (23).

Factors Affecting Interaction   A large number of factors govern the ability of proteins and liposomes to interact. Perhaps the simplest is surface charge. Many types of proteins interact most strongly with lipid vesicles if the vesicle and protein bear opposite charges; thus, cytochrome *c* (positive) interacts with phosphatidylserine liposomes (negative) but not with neutral liposomes (19).

For proteins that are negatively charged under physiological conditions, reducing the pH to below their isoelectric point tends to promote liposome interactions (24). However, one should also bear in mind that pH changes can alter the conformation of proteins, thus perhaps exposing regions suitable for hydrophobic interaction with lipids.

A second factor affecting protein interactions is the fluidity and packing of the bilayer. It is well known from surface balance studies of monolayer films that proteins most readily penetrate when the film pressure is low (10,25). The fluid-to-solid phase transition increases the packing of lipid molecules, thus tending to exclude protein from penetrating the lipid film. Similarly, the addition of cholesterol to monolayer or bilayer membranes (above the critical temperature) has a condensing effect and tends to retard protein penetration (19,22). We shall see that the inclusion of cholesterol is a very important factor for obtaining liposome formulations which are stable in the biological environment and which are suitable as drug carriers. For example, the inclusion of cholesterol in liposomes largely prevents the increase in transmembrane permeability caused by the interaction of the liposomes with soluble or peripheral membrane proteins (22). Interestingly, however, the permeability changes caused by myelin proteolipid, or by intrinsic proteins, are only minimally affected by cholesterol (19).

Another aspect of protein lipid-interactions and fluid-solid phase transitions concerns the tendency of proteins to undergo lateral segregation in the membrane during such transitions (9,26,27). Intrinsic membrane proteins seem to prefer to reside in fluid membrane domains; thus, when mixed solid-fluid domains exist (as during the phase transition) the proteins are markedly enriched in the fluid regions.

Consequences of Protein-Liposome Interaction   The binding and penetration of protein into the lipid bilayer can markedly alter the biophysical properties of liposomes (9,10,19,21). Observed changes included expansion and fluidization of the bilayer, alteration of the enthalpy and critical temperature of the phase transition, and increases in permeability to ions and small molecules. These effects are largely antagonized by the presence of a substantial (30-50 mol%) amount of cholesterol in the bilayer, or by keeping the lipid below its gel-fluid phase transition.

Rationale for Effects Observed in Protein-Liposome Interactions   The observations on lipid-protein interaction in liposomes can at least partially be rationalized in a fairly simple way. Certain soluble proteins (e.g., ribonuclease) have a very tight, compact structure (perhaps stabilized by S-S bonds). These proteins can bind to lipid surfaces via charge interactions but are incapable of undergoing conformational changes to expose hydrophobic regions for further interaction. Other soluble proteins (albumin) and peripheral membrane proteins (spectrin, cytochrome *c*, myelin basic protein) initially bind to lipid surfaces by charge interactions, but then undergo a denaturation event, unfold and expose hydrophobic regions which can interact extensively with fatty acid side chains, and cause major changes in bilayer organization, phase transitions, and permeability. Intrinsic membrane proteins (e.g., myelin proteolipid) are designed by nature to fit into a hydrophobic environment. They can insert themselves into lipid membranes, interact with a "halo" of bound lipid and leave the overall structure of the bilayer relatively unperturbed. It seems likely that some proteins may be able to undergo a conformational change from a soluble form to a lipophilic or intrinsic membrane form (12) and insert into membranes, or even through membranes, without causing gross alteration of structure. The interaction of intrinsic membrane proteins such as bacterial rhodopsin (28) and M13 phage coat protein (29) with lipid bilayers, and effects of cholesterol, lipid fluidity, and phase separation on such interactions are now starting to be examined using very sophisticated biophysical techniques. Further refinement of our basic concepts on lipid-protein interaction should emerge from these studies and others of equal sophistication (30-32).

Lipid-Protein Reconstitution   The subject of reconstituting functional membrane enzyme and transport systems from membrane proteins and bilayer vesicles is really beyond the scope of this chapter and has been reviewed elsewhere (9,10, 19; Malathi, Chap. 4). However, it seems worthwhile mentioning a few aspects of this interesting area.

The most successful approach to reconstitution of functional membrane systems was devised some time ago by Racker and his colleagues and involves the use of detergents to disperse both lipid and protein components followed by removal of the detergent allowing reassociation of the membrane (33). This general approach has now achieved a high level of sophistication in terms of reconstitution of membrane enzymes, transport systems, pharmacological receptors, and antigens; a few examples chosen out of many excellent studies are cited here (34-38).

An interesting finding coming out of reconstitution studies is that functional membrane proteins (enzymes) have requirements not only for a lipid milieu, but, in some cases, may require specific lipids (9,10). For example, the membrane (Na^+,K^+)-ATPase requires anionic lipids such as phosphatidylserine (39). By contrast, other enzymes, such as mitochondrial β-hydroxybuterate dehydrogenase (40) and pancreatic signal peptidase (41), seem to specifically require phosphatidylcholines.

Specific Cases of Protein-Liposome Interaction

Serum Lipoproteins

It has been known for some time that serum lipoproteins and their apolipoprotein subunits can interact with phospholipid vesicles. Much of the earlier work

in this area has been reviewed by Morisett et al. (42); some of the major findings
of these studies include the following:

1. Small unilamellar vesicles interact more extensively with lipoproteins or apo-
    lipoproteins than do large multilamellar vesicles of the same composition.
2. For homogenous phosphatidylchloines [e.g., dimyristoylphosphatidylcholine
    (DMPC)] interaction occurs extensively above the critical temperature when
    the bilayer membrane is in a fluid state. For lipids with heterogeneity in fatty
    acid composition [e.g., egg phosphatidylcholine (PC)] interaction is weaker
    than with homogenous synthetic phospholipids, even though both lipids are
    in a fluid state.
3. Different lipoprotein subunits have markedly different energetics for inter-
    action with vesicles.
4. Vesicle-apoprotein interactions result in alterations of vesicle bilayer proper-
    ties, including decreased motional freedom of fatty acid side chains, changes
    in packing, increases in permeability, and eventually destruction of the vesicles
    and formation of new apolipoprotein-lipid complexes (these are often discoidal
    in structure by electron microscopic observation). Transfer of lipid from the
    vesicle to lipoprotein can also take place.
5. Apolipoproteins undergo conformational changes upon binding to vesicles;
    this can be studied by spectroscopic techniques. Certain regions of apopro-
    teins (e.g., carboxyl terminal of apoprotein CIII) interact preferentially with
    the lipid vesicles.

More recently a number of investigators have studied liposome-lipoprotein
interaction in great detail. For example, the structures formed by interaction of
apoprotein CIII (43), high-density lipoprotein (HDL), or apoprotein AI (44,45)
with liposomes have been characterized by ultracentrifugation, column chromatography, and electron microscopy.

In a paper that caused considerable concern among those interested in liposomal drug delivery systems, Scherphof et al. (46) showed that egg PC vesicles
of multilamellar or unilamellar structure, when incubated with plasma or purified
lipoprotein, could transfer lipids to the lipoprotein and that the structure of the
vesicles could be disrupted. More recently, it has become clear that lipoproteins
can disrupt certain types of vesicles, but may not affect others. Thus, the presence of cholesterol has a marked stabilizing effect on vesicle structure. Guo et al.
(47), working with small unilamellar egg PC vesicles, have shown that the contents of the vesicle can be released via interaction with serum lipoproteins. The
order of potency is apolipoprotein > lipoprotein > serum. However, if cholesterol
is included in the vesicles, the ability of the apolipoproteins to interact and to release the vesicle contents is markedly diminished and is completely inhibited
above 37 mol% cholesterol. In a similar vein, Tall (48) has shown that the inclu-

sion of cholesterol inhibits release of [^{3}H] inulin from egg PC unilamellar vesicles in vivo.

Recent studies on the role of phospholipid composition and phase transition behavior in liposome-lipoprotein interactions have given some novel insights into the phenomenon. Scherphof et al. (49), working with whole plasma, and Swaney (50) and Van Tornout et al. (51), working with apoprotein AI and apoprotein AII, have demonstrated that lipoproteins interact best with multilamellar vesicles at the onset of the gel-fluid phase transition. The lipoprotein seems to require defects or discontinuities in the bilayer membrane in order to penetrate and to affect vesicle structure. This observation rationalizes previous findings that unilamellar vesicles interact more readily with lipoproteins than multilamellar vesicles; thus, the radius of curvature of the small unilamellar vesicles probably introduces defects in packing into the bilayer structure and potentiates interactions with lipoproteins. The new observation also rationalizes the role of cholesterol, which (as discussed above) tends to increase packing and to minimize discontinuities in the bilayer.

Recently, a good deal of attention has been paid to the interaction of model peptides with lipid bilayers. Segrest and his colleagues (52) have extended their studies of "amphipathic helix"-type peptides and have explored the role of charged and hydrophobic residues in promoting protein-lipid interactions. Gotto and his colleagues (53) have also extensively studied the interactions between model apolipoprotein peptides and lipids and have determined critical lengths and hydrophobicity for amphipathic helix-type structures. In addition to the amphipathic helix, other structures for lipophilic regions of proteins, including β-pleated sheats and the "β helix," have been proposed [reviewed by Kennedy (54)] .

Liposome Interaction with Lectins and Toxins

Lectins bind to oligosaccharide residues and thus may interact with glycolipids or glycoproteins embedded in membranes. Some toxins, such as cholera toxin, also interact with carbohydrate residues, whereas others, such as diphtheria toxin and mellitin, seem to interact with phospholipids directly. Therefore, study of the interaction of toxins and lectins with liposomes provides a valuable, easily manipulated model system for understanding the behavior of these important macromolecules in biological systems. Generally, lectins bind to cell surfaces, causing receptor redistribution and cell agglutination and are then subsequently internalized (55). Toxins of the type epitomized by diphtheria toxin and cholera toxin are composed of two types of subunits; one subunit (usually termed B) binds to cell surfaces and seems to assist in the translocation of the second, enzymatically active subunit (termed A) across the membrane. The second subunit perturbs important biological processes, (protein synthesis, adenylate cyclase), usually via adenosine diphosphate (ADP) ribosylation of critical components of those processes (56). Toxins of the type epitomized by bee mellitin have direct membrane-perturbing effects, leading to increased transmembrane fluxes and to cell lysis (57). In this section we deal with (a) the factors governing the binding of lectins and toxins to lipid vesicles, (b) the mechanisms of transmembrane penetration of the enzymatically active subunit of diphtheria-toxin-type molecules, and (c) the nature of the membrane perturbing effects of mellitin, colicins, and other toxins of this type.

Lectin-Binding Characteristics The binding of lectins to liposomes containing incorporated glycolipids (58) or glycoproteins (59,60) was first studied in a rather elementary way several years ago, usually by observing lectin-mediated agglutination of these particles. In addition, Juliano and Stamp (60) demonstrated that lectins could be used to bind glycoprotein containing liposomes to cells.

More recently, binding studies between lectins and vesicles have grown much more sophisticated. For example, Grant and his colleagues, using liposomes containing erythrocyte glycophorin (a receptor for wheat germ agglutinin and for M/N antibodies) and/or erythrocyte band III (a receptor for concanavalin A) have studied the nature of cooperativity in lectin binding (61). They find that highly cooperative behavior of lectin binding and agglutination can occur in a simple liposome-glycoprotein system, indicating that it is not necessary to postulate a role for complex cytoskeletal actions in explaining cooperativity. In addition, they have examined interactions between lectin receptor (band III) and antibody binding site (glycophorin) within the plane of the membrane, demonstrating that the presence of band III can interfere with agglutination mediated via anti-M/N antibody and glycophorin (62). These studies delineate some of the complexities of studying ligand binding to macromolecular receptors in membranes.

Recent studies on the lectin-mediated agglutination of glycolipid containing liposomes have shown that physical factors such as haptene density, haptene exposure, membrane fluidity, and liposome charge can be critical determinants of the aggregation process. For example, small unilamellar liposomes containing a mannosyl-terminated glycolipid were readily agglutinated by concanavalin A but only when the mole fraction of glycolipid was above 5%; thus, a critical density of membrane-bound haptene is necessary for aggregation to occur (63a). In another study (63b), the same group emphasized the importance of appropriate spacing between the bilayer and the haptenic groups for lectin-induced aggregation of liposomes to take place. In other investigations (64) both threshold and charge effects were evident. Concanavalin A was able to agglutinate liposomes containing a glucosyl lipid and bearing a negative charge but not similar liposomes which were uncharged. Ionic strength and divalent cation effects on lectin-vesicle interaction were also noted. The study of Curatalo et al. (65) demonstrated that the closeness of the haptene group to the bilayer surface is an important determinant of interaction. *Ricinus communis* agglutinin could aggregate small liposomes containing lactosylceramide, but not those containing galactosyl ceramide, although

both have an appropriate haptene residue (galactose); presumably the presence of
a subterminal glucose residue in the lactosyl ceramide acts as a "spacer arm" and
enhances the availability of the terminal galactose residue.

Most available evidence indicates that lectins bind to oligosaccharide-contain-
ing liposomes but do not perturb the bilayer or disrupt liposome structure. Lectin binding does not cause changes in the release of entrapped contents from
liposomes (60) since intact liposomes can be recovered from lectin-liposome aggregates by treatment with haptene sugars (65) (i.e., no fusion has occurred). Finally,
lectin binding does not perturb bilayer fluidity as measured by fluorescence polarization of diphenyl hexatriene (66). However, one should note that studies in
lipid monolayers have suggested that concanavalin A alone can penetrate lipid
films (67).

Binding and Penetration of Subunit Toxins   In cells, the action of subunit
toxins seems to involve three distinct steps (56). The first is binding via the B
subunit to the cell surface. The next step seems to involve the insertion into the
membrane of a hydrophobic segment of the B subunit and a consequent penetration of the membrane by the A subunit. Finally, the enzymatically active A subunit catalyzes some intracytoplasmic event which leads to cytotoxicity. Lipid
membranes have been exceedingly useful in studying the first two steps of toxin
action.

For example, lipid vesicle studies have been used to clarify the nature of the
membrane-binding site for toxins. Thus, cholera toxin can bind to and agglutinate liposomes containing ganglioside GM_1, its putative membrane receptor, but
does not agglutinate liposomes containing GM_2, GO_{1a}, or GM_3 (68). Tetanus
toxin has also been shown to interact with ganglioside-containing membranes,
but the precise identity of the glycolipid involved is not clear (69). In contrast
to the requirement for oligosaccharides found with cholera and tetanus, diphtheria
toxin seems to bind directly to the phospholipids themselves (70). Recent evidence suggests that diphtheria toxin binds specifically to phospholipids containing free phosphate groups (e.g., cardiolipin, phosphatidylinositol phosphate).
The interaction with the toxin, however, is also modulated by the nature of the
other adjacent phospholipids, which do not directly bind to the ligand (71).

In contrast to the case of lectins, the binding of protein toxins to lipid membranes causes profound effects on membrane permeability and organization.
Thus, choleragen caused the release of entrapped glucose from ganglioside liposomes, as did the B protomer; the A protomer was ineffective in this regard. Interestingly, neither choleragen nor the B protomer released glucose from ganglio-
side-free liposomes, indicating that a conformational change in the B subunit
probably takes place subsequent to binding to ganglioside residues on the vesicle
surface (72). Several groups have reported that diphtheria toxin can induce ion-
conducting channels in bilayer membranes (73,74). The establishment of such

channels presumably indicates penetration of part of the molecule into and through the bilayer. Binding of the diphtheria toxin to the bilayer is accentuated by low pH, but penetration requires both low pH and a transmembrane potential difference (73). These observations have a number of interesting implications, not only for toxin action but also for the general question of the assembly of proteins into membranes.

Membrane-Active Toxins   There is a large and diverse group of animal, plant, and microbial toxins whose major action seems to be directly on the stability and permeation properties of lipid membranes. In many cases, however, the precise details of toxin action remains obscure or tentative.

Mellitin is a toxic polypeptide (26 amino acids) which is an abundant component of bee venom. Mellitin is surface active and is strongly hemolytic, suggesting a direct action on membranes (75). In water, mellitin is in an extended configuration, whereas in a membrane-type environment the α-helical character increases (76). Mellitin has dramatic effects on planar lipid membranes, rapidly causing disruption of the film (77). It has been postulated that mellitin is composed of hydrophobic and amphiphilic domains and that it inserts into membrane in a wedgelike manner (77). When mellitin interacts with dimyristoyl phosphatidylcholine liposomes, the gel-fluid phase transition of the bulk lipid is broadened and shifted to lower temperatures (22.5°C to 17°C) (78). In addition, a minor transition is observed at 29°C and seems to be associated with the melting behavior of a halo of boundary lipid ($\sim$7 mol per mole of mellitin) which surrounds the hydrophobic segment of the polypeptide. Thus, the toxic and lytic actions of mellitin are probably readily understood in terms of the molecule's strong interaction with lipids and its tendency to insert into and alter the geometry of bilayer membranes.

Another toxin which interacts strongly with lipids is that from the sea anemone *Stoichatis heliantus*. This component (probably a 17,500-dalton polypeptide) binds most strongly to sphingomyelin (79). The toxin can release entrapped markers such as $^{86}Rb^+$ from the sphingomyelin-containing liposomes but is less effective with liposomes composed of phosphatidylcholines (80). In planar bilayer membranes the toxin forms channels which primarily conduct monovalent cations. The conduction characteristics are voltage and polarity dependent, indicating that insertion of the toxin into the membrane is dependent on an appropriate transmembrane potential (81). This is reminiscent of the case with diphtheria toxin described above.

Liposomes have also been used to study the mechanism of action of streptolysin O (82). This toxin preferentially binds cholesterol and can release trapped markers from cholesterol-containing small unilamellar vesicles. Interestingly, the toxin fails to promote release from large multilamellar vesicles, suggesting that the packing of the vesicle membrane is an important determinant of toxin penetration. Streptolysin O is not a lipase (82).

Some bacterial colicins, particularly those termed K, Ia, and E_1, are thought to exert their toxic effects via a direct action on lipid membranes (83,84). These molecules apparently insert into the bilayer and form conducting channels in planar membranes or liposomes which are rather nonselective to low molecular weight substances but which can discriminate between low and high molecular weight components (e.g., choline versus inulin). The channels formed by colicin K have been characterized as being voltage dependent, whereas the actions of I_2 and E_1 on lipid membranes seem to be independent of a transmembrane voltage. Possible lipid specificities of these toxins have not yet been analyzed in detail.

A number of lipase-type toxins have also been described. These include cobra cardiotoxin (85) and β-bungarotoxin (86), both of which have been studied in liposome systems.

Liposome Interaction with Clotting Components

The role of phospholipids in blood clotting has been recognized for a very long time. However, until recently, the biochemical details of this involvement were rather poorly understood. An overview of lipid involvement in coagulation is given in the review by Zwaal (87). Basically, there are three processes in the coagulation cascade which involve phospholipid surfaces: (a) the conversion of factor X to Xa by a complex of IXa, VIIIa, calcium, and phospholipid in the intrinsic pathway; (b) the conversion of X to Xa by a complex of VIIa, III, calcium, and phospholipid in the extrinsic pathway; and (c) the conversion of prothrombin (II) to thrombin (IIa) by a complex of Xa, Va, calcium, and phospholipid. The richest store of biochemical information currently available concerns the third reaction, the conversion of prothrombin to thrombin by the prothrombinase complex.

There is now good evidence to suggest that the prothrombinase complex is comprised of Xa, which is bound to a phospholipid surface via calcium bridges involving γ-carboxyglutamic residues, and Va, which binds to lipid via hydrophobic interactions (88). Prothrombin interacts with this complex, primarily via calcium-carboxyglutamic acid bridges concentrated near the N terminal of the molecule. The nature of the lipid surface required for an effective prothrombinase complex has been explored in some detail. Although negative charge is required, an excessive amount of charge is inhibiting. An optimal charge density corresponds to a zeta potential of $-1.6 \ S^{-1} \ V^{-1} \ cm^{-1}$. In addition, the negative charge must be in a fluid membrane and be uniformly distributed. Thus, when the negatively charged lipid undergoes a phase separation, the activity of the complex may be reduced (87). Phosphatidylserine (PS) seems to be the most effective lipid in terms of binding of prothrombin and factor Xa to the complex. Binding of both factors increases with phosphatidylserine content up to $\sim 15\%$, with a binding stoichiometry of 9 PS residues per prothrombin or 5 per factor Xa. Above 20% PS the limiting factor in binding seems to be the protein packing den-

sity on the membrane surface (89). Recently, the role of calcium in the binding
of prothrombin to phospholipids has been studied in detail (90,91).

Recent studies have examined the interaction of other vitamin K-dependent
clotting factors with phospholipids. One group has measured the calcium-dependent binding to lipid of six vitamin K-dependent factors (92). Another group
has found that factor III must interact with phospholipid bilayers in order to display activity and that cadmium promotes the association of this factor with
lipids (93).

Liposome Interaction with Other Blood Proteins

Recently, there have been a number of studies on the interaction of liposomes
with complex mixtures of proteins such as those found in whole serum and in the
membranes of cells. The intent of these studies has been (a) to examine the proteins capable of binding to liposomes and (b) to assess the stability of liposomes in
the biological environment. Allen and Clelend (94) have shown that serum proteins promote the efflux of the fluorescent dye calcein from small unilamellar
vesicles. As found by others, the vesicles were stabilized against protein disruption by the inclusion of cholesterol. Interestingly, the presence of an osmotic
gradient across the vesicle membrane enhanced the protein-mediated efflux of
the dye.

The identity of proteins binding to liposomes has been investigated in detail
by Juliano and Lin (95) using gel electrophoresis techniques. In contrast to a
previous report (96), which suggested that α_2-macroglobulin was the only blood
protein binding to liposomes, Juliano and Lin found a variety of serum components which bound tightly to the liposome surface. Some of these bound proteins
could be readily identified as major blood components (albumin, IgG subunits,
aprprotein AI, α_2-macroglobulin), whereas other bound proteins could not be
readily identified and probably represented minor serum components. The patterns of bound proteins were highly dependent on the chemical and physical
properties of the liposomes. For example, while negatively charged liposomes
bound an apparently random sample of serum proteins, positive and neutral liposomes tended to bind selectively to a group of high molecular weight ($>$200,000
daltons) protein components. Binding of these high molecular weight components was very rapid (1 min at 37°C), essentially irreversible, and the components
remained at the outer surface of the liposome and did not disrupt the vesicle
structure. An implication of these studies is that, upon injection into the blood,
liposomes rapidly become irreversibly coated with a layer of serum proteins. It
seems possible that the nature of this layer (which differs for different types of
liposomes) is a main determinant of the clearance kinetics and tissue distribution
of injected liposomes.

In a novel and interesting study, Bouma et al. (97) found that liposomes
could extract certain proteins (particularly acetylcholinesterase) from erythro-

cytes. It is not clear if the same is true of other cell types, but one would suspect
that this might be so.

INTERACTIONS OF DRUGS WITH LIPOSOMES

Basic Aspects

Liposomes have attracted a good deal of attention as a potentially valuable system for the controlled delivery of drugs. In this regard a wide variety of pharma-
ceutical substances have been entrapped or encapsulated in liposomes, including
antitumor agents, antibiotics, anti-inflammatory substances, and drugs acting on
the central nervous system (CNS). It is not the intent of this chapter to review
the pharmacological or therapeutic consequences of these studies. Rather, we
shall confine our attention to problems of drug incorporation or formulation in
the liposome system and to the effects of drugs on the properties and stability of
liposomes.

It seems self-evident that polar or hydrophilic drugs will be found in the internal aqueous compartments of liposomes, while hydrophobic or amphipathic
drugs can partition into the liposome membrane. This has been demonstrated
directly by Stamp and Juliano (98). Thus, brief sonication of liposomes containing water-soluble drugs causes rapid release of all the entrapped material. By contrast, sonication of liposomes containing lipophilic drugs results in slow and partial
release, indicating that the drug is stably associated with the liposome membrane.

The amount of polar drug capable of being entrapped per unit weight of
liposomal material depends on the solubility of the drug and on the volume of
water encapsulated per mass of lipid. The latter parameter differs drastically for
different types of liposomes. For example, Szoka and Papahadjopoulas (99) have
examined the encapsulation properties of small unilamellar vesicles (SUVs), large
multilamellar vesicles (MLVs), and large unilamellar vesicles (LUVs). The aqueous-
phase trapping (microliters per micromole) and encapsulation efficiencies (percent of drug encapsulated) were as follows: SUVs ($0.5\ \mu l/\mu mol$, 0.5-1.0%), MLVs
($1-5\ \mu l/\mu mol$, 5-15%), LUVs ($15\ \mu l/\mu mol$, 35-65%). Thus, at first glance, the
LUV preparation seems most ideal for efficient encapsulation of water-soluble
drugs. In many cases this is indeed true; however, sometimes LUV preparations
are less desirable than MLVs since the LUVs are less stable in protein-containing
biological fluids. For example, efflux rates of drugs are higher from LUVs than
from MLVs. In some cases the use of SUVs may be desirable since they tend to
remain in the circulation longer than other types of vesicles. Although MLVs are
less efficient than LUVs in drug encapsulation, the water content and the encapsulation efficiency of MLVs can be manipulated by changing the charge on these
vesicles. Increasing the charge density increases the spacing between the bilayers
and thus the amount of entrapped aqueous volume. Generally, the use of nega-

tively charged lipid is preferred since positively charged lipids are considered to be toxic. For use in drug delivery one must formulate liposomes which are stable in vivo. As discussed in the section on liposome-protein interactions, this requires the use of lipids that are in the gel phase at physiologic temperature or the inclusion of substantial fractions of cholesterol in the liposome mixture. Thus, there is no clear choice for an optimal vesicle. The choice really depends on the intended application.

In the cases of lipophilic or amphipathic drugs, these molecules clearly have the capacity to interact directly with the bilayer membrane of liposomes. Several groups have now demonstrated that lipophilic or amphipathic drugs can affect the physical properties of the bilayer membranes in very profound ways (100-103). For example, Juliano and Stamp (100) have shown that the lipophilic antitumor drugs actimomycin D and vinblastine can markedly affect the phase transition behavior of DMPC liposomes; hydrophilic drugs such as fluorodeoxyuridine had no such effect. In a similar vein, anthracycline drugs were found to have strongly perturbing effects on lipid membranes, particularly those containing negatively charged lipids (101,103). Therefore, particularly with nonpolar drugs, the drug-liposome combination must be considered as an interacting system, not in isolation from each other.

Since nonpolar drugs reside in the membrane environment, the details of membrane structure affect the efficiency of drug incorporation. Nonpolar drugs tend to prefer more fluid membrane environments rather than more rigid ones (98). In addition, the presence of drug molecules in the bilayer has a negative effect on the incorporation of additional drug molecules. It has been found that the encapsulation efficiency tends to go down as the total amount of nonpolar drug per weight of lipid increases (98,104), although this is not necessarily true for polar drugs.

The efflux rates of both polar and nonpolar drugs from liposomes seem to be controlled by similar factors, with the absolute rates of efflux of nonpolar drugs being higher for most liposome types at most temperatures (100). For both polar and nonpolar drugs, the permeation rate through membranes composed of pure phospholipid shows a marked discontinuity at the phase transition, where a distinct maximum in drug efflux is observed (100,105). This may be due to the presence of membrane discontinuities at the junction of fluid- and gel-state domains as proposed by Papahadjopoulos (106). These interesting characteristics of liposomes have been exploited by Yatvin et al. in constructing liposomes for local release of drugs by hyperthermia (105).

Liposomes and Specific Drug Classes

In the following sections we will examine in detail the interaction of liposomes with specific types of drugs. Our concern will not be with the therapeutic utility of liposomal drug delivery approaches but rather with the factors leading to efficient and stable incorporation of drugs into liposomes. In other instances we will survey the use of liposomes as a model for elucidating the mechanisms of action of membrane-active drugs.

Antitumor Agents

Cytosine Arabinoside This drug is a potent inhibitor of DNA polymerase and is used primarily in the treatment of adult myelogenous leukemia. Several groups have worked extensively with liposome-encapsulated cytosine arabinoside in terms of exploring alterations in pharmacokinetics (104) and therapeutic efficacy (107,108). Cytosine arabinoside, a highly polar drug, can be readily encapsulated in both multilamellar vesicles and large unilamellar vesicles, with the encapsulation efficiencies being substantially larger in the latter. It seems likely that the basic mode of action of the liposomal carrier for this drug is to act as a sustained-release system (109; see Chap. 9). Thus, factors that contribute to the stability and longevity of the liposomes in vivo will contribute to effective pharmacological behavior. This has been examined in detail by Mayhew et al. (110), who studied the effects of cholesterol content on the stability and antileukemic effects of liposomes containing cytosine arabinoside. The use of MLVs containing cytosine arabinoside as a localized drug delivery system, with the lung as a target organ, has been explored by Juliano and McCullough (111). These studies emphasized pharmacological effects rather than drug formulation in liposomes.

Anthracyclines Adriamycin, the best known anthracycline drug, is an antitumor drug which is effective against a wide variety of solid tumors and leukemias. There has been a lively interest in liposome-encapsulated anthracycline drugs for two reasons. On the one hand, it seems possible that membranes may be one of the sites of action of such drugs and thus the liposomal system offers an interesting model (112). On the other hand, since uptake of liposomes by the heart is low, liposomal encapsulation of adriamycin may offer a way of circumventing the well-known myocardial toxicity of this powerful antitumor agent (113). Adriamycin is an amphipathic drug with a bulky ring system and an octanol-water partition coefficient near 1.0. It also has an amino function which normally imparts a positive charge to the molecule.

The therapeutic effects of encapsulated adriamycin have been studied by Rahman et al. (113) and by Tökes and his colleagues (114,115). Both of these groups seemed to be able to reduce the myocardial toxicity of adriamycin while maintaining good antitumor potency. There is some concern about the mechanisms of encapsulation of adriamycin in these studies. Despite the value of the octanol-water partition coefficient of adriamycin (which would favor easy passage across membranes), both groups claim to get highly efficient incorporation of adriamycin into liposomes of either positive or negative charge. Complex formation of adriamycin with negatively charged lipids would be easy to understand, but the ability of positively charged lipids to bind the drug is perplexing. Tökes' group (114,115) has evolved a novel two-step procedure for the encapsulation of adriamycin.

There has been considerable interest in the hypothesis that membrane lipids, rather than DNA, might be the major intracellular target of anthracycline-type drugs (112). Tritton and his colleagues (116) as well as others (117,118) have studied the physical chemistry of adriamycin-lipid interaction in some detail. The drug can readily bind to negatively charged lipids such as cardiolipin (2/1 complex) or phosphatidylserine (1/1). In gel-state membranes, the drug can bind but not intercalate into the bilayer. By contrast, in fluid membranes the drug can penetrate partially into the bilayer in the presence or absence of negatively charged lipids (116); however, the binding properties of adriamycin and its effect on the bilayer are different when a small amount of a negatively charged lipid, such as cardiolipin, is present (116). It is interesting to note that the interaction energies of anthracyclines with lipids (118) and the effects of lipid binding on the anthracycline spectrum (117) are similar to the same parameters when anthracyclines interact with DNA. This supports the concept of membrane lipids as an alternative target for anthracycline action.

Antimetabolites   There has been a diverse set of studies with liposome-encapsulated antimetabolite drugs such as methotrexate and purine and pyrimidine analogs. A number of workers have studied pharmacokinetics (119), cytotoxicity (120), and antitumor effects (121) of liposomal-entrapped methotrexate. Methotrexate is a charged, water-soluble drug; its encapsulation behavior is similar to that of cytosine arabinoside in that it is entrapped in proportion to the fraction of the aqueous space that is enclosed by the vesicles. The studies mentioned above do not provide many details on formulation aspects of the encapsulation of methotrexate. In an interesting study, Weinstein and his colleagues (122) made use of the phenomenon mentioned above that pure phospholipid liposomes tend to leak their contents most rapidly at the phase-transition temperature. This effect was exploited in the treatment of solid L1210 tumors with a combination of local hyperthermia and temperature-sensitive methotrexate-containing liposomes. High levels of drug release were obtained within the tumor and the therapeutic results seemed quite encouraging.

Other antimetabolites, such as floxuridine (123) and 6-mercaptopurine and 8-azaguanine (124), have also been encapsulated in lipid vesicles. It was claimed that the presence of sphingomyelin in the vesicles was an important determinant for the successful encapsulation of floxuridine (123). A group of Japanese workers has examined the encapsulation of illudin S (125) and of neocarzinostatin (126) and have demonstrated, in the later case, a correlation between in vitro stability of the drug-liposome complex and the antitumor effect.

Lipophilic Antitumor Agents   The lipophilic antitumor drug actinomycin D
has been studied in terms of its pharmacokinetics (104) and in terms of its antitumor potencies (127,128). As mentioned above, this drug can intercalate into
the bilayer and perturb its structure (100). The encapsulation efficiencies of
actinomycin D decline as the total dose increases (104), a characteristic typical
of highly lipophilic drugs. The lipophilic antimitotic vinca alkaloid drugs have
also been investigated in terms of pharmacokinetics (104). In addition, a careful
study has been made of the uptake of *Vinca* alkaloid-containing liposomes by
cultured fibroblasts (129) and of the therapeutic effects of encapsulated *Vinca*
alkaloids in leukemic mice (130). In both of these studies very substantial levels
of drug incorporation were achieved (>10%). This is more than would be expected using the same encapsulation procedures with a water-soluble drug; once
again this emphasizes the different modes of encapsulation of hydrophilic and
lipophilic drugs. Other studies have examined the utilization of highly unstable,
highly lipophilic nitrosourea compounds in liposomes (131), although not a great
deal of information on the formulation of the drug-liposome combination is available in these studies.

Antimicrobial Drugs

Once again, as in the case of antitumor agents, liposomes have been investigated
both as a potential drug delivery system and as a model for elucidating therapeutic
or toxic actions of drugs which involve membrane components.

Aminoglycoside and β-Lactam-Type Antibiotics   There is a rapidly growing
literature on the interaction of these "conventional"-type antibiotics with liposomes. Kimura et al. (132) studied the transmembrane efflux rate of β-lactam
antibiotics from liposomes as a model for passive intestinal transport of these
drugs. They found good correlation between the efflux rates and gastrointestinal
uptake, especially when intestinal lipids were used to form the liposomes.

There have been a few preliminary explorations of the therapeutic uses of
liposome-encapsulated antibiotics, mainly in connection with aminoglycoside-
type drugs. Bonventre and Gregoriadis (133) found that liposomal dihydrostreptomycin was effective in the killing of intraphagocytic *Staphylococcus aureus,*
whereas the free drug was not. Morgan and Williams (134) examined the encapsulation, stability, and pharmaceutic properties of liposomal gentamicin, finding
that this positively charged drug can be stably encapsulated, particularly in negatively charged liposomes.

One of the most interesting developments in this area concerns the use of
lipid monolayers and bilayers as a model system to evaluate the ototoxicity of
aminoglycoside antibiotics. It is well known that although the therapeutic effects
of aminoglycosides are due primarily to action at the level of the bacterial ribosome, these drugs also have serious toxic effects in the mammalian auditory system. It has been suggested that the ototoxicity of the positively charged amino-

glycoside drugs is due to their propensity to bind to polyphosphoinositol lipids, which are known to be enriched in the auditory apparatus (135). There is a good deal of data to indicate that aminoglycoside drugs can bind to negatively charged lipids, the order of potency being neomycin > gentamycin > tobramycin > amikacin > kanamycin > streptomycin (136). In addition, there is a specificity in terms of the lipids; thus surface balance studies of neomycin interactions with lipid films indicated that the strongest interaction is with polyphosphoinositides > cardiolipin, sulfatides > phosphatidylserine, phosphatidic acid, phosphatidyl-inositol (137). These observations strongly support the concept that the ototoxic action of aminoglycosides is explained by their selective interaction with poly-phosphoinositides (138).

Polymixin   Polymixins are a class of bacterially derived amphipathic poly-peptides. Acting primarily on gram-negative organisms, they are a valuable tool for topical and parenteral therapy. The biochemistry and pharmacology of the polymixins has been reviewed recently in detail (139). Our concern is with the physical-chemical details of the polymixin-lipid interactions. There have been several interesting studies recently along these lines. Experiments with polymixins in planar bilayer membranes (i.e., black membranes) suggest that the absorption of polymixins to one side of a bilayer of negatively charged lipids causes destabili-zation, as reflected by an increase in conductance (140). This is very similar to the situation with calcium ions and negative bilayers. The effect of polymixin on phase transition behavior and membrane fluidity as assessed by diphenylhexa-triene fluorescence polarization has been investigated in detail by Galla and col-leagues (141,142). These workers find that when polymixin binds to a negatively charged lipid layer, the phase transition is split; the lower transition they inter-pret as being due to lipid directly bound to the polymixin molecules, while the upper transition is the unperturbed lipid; thus, a lateral phase separation seems to take place. The binding properties of polymixin are highly dependent on pH and ionic strength (142), as one might expect from its polypeptide nature. Mono-layer studies (142) reveal that one cationic polymixin molecule interacts with five anionic phospholipids to produce a total charge neutralization. Calcium, but not monovalent cations, can partially compete with the polymixin-anionic lipid interaction. It has been suggested that the entire polymixin molecule binds to one half the lipid bilayer and partially penetrates, reminiscent of the situation with bee mellitin, another surface-active polypeptide. In fact, discussion of the behavior of polymixin involves aspects of both drug-liposome interaction and protein-liposome interaction.

Antiparasitic and Antiviral Agents   There is a diverse literature on encapsula-tion of antiviral and antiparasitic compounds in lipid vesicles. Although this is an area of considerable therapeutic potential, we will deal with it only passingly since there is not a great deal of information on drug-lipid interaction with regard to these substances.

Several groups have successfully used liposome-encapsulated antimonial drugs such as Pentostam in the treatment of leishmaniasis, a tropical protozoan infection which affects cells of the reticuloendothelial system. Since both parasite and drug-liposome complexes tend to accumulate at the same site (i.e., in the endocytotic vacuoles of macrophages), the therapeutic index of the drug is markedly enhanced via encapsulation (144-146). Other drugs, including 8-aminoquinolines, were also effective antileishmanial compounds when given entrapped in liposomes (147). In these studies the precise composition of the liposomes did not seem to matter a great deal, as long as the formulation was capable of surviving in vivo (144,147). In a related vein, primaquine-containing liposomes have been used for chemotherapy of experimental malaria (148).

There is one report on the use of an antiviral drug in liposomes. This concerns the therapy of herpes simplex keratitis with Idoxuridine (149). It would seem that both the uses of antiviral drugs and the exploration of possibilities for therapy of disease of the eye would be promising future areas for experimentation with liposomal drug delivery systems.

Antifungal Agents   Amphotericin B, a polyene antibiotic, is a potent but dangerous drug used for the therapy of systemic mycoses. Although it displays considerable toxicity, it is one of the few weapons available for the treatment of these diseases (150). Amphotericin B is a membrane-active drug that binds to sterols. The drug has a higher affinity for fungal sterols such as ergosterol but also binds to cholesterol, which may account for its toxicity in humans (151). The mechanism of action of amphotericin involves formation of a complex with sterols which alters membrane permeability properties and leads to osmotic destabilization of the cells (152). Amphotericin does not alter the permeability of sterol-free membranes.

Recently, a number of laboratories have explored in detail the mechanism of action of amphotericin B. The drug apparently forms channels 6-8 Å in diameter in sterol-containing membranes, which permit passage of small inorganic ions (153). Larger molecules such as glucose or dextran do not pass through the amphotericin channels and are released only if there is osmotic lysis (154). By contrast, the related antibiotic filipin caused major disruption of membrane structure, allowing direct release of large molecules. Amphotericin does not cause as drastic a change in membrane physical properties (e.g., bilayer fluidity) as does Filipin (155). However, at amphotericin B/sterol ratios greater than 0.7, some acyl chain disordering is observed (156). The ability of amphotericin and related antibiotics such as pimariacin and etruscomycin to release ions from vesicles is markedly dependent on their sterol content. Amphotericin B shows a maximal effect at a cholesterol content of 20 mol% (157).

Other Antimicrobials   Recently, there have been a number of studies exploring the question of possible direct effects of antimicrobial agents on membranes using liposomes as a model system. These studies include a very sophisticated analysis of the action of the germacide 3,3′,4′,5-tetrachlorosalicylanalide with liposomes (158) and studies of the actions of chloroquine, an antimalarial (159), and citrinin, a mycotoxin (160), on lipids.

Anti-Inflammatory and Immunomodulatory Agents

There has been a recent burst of interest in the use of liposomes as a vector to deliver drugs selectively to macrophages and other phagocytic cells. These studies have involved both agents that tend to suppress macrophage activity, such as glucocorticoids, and agents that tend to activate macrophages, such as lymphokines and muramyl dipeptide.

Dingle and his colleagues have studied the effect of liposome-encapsulated corticosteroids on joint inflammation (161). From the viewpoint of this article, one of the interesting chemical findings was that long-chain steroid esters could be stably incorporated in liposomes, while the corresponding unesterified drugs were not (162). Detailed physical studies of the interaction of cortisol (163) and of long-chain cortisol esters (164) with phospholipids have been carried out. These studies show that cortisol interacts primarily with the polar head group of phospholipids. When cortisol esters are used, the fatty acid side chains can penetrate the interior of the bilayer and help to anchor the corticosteroid. Because of limitations due to the packing of the bulky cortisol entities at the bilayer surface, only a limited amount of material (12-20 mol%) can be successfully incorporated into liposome membranes.

Liposomes containing trace amounts of [^{3}H] prednisolone have been studied in terms of use as an intramuscular sustained-release preparation (165). Few details on the formulation or stability of the vesicles were given in this study. Prostaglandins have also been incorporated into lipid vesicles for the purpose of testing possible ionophoretic activity (166). Little has been done with the utilization of prostaglandins in vesicles for therapeutic purposes, although these highly potent lipophilic drugs would seem a natural candidate for membrane insertion, provided that problems of stability can be overcome.

In an extremely interesting series of papers, Fidler, Poste, and their colleagues have shown that liposomal encapsulation can potentiate the action of immunomodulating substances such as T-cell-derived lymphokines (MAF) and muramyl dipeptide, on macrophages (167-169). Stimulation of macrophages to become tumorcidal with liposomal immunomodulators can occur not only in vitro but in vivo as well. Intravenous injection of appropriate liposomes containing MAF into rats resulted in the activation of rat alveolar macrophages (170). Studies of the effect of liposome composition on this process indicated that large, negatively charged vesicles are most appropriate, presumably because they readily lodge in the lung capillaries. Recently, these studies have been extended to the use of liposomal immunomodulators to reduce the incidence of pulmonary

metastasis in an animal tumor system (171). Because of the lack of radiolabeled
immunomodulating substances, these studies contain relatively little information
on the formulation characteristics or stability of the liposome-modulator complexes used.

CNS-Acting Drugs

There is only a limited literature on the use of liposomes as a delivery system for
CNS-acting drugs, or as models to aid in the understanding of the mechanisms of
action of such drugs. There is one report on the pharmacologic effects of liposome-encapsulated lithium, an agent used in manic-depressive disease (172).
Encapsulation changed the lithium-induced polydipsic effect but did not result
in more lithium accumulation in the brains of rats. More extensive literature
exists on the use of lipid bilayers to elucidate the possible membrane mechanism
of drugs that act in the CNS.

Ethanol   Johnson et al. (173,174) have studied the effects of ethanol in
liposomes made from the brain lipids of normal or ethanol-tolerant mice. Using
fluorescence polarization techniques, they have found that liposomes from alco-
hol-tolerant animals are less susceptible to the membrane fluidizing effects of
ethanol than are liposomes made from lipids of normal mice. They relate this
to the changes in brain lipid composition, particularly cholesterol levels.

Neuroleptics   Neuroleptic drugs such as chloropromazine, in addition to
binding to specific receptors in the nanomolar concentration range, also bind to
lipids in a relatively nonspecific manner. The interaction of chloropromazine
with liposomes has been studied by equilibrium dialysis techniques (175) and by
fluorimetry (176). The drug bound better to fluid membranes than to gel-state
membranes and cholesterol reduced drug binding. Chloropromazine had a fluidizing effect on bilayers as measured by perylene fluorescence depolarization.
From these studies it seems that chloropromazine binds to the surface of membranes and also penetrates and fluidizes the interior of the bilayer. It is not
clear if any of these effects are related to the toxic or therapeutic action of
drugs of this type.

Anesthetics   Since the primary site of action of local and general anesthetics
is at the membrane level, there have been an enormous number of studies on the
effects of anesthetics on the physical properties of various types of bilayers. It
is really beyond the scope of this review to deal with this large body of literature.
Rather we refer the reader to several timely and thoughtful analyses on the topic
of mechanisms of anesthetic action (177-181).

Other Drugs   There are a few other reports on the interactions of various
other types of drugs with lipid bilayers; these include studies on antihistamines
(182), colchicine (used in the treatment of gout) (183), and the amantadines,

which are used as antiviral or antiplatelet drugs (184). This last report contains
a careful correlation between the effects of amantadine derivatives on phase transitions in lipid bilayers and effects on platelet aggregation, suggesting that the
basic action of these interesting drugs, with their cagelike structures, is at the
level of the bilayer membrane.

REFERENCES

1.   Bangham, A. D. 1972. Liposomes. Annu. Rev. Biochem. 41:753.
2.   Juliano, R. L. 1981. Liposomes as a drug delivery system. Trends Phar-
     macol. Sci. 2(2):34–41.
3.   Zilversmit, D. B., L. B. Hughes, and J. Balmer. 1978. Stimulation of cho-
     lesterol ester exchange by lipoprotein free rabbit plasma. Biochim. Biophys.
     Acta 409:343.
4.   Racker, E. 1975. Reconstitution, mechanism of action and control of ion
     pumps. Biochem. Soc. Trans. 3:785.
5.   Shamoo, A. C. 1975. *Carriers and Channels in Biological Membranes.* New
     York Academy of Sciences, New York.
6.   Singer, S. J. and G. L. Nicolson. 1972. The fluid mosaic model of the struc-
     ture of cell membranes. Science 175:720.
7.   Rothman, J. E. and J. Lenard. 1977. Membrane asymmetry. Science 195:
     743.
8.   Nicolson, G. E. 1976. Transmembrane control of the receptors on normal
     and tumor cells. Biochim. Biophys. Acta 475:57.
9.   Lenaz, G. 1980. Role of lipids in the structure and function of membranes.
     Subcell. Biochem. 6:233.
10. Quinn, P. J. and D. Chapman. 1980. The dynamics of membrane struc-
     ture. CRC Crit. Rev. Biochem. 8:1.
11. Blobel, G. and B. Dobberstein. 1975. Transfer of proteins across mem-
     branes. J. Cell Biol. 67:835.
12. Wickner, W. 1979. The assembly of proteins in the biological membranes:
     the membrane trigger hypothesis. Annu. Rev. Biochem. 48:23.
13. Tanner, M. J. A. 1978. Erythrocyte glycoproteins. Curr. Top. Membr.
     Transp. 11:279.
14. Compans, R. W. and M. C. Kemp. 1978. Membrane glycoproteins of
     enveloped viruses. Curr. Top. Membr. Transp. 11:233.
15. Rothstein, A. Cabantchik, Z., and Knauf, R. 1976. Mechanism of anion
     transport in red blood cells. Fed. Proc. 35:3.
16. Demel, R. A. and F. B. DeKruyff. 1976. Functions of sterols in mem-
     branes. Biochim. Biophys. Acta 457:109.
17. Shimshick, B. J. and H. M. McConnell. 1973. Lateral phase separation in
     phospholipid membranes. Biochemistry 12:2351.
18. Jacobson, K. and D. Papahadjopoulos. 1975. Phase transitions and phase
     separation in phospholipid membranes induced by changes in temperature,
     pH and the concentrations of divalent cations. Biochemistry 14:152.

19. Kimelberg, H. K. 1976. Protein-liposome interactions and their relevance to the structure and function of cell membranes. Mol. Cell. Biochem. 10: 171.

20. Kimelberg, H. K. and D. Papahadjopoulos. 1974. Effects of phospholipid chain fluidity phase transitions and cholesterol on the $(Na^+ + K^+)$ stimulated adenosine triphosphatase. J. Biol. Chem. 249:1071.

21. Papahadjopoulos, D., M. Moscarello, E. M. Eyler, and T. Isaac. 1975. Effects of proteins on the thermotropic phase transition of phospholipid membranes. Biochim. Biophys. Acta 40:317.

22. Papahadjopoulos, D., M. Cowden, and H. K. Kimelberg. 1973. Role of cholesterol in membranes. Biochim. Biophys. Acta 330:8.

23. Jost, P. C. and O. H. Griffith. 1980. The lipid-protein interface in biological membranes. In *Annals of the New York Academy of Sciences*, Vol. 348 (*Lipoprotein Structure*). New York Academy of Sciences, New York, p.391.

24. Juliano, R. L., H. K. Kimelberg, and D. Papahadjopoulos. 1971. Synergistic effects of membrane protein (Spectrin) and Ca^{2+} on the Na^+ permeability of phospholipid vesicles. Biochim. Biophys. Acta 241:894.

25. Van Deenen, L. L. M., R. A. Demel, W. S. M. Geurts van Kessel, H. H. Kamp, B. Roelofsen, A. J. Verkleij, K. W. Wirtz, and R. F. A. Zwaal. 1976. Phospholipases and monolayers as tools in studies of membrane structure. In *The Structural Basis of Membrane Function*, Y. Hatefi and L. D. Ohaniace (Eds.). Academic Press, New York, p. 21.

26. Brulet, P. and H. M. McConnell. 1976. Protein and lipid interactions: glycophorin and dipalmitoyl phosphatidyl choline. Biochem. Biophys. Res. Commun. 68:363.

27. Schinitzky, M. and P. Henkart. 1980. Fluidity of cell membranes—current concepts and trends. Int. Rev. Cytol. 60:121.

28. Cherry, R. J., U. Müller, C. Holenstein, and M. P. Heyn. 1980. Lateral segregation of proteins induced by cholesterol in bacteriorhodopsin-phospholipid vesicles. Biochim. Biophys. Acta 596(1):145.

29. Smith, L. M., J. L. Rubenstein, J. W. Parce, and H. H. McConnell. 1980. Lateral diffusion of M-13 coat protein in mixtures of phosphatidylcholine and cholesterol. Biochemistry 19(25):5907.

30. Pink, D. A. and D. Chapman. 1979. Protein-lipid interactions in bilayer membranes: a lattice model. Proc. Natl. Acad. Sci. USA 76(4):1542.

31. D. E. Wolf, J. Schlessinger, E. L. Elson, W. W. Webb, R. Blumenthal, and P. Henkart. 1977. Diffusion and patching of macromolecules on planar lipid bilayer membranes. Biochemistry 16(15):3476.

32. Owicki, J. C. and H. M. McConnell. 1979. Theory of protein-lipid and protein-protein interactions in bilayer membranes. Proc. Natl. Acad. Sci. USA 76(10):4750.

33. Racker, E. 1973. A new procedure for the reconstitution of biologically active phospholipid vesicles. Biochem. Biophys. Res. Commun. 55:224.

34. Frenkel, E. J., B. Roelofsen, U. Brodbeck, and L. L. Van Deenen. 1980. Lipid protein interactions in human erythrocyte-membrane acetylcholinesterase. Eur. J. Biochem. 109(2):377.

35. Niggli, V., E. S. Adunyah, J. T. Penniston, and E. Carafoli. 1981. Purified (Ca^{2+}-Mg^{2+})-ATPase of the erythrocyte membrane. J. Biol. Chem. 256(1): 395.

36. Madden, T. D., D. Chapman, and P. J. Quinn. 1979. Cholesterol modulates activity of calcium-dependent ATPase of the sarcoplasmic reticulum. Nature 279(5713):538.

37. Curman, B., L. Klareskog, and P. A. Peterson. 1980. On the mode of incorporation of human transplantation antigens into lipid vesicles. J. Biol. Chem. 255(16):7820.

38. Wu, W. C. S., H. P. H. Moore, and M. A. Raftery. 1981. Quantitation of cation transport by reconstituted membrane vesicles containing purified acetylcholine receptor. Proc. Natl. Acad. Sci. USA 78:775.

39. Kimelberg, H. K. and D. Paphadjopoulos. 1972. Biochim. Biophys. Acta 282:277.

40. Gazzotti, P., H. G. Bock, and S. Fleischer. 1975. Interaction of D-β-hydroxy-butyrate dehydrogenase with phospholipids. J. Biol. Chem. 250: 5782.

41. Jackson, R. C. and W. R. White. 1981. Phospholipid is required for the processing of presecretory proteins by detergent-solubilized canine pancreatic signal peptides. J. Biol. Chem. 256(5):2545.

42. Morrisett, J. D., R. L. Jackson, and A. M. Gotto. 1977. Lipid protein interaction in the plasma lipoproteins. Biochim. Biophys. Acta 472:93.

43. Aune, K. C., J. G. Gallagher, A. M. Gotto, Jr., and J. D. Morrisett. 1977. Physical properties of the dimyristoylphosphatidylcholine vesicle and of complexes formed by its interaction with apolipoprotein C-III. Biochemistry 16(10):2151.

44. Tall, A. R., V. Hogen, L. Askinazi, and D. M. Small. 1978. Interaction of high density lipoproteins with DMPC multilamellar liposomes. Biochemistry 17:322.

45. Tall, A. R. and P. H. Green. 1981. Incorporation of phosphatidylcholine into spherical and discoidal lipoproteins during incubation of egg phosphatidylcholine vesicles with isolated high density lipoproteins or with plasma. J. Biol. Chem. 256(4):2035.

46. Scherphof, G., F. Roerdink, M. Waite, and J. Parks. 1978. Disintegration of phosphatidylcholine liposomes in plasma as result of interaction with high-density lipoproteins. Biochim. Biophys. Acta 542(2):296.

47. Guo, L. S., R. L. Hamilton, J. Goerke, J. N. Weinstein, and R. J. Havel. 1980. Interaction of unilamellar liposomes with serum lipoproteins and apolipo-proteins. J. Lipid Res. 21(8):993.

48. Tall, A. R. 1980. Studies on the transfer of phosphatidylcholine from unilamellar vesicles into plasma high density lipoproteins in the rat. J. Lipid Res. 21(3):354.

49. Scherphof, G., H. Morselt, J. Regts, and J. C. Wilschut. 1979. The involvement of the lipid phase transition in the plasma-induced dissolution of multilamellar phosphatidylcholine vesicles. Biochim. Biophys. Acta 556 (2):196.

50. Swaney, J. B. 1980. Mechanisms of protein-lipid interaction. Association of apolipoproteins A-I and A-II with binary phospholipid mixtures. J. Biol. Chem. 255(18):8791.

51. van Tornout, P., R. Vercaemst, M. J. Lievens, H. Caster, H. Rosseneu, and G. Assmann. 1980. Reassembly of human apoproteins A-I and A-II with unilamellar phosphatidylcholine-cholesterol liposomes. Association kinetics and characterization of the complexes. Biochim. Biophys. Acta 691(3): 509.

52. Kanellis, P., A. Y. Romans, B. J. Johnson, H. Kercret, R. Chiovetti, Jr., T. M. Allen, and J. P. Segrest. 1980. Studies of synthetic peptide analogs of the amphipathic helix. Effect of charged amino acid residue topography on lipid affinity. J. Biol. Chem. 255(23):11464.

53. Sparro, J. T. and A. M. Gotto. 1980. Phospholipid binding studies with synthetic apolipoprotein fragments. Ann. N.Y. Acad. Sci. 348:187.

54. Kennedy, J. J. 1978. Structure of membrane proteins. J. Membr. Biol. 42:265.

55. Goldstein, I. J. and C. E. Hayes. 1978. Lectins: carbohydrate binding proteins of plants and animals. Adv. Carbohyd. Chem. Biochem. 35:127.

56. Pappenheimer, A. M. 1979. Interactions of protein toxins with mammalian cell membranes. In *Microbiology*, P. Schlessinger (Ed.). American Society for Microbiology, Washington, D.C., pp. 187–192.

57. Sessa, G., T. H. Freer, C. Colaciccio, and C. Weissmann. 1969. J. Biol. Chem. 244:275.

58. Ghosh, P. and B. K. Bachawat. 1980. Grafting of different glycosides on the surface of liposomes and its effect on the tissue distribution of ^{125}I-IgG encapsulated in liposomes. Biochim. Biophys. Acta 632:562.

59. Redwood, W. R., V. Jansons, and B. C. Patel. 1975. Biochim. Biophys. Acta 406:347.

60. Juliano, R. L. and D. Stamp. 1976. Lectin mediated attachment of glycoprotein bearing liposomes to cells. Nature 261:235.

61. Ketis, N. V., J. Girdleston, and C. W. Grant. 1980. Positive cooperativity in a (dissected) lectin-membrane glycoprotein binding event. Proc. Natl. Acad. Sci. USA 77(7):3788.

62. Sharom, F. J., D. G. Barratt, and C. W. Grant. 1977. Glycophorin and the concanavalin A receptor of hyman erythrocytes: their receptor function in lipid bilayers. Proc. Natl. Acad. Sci. USA 74(7):2751.

63a. Rando, R. R. and F. W. Bangerter. 1979. Threshold effects on the lectin mediated aggregation of synthetic glycolipid-containing liposomes. J. Supramol. Struct. 11(3):295.

63b. Slama, J. S. and R. R. Rando. 1980. Lectin-mediated aggregation of liposomes containing glycolipids with variable hydrophilic spacer arms. Biochemistry 19:4595.

64. Hampton, P. Y., R. W. Holz, and I. J. Goldstein. 1980. Phospholipid, glycolipid and ion dependencies of concanavalin A and *Ricinus communis* agglutin I induced aggregation of lipid vesicles. J. Biol. Chem. 255:6760.

65. Curatolo, W., A. O. Yau, D. M. Small, and B. Sears. 1978. Lectin-induced agglutination of phospholipid/glycolipid vesicles. Biochemistry 17(26):5740.

66. Roche, A. C., R. Maget-Dana, A. Obrenovitch, K. Hildenbrand, C. Nocolau, and M. Monsigny. 1977. Interaction between vesicles containing gangliosides and limulin (*Limulus polyphemus* lectin). Fluorescence polarization of 1,6-diphenyl-1,3,5-hexatriene. FEBS Lett. 93(1):91.

67. Read, B. D., R. A. Demel, H. Weingandt, and L. L. van Deenen. 1977. Specific interaction of concanavalin A with glycolipid monolayers. Biochim. Biophys. Acta 470(2):325.

68. Richards, R. L., J. Moss, C. R. Alving, P. H. Fishman, and R. O. Brady. 1979. Choleragen (cholera toxin): a bacterial lectin. Proc. Natl. Acad. Sci. USA 76(4):1673.

69. Morris, N. P., E. Consiglio, L. D. Kohn, W. H. Habig, M. C. Hardegree, and T. B. Helting. 1980. Interaction of fragments B and C of tetanus toxin with neural and thyroid membranes and with gangliosides. J. Biol. Chem. 255 (13):6071.

70. Boquet, P. 1979. Interaction of diphtheria toxin fragments A, B and protein crm 45 with liposomes. Eur. J. Biochem. 100(2):483.

71. Alving, C. R., B. H. Iglewski, K. A. Urban, J. Moss, R. L. Richards, and J. C. Sandoff. 1979. Binding of diphtheria toxin to phospholipids in liposomes. Proc. Natl. Acad. Sci. USA 77(4):1986.

72. Moss, J., R. L. Richards, C. R. Alving, and P. H. Fishman. 1977. Effect of the A and B protomers of choleragen on release of trapped glucose from liposomes containing or lacking ganglioside GM1. Unpublished data.

73. Donovan, J. J., M. I. Siman, D. K. Draper, and M. Montal. 1981. Diphtheria toxin forms transmembrane channels in planar lipid bilayers. Proc. Natl. Acad. Sci. USA 78:172.

74. Boquet, P. 1979. Eur. J. Biochem. 100:483.

75. Haberman, E. 1972. Science 177:314.

76. Drake, A. F. and R. C. Hider. 1979. The structure of melittin in lipid bilayer membranes. Biochim. Biophys. Acta 555(2):371.

77. Dawson, C. R., A. F. Drake, J. Helliwell, and R. C. Hider. 1978. The interaction of bee melittin with lipid bilayer membranes. Biochim. Biophys. Acta 510(1):75.

78. Lavialle, F., I. W. Levin, and C. Mollay. 1980. Interaction of melittin with dimyristoyl phosphatidylcholine liposomes: evidence for boundary lipid by Raman spectroscopy. Biochim. Biophys. Acta 600(1):62.

79. R. Linder, A. W. Bernheimer, and K. S. Kim. 1977. Interaction between sphingomyelin and a cytolysin from the sea anemone *Stoichactis helianthus*. Biochim. Biophys. Acta 467(3):290.

80. Shin, M. L., D. W. Michaels, and M. M. Mayer. 1979. Membrane damage by a toxin from the sea anemone *Stoichactis helianthus*. II. Effect of membrane lipid composition in a liposome system. Biochim. Biophys. Acta 555(1):79.

81. Michaels, D. W. 1979. Membrane damage by a toxin from the sea anemone

Stoichactis helianthus. I. Formation of transmembrane channels in lipid bilayers. Biochim. Biophys. Acta 555(1):67.

82. Fehrenbach, F. J. and M. Eibl. 1977. Interaction of streptolysin-O with natural and artificial membranes. Z. Naturforsch. 32(1-2):101.

83. Schein, S. J., B. L. Kagan, and A. Finkelstein. 1978. Colicin K acts by forming voltage-dependent channels in phospholipid bilayer membranes. Nature 276(5684):159.

84. Tokuda, H. and J. Konisky. 1979. Effect of colicins Ia and E1 on ion permeability of liposomes. Proc. Natl. Acad. Sci. USA 76(12):6167.

85. Hsia, J. C., S. S. Er. and C. Y. Lee. 1978. Effects of Ca^{2+} and membrane surface charge on the direct lytic activity of cobra cardiotoxin—a membrane spin assay. Biochem. Biophys. Res. Commun. 80(2):472.

86. Strong, P. N. and R. B. Kelley. 1977. Membranes undergoind phase transitions are preferentially hydrolyzed by beta-bungarotoxin. Biochim. Biophys. Acta 469(2):231.

87. Zwaal, R. F. A. 1978. Membrane and lipid involvement in blood coagulation. Biochim. Biophys. Acta 515:163.

88. Nesheim, M. E., J. B. Taswell, and K. G. Mann. 1979. The contribution of bovine factor V and factor Va to the activity of prothrombinase. J. Biol. Chem. 254(21):10952.

89. Nelsesteun, G. L. and M. Broderius. 1977. Interaction of prothrombin and blood-clotting factor X with membranes of varying composition. Biochemistry 16(19):4172.

90. Dombrose, F. A., S. N. Gitel, K. Zawalich, and C. M. Jackson. 1979. The association of bovine prothrombin fragment 1 with phospholipid. J. Biol. Chem. 254(12):5027.

91. Resnick, R. M. and G. I. Nelsestuen. 1980. Prothrombin-membrane interaction. Effects of ionic strength, pH, and temperature. Biochemistry 19 (13):3028.

92. Nelsestuen, G. L., W. Kisiel, R. G. Di Scipio. 1978. Interaction of vitamin K dependent proteins with membranes. Biochemistry 17(11):2134.

93. Carson, S. D. and W. H. Konigsberg. 1980. Cadmium increases tissue factor (coagulation factor III) activity by facilitating its reassociation with lipids. Science 208:307.

94. Allen, T. M. and L. G. Clelend. 1980. Serum-induced leakage of liposome contents. Biochim. Biophys. Acta 597(2):418.

95. Juliano, R. L. and G. Lin. 1980. The interaction of plasma proteins with liposomes: protein binding and effects on the clotting and complement systems. In *Liposomes and Immunobiology,* B. Tom and H. Six (Eds.). Elsevier, New York, pp. 49-66.

96. Black, C. P. V. and G. Gregoriadis. 1976. Interaction of liposomes with blood plasma proteins. Biochem. Soc. Trans. 4:253.

97. Bouma, S. R., F. W. Drislane, and W. H. Huestis. 1977. Selective extraction of membrane-bound proteins by phospholipid vesicles. J. Biol. Chem. 252(19):6759.

98. Stamp, D. and R. L. Juliano. 1979. Factors affecting the encapsulation of drugs within liposomes. Can. J. Physiol. Pharmacol. 57(5):535.

99. Szoka, F. and D. Papahadjopoulos. 1978. A new procedure for the preparation of liposomes with large internal aqueous space and high capture by reverse phase evaporation. Proc. Natl. Acad. Sci. USA 75: 4194.

100. Juliano, R. L. and D. Stamp. 1979. Interactions of drugs with lipid membranes: characteristics of liposomes containing polar or non-polar anti-tumor drugs. Biochim. Biophys. Acta 586:137.

101. Tritton, T. R., S. A. Murphree, and A. C. Sarotorelli. 1977. Characterization of drug-membrane interactions using the liposome system. Biochem. Pharmacol. 26(24):2319.

102. Lee, A. G. 1978. Effects of charged drugs on the phase transition temperatures of phospholipid bilayers. Biochim. Biophys. Acta 514:95.

103. Goldman, R., T. Facchinetti, D. Bach, A. Raz, and M. Shinitzky. 1980. A differential interaction of daunomycin, adriamycin and their derivatives with human erythrocytes and phospholipid bilayers. Biochim. Biophys. Acta 512(2):254.

104. Juliano, R. L. and D. Stamp. 1978. Pharmacokinetics of liposome-encapsulated anti-tumor drugs. Biochem. Pharmacol. 27(1):21.

105. Yatvin, M. B., J. N. Weinstein, W. H. Dennis, and R. Blumenthal. 1978. Design of liposomes for enhanced local release of drugs by hyperthermia. Science 202(4374):1290.

106. Papahadjopoulos, D., K. Jacobsen, S. Nir, and T. Isaac. 1973. Biochim. Biophys. Acta 311:330.

107. Mayhew, E., D. Papahadjopoulos, Y. M. Rustum, and C. Dave. 1978. Use of liposomes for the enhancement of the cytotoxic effects of cytosine arabinoside. Ann. N.Y. Acad. Sci. 308:371.

108. Kataoka, T. and T. Kobayashi. 1978. Enhancement of chemotherapeutic effect by entrapping 1-β-D-arabinofuranosyl cytosine in lipid vesicles and its mode of action. Ann. N.Y. Acad. Sci. 308:387.

109. Mayhew, E. 1981. Proc. NATO Inst. Drug Targeting, Sounion, Greece.

110. Mayhew, E., Y. M. Rustum, F. Szoka, and D. Papahadjopoulos. 1979. Role of cholesterol in enhancing the antitumor activity of cytosine arabinoside entrapped in liposomes. Cancer Treat. Rep. 63:1923.

111. Juliano, R. L. and H. W. McCullough. 1980. Controlled delivery of an anti-tumor drug: localized action of liposome encapsulated cytosine arabinoside administered via the respiratory system. J. Pharmacol. Exp. Ther. 214(2):381.

112. Tritton, R. T., S. A. Murphree, and A. C. Sartorelli. 19xx. Adriamycin: a proposal on the specificity of drug action. Biochem. Biophys. Res. Commun. 84(3):802.

113. Rahman, A., A. Kessler, N. More, B. Sikic, G. Rowden, P. Woolley, P. S. Schein. 1980. Liposomal protection of adriamycin-induced cardiotoxicity in mice. Cancer. Res. 40(5):1532.

114. Forssen, E. A. and Z. A. Tökes. 1979. In vitro and in vivo studies with adriamycin liposomes. Biochem. Biophys. Res. Commun. 91:1295.

115. Forssen, E. A. and Z. A. Tökes. 1981. Use of anionic liposomes for the reduction of chronic doxorubicin induced cardiotoxicity. Proc. Natl. Acad. Sci. USA 78:1873.

116. Karczmar, G. S. and T. R. Tritton. 1979. The interaction of adriamycin with small unilamellar vesicle liposomes. A fluorescence study. Biochim. Biophys. Acta 557(2):306.

117. Goormaghtigh, E., P. Chatelain, J. Caspers, and J. M. Ruysschaert. 1980. Evidence of a specific complex between adriamycin and negatively-charged phospholipids. Biochim. Biophys. Acta 597(1):1.

118. Vilallonga, F. A. and F. W. Phillips. 1978. Interaction of doxorubicin with phospholipid monolayers. J. Pharm. Sci. 67(6):773.

119. Kimelberg, H. K. and M. L. Atchison. 1978. Effects of entrapment in liposomes and the distribution, degradation and effectiveness of methotrexate in vivo. Ann. N.Y. Acad. Sci. 308:395.

120. Todd, J. A., A. M. Levine, and Z. A. Tökes. 1980. Liposome-encapsulated methotrexate interactions with human chronic lymphocytic leukemia cells. J. Natl. Cancer Inst. 64(4):715.

121. Kosloski, M. J., F. Rosen, R. J. Milholland, and D. Papahadjopoulos. 1978. Effect of lipid vesicle (liposome) encapsulation of methotrexate on its chemotherapeutic efficacy in solid rodent tumors. Cancer. Res. 38(9): 2843.

122. Weinstein, J. N., R. L. Magin, R. L. Cysyk, and D. S. Zaharko. 1980. Treatment of solid L1210 murine tumors with local hyperthermia and temperature-sensitive liposomes containing methotrexate. Cancer Res. (40(5):1388.

123. Simmons, S. P. and P. A. Kramer. 1977. Liposomal entrapment of floxuridine. J. Pharm. Sci. 66(7):1729.

124. Kano, K. and J. H. Fendler. 1977. Enhanced uptake of drugs in liposomes: use of labile vitamin B12 complexes of 6-mercaptopurine and 8-azaguanine. Life Sci. 20(10):1729.

125. Shinozawa, S., K. Tsutsui, and T. Oda. 1979. Enhancement of the antitumor effect of illudin S by including it into liposomes. Experientia 35 (8):1102.

126. Shinozawa, S., Y. Araki, and T. Oda. 1980. Antitumor effect of neocarzinostatin entrapped in liposomes. Gann 71(1):107.

127. Rahman, Y. E., W. R. Hansen, J. Bharwada, E. J. Ainsworth, and B. N. Jaroslav. 1978. Mechanism of reduction of antitumor drug toxicity by liposome encapsulation. Ann. N.Y. Acad. Sci. 308:325.

128. Kedar, A., E. Mayhew, R. H. Moore, P. Williams, and G. P. Murphy. 1980. Effect of actinomycin D-containing lipid vesicles on murine renal adenocarcinoma. J. Surg. Oncol. 15(4):363.

129. Layton, D., J. de Meyere, M. P. Collard, and A. Trouet. 1979. The accumulation by fibroblasts of liposomally encapsulated vinblastine. Eur. J. Cancer 15(12):945.

130. Layton, D. and A. Trouet. 1980. A comparison of the therapeutic effects of free and liposomally encapsulated vincristine in leukemic mice. Eur. J. Cancer 16(7):945.

131. Ritter, C. and R. J. Rutman. 1978. Relative enhancement by various liposomes of BCNU effectiveness against L-1210 leukemia in vivo. Rex. Commun. Chem. Pathol. Pharmacol. 30(1):123.

132. Kimura, T., M. Yoshikawa, M. Yasuhara, and H. Sezaki. 1980. The use of liposomes as a model for drug absorption: beta-lactam antibiotics. J. Pharm. Pharmacol. 32(6):394.

133. Bonventre, P. F. and G. Gregoriadis. 1978. Killing of intraphagocytic *Staphylococcus aureus* by dihydrostreptomycin entrapped within liposomes. Antimicrob. Agents Chemother. 13(6):1049.

134. Morgan, J. R. and K. E. Williams. 1980. Preparation and properties of liposome associated gentamicin. Antimicrob. Agents Chemother. 17(4): 544.

135. Schacht, J., S. Lodhl, and N. D. Weiner. 1977. Effects of neomycin on polyphosphoinositides in inner ear tissues and monomolecular films. Adv. Exp. Med. Biol. 84:191.

136. Alexander, A. M., I. Gonda, E. S. Harpur, and J. B. Kayes. 1979. Interaction of aminoglycoside antibiotics with phospholipid liposomes studied by microelectrophoresis. J. Antibiot. (Tokyo) 32(5):504.

137. Lodhi, S., N. D. Weiner, and J. Schacht. 1979. Interactions of neomycin with monomolecular films of polyphosphoinositides and other lipids. Biochim. Biophys. Acta 557(1):1.

138. Lodhi, S., N. D. Weiner, I. Mechigan, and J. Schacht. 1980. Ototoxicity of aminoglycosides correlated with their action on monomolecular films of polyphosphoinositides. Biochem. Pharmacol. 29(4):597.

139. Storm, D. R., K. S. Rosenthal, and P. E. Swanson. 1977. Polymyxin and related peptide antibiotics. Annu. Rev. Biochem. 46:723.

140. Miller, I. R., D. Bach, and M. Teuber. 1978. Effect of polymyxin B on the structure and the stability of lipid layers. J. Membr. Biol. 39(1):49.

141. Hartman, W., H. J. Galla, and E. Sackmann. 1978. Polymyxin binding to charged lipid membranes. An example of cooperative lipid-protein interaction. Biochim. Biophys. Acta 510(1):124.

142. Six, F., and H. J. Galla. 1979. Cooperative lipid-protein interaction. Effect of pH and ionic strength on polymyxin binding to phosphatidic acid membranes. Biochim. Biophys. Acta 557(2):320.

143. El Mashak, E. M. and J. F. Tocanne. 1980. Polymyxin B-phosphatidylglycerol interactions. A monolayer (pi, delta V) study. Biochim. Biophys. Acta 596(2):165.

144. Alving, C. R., E. A. Steck, W. J. Chapman, Jr., V. B. Waits, L. D. Hendricks, G. M. Swartz, Jr., and W. L. Hanson. 1978. Therapy of leishmaniasis: superior efficacies of liposome-encapsulated drugs. Proc. Natl. Acad. Sci. USA 75(6):2959).

145. Black, C. D., G. J. Watson, and R. J. Ward. 1977. The use of pentostam

liposomes in the chemotherapy of experimental leishmaniasis. Trans. R. Soc. Trop. Med. Hyg. 76(1):550.

146. New, R. R. and M. L. Chance. 1980. Treatment of experimenta cutaneous leishmaniasis by liposome-entrapped Pentostam. Acta Trop. (Basel) 37(3): 253.

147. Alving, C. R., E. A. Steck, W. L. Chapman, Jr., V. B. Waits, H. D. Hendricks, G. M. Schwartz, Jr., and W. L. Hanson. 1980. Liposomes in leishmaniasis: therapeutic effects of antimonial drugs, 8-aminoquinolines, and tetracycline. Life Sci. 26(26):2231.

148. Prison, P., R. F. Steiger, A. Trouet, J. Gillet, and F. Herman. 1980. Primaquine liposomes in the chemotherapy of experimental murine malaria. Ann. Trop. Med. Parasitol. 74(4):383.

149. Smolin, G., M. Okumoto, S. Feiler, and D. Condon. 1981. Idoxuridine-liposome therapy for herpes simplex keratitis. Am. J. Ophthalmol. 91(2); 220.

150. Bennett, J. E. 1974. Chemotherapy of systemic mycoses. N. Engl. J. Med. 290:30.

151. Hamilton-Miller, J. M. T. 1974. Fungal steroids and the mode of action of polyene antibiotics. Adv. Appl. Microbiol. 17:109.

152. Gale, E. F. 1974. Release of potassium ions from *Candida albicans* in the presence of polyene antibiotics. J. Gen. Microbiol. 80:451.

153. Borisova, M. P., L. N. Ermishkin, and A. Y. Silberstein. 1979. Mechanism of blockage of amphotericin B channels in a lipid bilayer. Biochim Biophys. Acta 553(3):450.

154. Oku, N., S. Nojima, and K. Inoue. 1980. Selective release of non-electrolytes from liposomes upon perturbation of bilayers by temperature change or polyene antibiotics. Biochim. Biophys. Acta 595:277.

155. Ohki, K., Y. Nozawa, and S. T. Ohnishi. 1979. Interaction of polyene antibiotics with sterols in phosphatidylcholine bilayer membranes as studied by spin probes. Biochim. Biophys. Acta 554(1):39.

156. Oehlschlager, A. C. and P. Laks. 1980. Nitroxide spin-probe study of amphotericin B-ergosterol interaction in egg phosphatidylcholine multi-layers. Can. J. Biochem. 58(10):978.

157. Teerlink, T., B. de Kruiiff, and R. A. Demel. 1980. The action of primaricin, etruscomycin and amphotericin B on liposomes with varying sterol content. Biochim. Biophys. Acta 599(2):484.

158. Barratt, M. D. and A. W. Weaver. 1979. The interaction of 3,3',4',5-tetra-chlorosalicylanilide with phosphatidylcholine bilayers. Biochim. Biophys. Acta 555(2):337.

159. Ganesan, M. G., M. Lakshmanan, K. V. Ravindran, and C. Damodaran. 1980. Action of citrinin on liposomes. Z. Naturforsch. 34C(5-6):397.

160. Harder, A., S. Kovatchev, and H. Debuch. 1980. Interactions of chloroquine with different glycerophospholipids. Hoppe Seylers Z. Physiol. Chem. 361(12):1847).

161. Dingle, J. T. 1978. Articular damage in arthritis and its control. Ann. Intern. Med. 88(6):821.

162. Shaw, I. H. and J. Dingle. 1980. Liposomes as steroid carriers in the intra articular therapy of rheumatoid arthritis. In *Liposomes in Biological Systems,* G. Gregoriadis (Ed.). Wiley, Chichester, England, pp. 299–324.

163. Cleary, G. W. and J. L. Zatz. 1977. Interaction of hydrocortisone with model membranes containing phospholipid and cholesterol. J. Pharm. Sci. 66(7):975.

164. Fildes, F. J. and J. E. Oliver. 1978. Interaction of cortisol-21-palmitate with liposomes examined by differential scanning calorimetry. J. Pharm. Pharmacol. 39(6):337.

165. Shinozawa, S., Y. Araki, and T. Oda. 1979. Distribution of [^{3}H] prednisolone entrapped in lipid layer of liposome after intramuscular administration in rats. Res. Commun. Chem. Pathol. Pharmacol. 24(2):223.

166. Weissman, G., P. Anderson, C. Serhan, E. Samuelsson, and E. Goodman. 1980. A general method, employing arsenazo III in liposomes for study of calcium ionophores: results with A23187 and prostaglandins. Proc. Natl. Acad. Sci. USA 77(3):1506.

167. Poste, G., R. Kirsch, W. E. Fogler, and I. J. Fidler. 1979. Cancer Res. 39:881.

168a. Sone, S. and I. J. Fidler. 1980. Synergistic activation by lymphokines and muramyl dipeptide of tumorcidal properties in rat alveolar macrophages. J. Immunol. 125(6):2454.

168b. Sone, S., G. Poste, and I. J. Fidler. 1980. Rat alveolar macrophages are susceptible to activation by free and liposome-encapsulated lymphokines. J. Immunol. 124(5):2197.

169. Fidler, I. J., A. Raz, W. E. Fogler, L. C. Hoyer, and G. Poste. 1981. The role of plasma membrane receptors and the kinetics of macrophage activation by lymphokines encapsulated in liposomes. Cancer Res. 41(2):495.

170. Fidler, I. J., A. Raz, W. E. Fogler, R. Kirsh, P. Bugelski, and G. Poste. 1980. Design of liposomes to improve delivery of macrophage-augmenting agents to alveolar macrophages. Cancer. Res. 40(12):4460.

171. Fidler, I. J. 1980. Therapy of spontaneous metastases by intravenous injection of liposomes containing lymphokines. Science 208:1469.

172. Ziblerman, Y., D. Lichtenberg, and Y. Gutman. 1979. The use of phospholipid liposomes for lithium administration. J. Pharmacol. 31(9):619.

173. Johnson, D. A., N. M. Lee, R. Cooke, and H. H. Loh. 1979. Ethanol-induced fluidization of brain lipid bilayers: required presence of cholesterol in membranes for the expression of tolerance. Mol. Pharmacol. 15(3):739.

174. Johnson, D. A., N. M. Lee, R. Cooke, and H. H. Loh. 1980. Adaptation to ethanol-induced fluidization of brain lipid bilayers; cross-tolerance and reversibility. Mol. Pharmacol. 17(1):52.

175. Schwendener, R. A. and H. G. Wader. 1978. The binding of chlorpromazine to bilayer liposomes. Evaluation of stoichiometric constants from equilibrium and steady state studies. Biochem. Pharmacol. 27(23):2721.

176. Römer, J. and M. H. Bickel. 1979. Interactions of chlorpromazine and imipramine with artificial membranes investigated by equilibrium dialy-

sis, dual-wavelength photometry, and fluorimetry. Biochem. Pharmacol. 28(6):799.
177. Lee, A. G. 1977. Local anesthesia: the interaction between phospholipids and chlorpromazine, propranolol, and practolol. Mol. Pharmacol. 13(3):474.
178. Hill, M. W. 1978. Interaction of lipid vesicles with anesthetics. Ann. N.Y. Acad. Sci. 308:101.
179. Bangham, A. D. and W. T. Mason. 1980. Anesthetics may act by collapsing pH gradients. Anesthesiology 53(2):135.
180. Franks, N. P. and W. R. Lieb. 1978. Where do general anesthetics act? Nature 274(5669):339.
181. Mountcastel, D. B., R. L. Biltonen, and M. J. Halsey. 1978. Effect of anesthetics and pressure on the thermotropic behavior of multilamellar dipalmitoylphosphatidylcholine liposomes. Proc. Natl. Acad. Sci. USA 759(10):4906.
182. Rooney, E. K., M. G. Gore, and A. G. Lee. 1979. Interaction of antihistamines with lipid bilayers. Biochem. Pharmacol. 28(14):2199.
183. Altstiel, L. D. and F. R. Landsberger. 1977. Interaction of colchicine with phosphatidylcholine membranes. Nature 269(5623):70.
184. Colman, R. W., J. Kuchibhotla, M. K. Jain, and R. K. Murray, Jr. 1977. Phase separation in phosphatidylcholine bilayers as predictor of inhibition of blood platelet aggregation by amantadines. Biochim. Biophys. Acta 467:273.

3

Liposome-Cell Interactions in Vitro

Leaf Huang / University of Tennessee, Knoxville, Tennessee

INTRODUCTION

In previous chapters, the formation and properties of liposomes have been discussed in detail. Although earlier liposome research centered around the use of liposomes as a model for biological membranes, more recent studies involve the use of liposomes as a carrier system for the delivery into cells of biologically active substances. For this reason, ample research activities have been focused on the mechanisms of liposome-cell interactions. In this chapter we summarize these studies and offer a unified view as to how liposomes interact with cells in vitro. Since some excellent reviews (1-5) were written not too long ago, this chapter is not intended to be exhaustive, but rather selective, with emphasis on some recent developments of the subject. These discussions begin with some descriptions of the various biochemical, biophysical, and cytological methods employed to elucidate the mechanisms of liposome-cell interaction, followed by a discussion of the various mechanisms of interaction.

METHODS FOR STUDYING LIPOSOME-CELL INTERACTIONS

Uptake of Radioactive Markers

This is the most convenient and widely used method for the quantitation of liposome uptake. The marker for the liposome bilayer could be phospholipid, cholesterol, or a tracer lipid. Markers for the trapped content have also been used, such as sucrose, glucose, $Cr_2O_7^{2-}$ and inulin. The data should be interpreted with caution since a number of pitfalls are associated with this type of experiment. It is known that phospholipids and cholesterol are free to move out of liposomes either as monomers or as part of an exchange reaction with other membranous systems or with lipoproteins (see below). Thus, the uptake of radioactive lipids by cells often overrepresents the actual uptake of liposomes by cells. Furthermore, the marker lipids may be sensitive to metabolic alterations by enzymes present in the medium or at the cell surface and the radioactive metabolites may not be faithful markers of liposomes. Lipid markers such as cholesteryl esters are sensitive to a transfer or exchange reaction catalyzed by serum proteins (6,7). Leakage of a trapped marker from the liposomes or from the cells after the marker is taken up intracellularly may further complicate the interpretation. The simultaneous use of more than one marker in the liposomes (e.g., a lipid-soluble marker and a trapped marker) usually improves the reliability of the data. Since charged molecules or macromolecules such as proteins or polysaccharides are less likely to diffuse through liposome membranes, they are better trapped markers than are small, uncharged ones such as sucrose and glucose. A serious problem in using trapped markers is that the trapping efficiency of liposomes is generally small (see Chap. 1) and it is therefore necessary to use a large quantity of the radioactive marker, which is often economically prohibitive. Radioiodinated low molecular weight proteins are more favorable markers in this regard. They can be highly water soluble, latent within liposomal and cellular membranes and amenable to be labeled economically to a very high specific activity. For example, ^{125}I-labeled lysozyme (molecular weight about 13,000) can be stably trapped in small unilamellar vesicles (SUVs), large unilamellar vesicles (LUVs), and multilamellar vesicles (MLVs) with an efficiency equal to that of [^{14}C] sucrose (8).

In search of a suitable lipid marker, radioactive acyl cholesteryl ether, such as octadecenyl cholesteryl ether, has been shown to be nonmetabolizable both in vitro and in vivo (9). In our laboratory, ^{3}H-labeled hexadecyl cholesteryl ether was used together with [^{14}C] phosphatidylcholine and [^{131}I] lysozyme in a triple-label experiment to study liposome uptake by cultured V79 cells. As shown in Table 1, at 37°C all three markers were taken up to the same extent, indicating the uptake of intact liposomes. At 4°C, however, only [^{131}I] lysozyme and [^{3}H]-hexadecyl cholesteryl ether were taken up together; [^{14}C] phosphatidylcholine uptake was about twice of the other two markers, probably as a result of rapid phospholipid exchange with cellular lipids at this temperature (see below). This

Table 1 Uptake of Radiolabeled Liposome Markers by Chinese Hamster V79 Cells[a]

Incubation temperature ($^\circ$C)	Observed uptake (cpm $\times$ 10^{-3})			Percent deviation from theoretical uptake		
	^{131}I	^{14}C	^{3}H	^{131}I	^{14}C	^{3}H
37	3.9	3.2	24.2	100	94	111
4	0.73	1.21	4.14	100	192	102

[a]SUVs of egg PC were prepared by sonication. They contained trace [^{14}C]DOPC and hexadecyl [^{3}H]cholesteryl ether in the lipid phase and [^{131}I]lysozyme in the entrapped aqueous phase. Incubation was for 1 hr with 10^7 monolayer cells and 1.46 mg of PC per milliliter in Gey's solution. The radioactivity ratio of the three markers in the original liposomes was ^{131}I:^{14}C:^{3}H = 1:0.87:5.58. The theoretical uptake of [^{14}C]DOPC and hexdecyl [^{3}H]cholesteryl ether with respect to [^{131}I]lysozyme uptake is calculated according to these ratios. Percent deviation of the observed uptake from the theoretical uptake is shown in the table.

Source: Unpublished data of M. E. French and L. Huang.

result indicated that [^{3}H] hexadecyl cholesteryl ether is as good a liposome marker as [^{131}I] lysozyme. Since incorporation of tracer quantity of acyl cholesteryl ether into liposomal membranes is 100%, only small amounts of the marker are needed for a given experiment. Unfortunately, this radiomarker is not yet commercially available, but the synthesis is relatively simple (10).

Carboxyfluorescein

Carboxyfluorescein (CF) is a water-soluble fluorophore which can be trapped in liposomes at various concentrations. At low concentrations ($<$1 mM), CF fluorescence is directly proportional to its concentration (11). At higher concentrations, self-quenching of fluorescence becomes increasingly significant. For example, liposomes containing 200 mM CF are only slightly fluorescent. However, upon dilution of the fluorophore (e.g., by lysis of liposomes with a detergent or by releasing the fluorophore into cell cytoplasm as a result of liposome-cell fusion), the fluorescence quantum yield of CF increases about 300-fold (12). Theoretically, under conditions in which endocytosis of intact liposomes and liposome leakage are excluded, the extent of liposome-cell fusion can be determined by measuring the relative fluorescence of the original liposome preparation and the fluorescence of the liposome-treated cells in the presence and absence of detergent. This approach has been used by several investigators (11-19,109). However, recent studies on this probe have revealed several problems. CF leakage from liposomes is not negligible, especially at acidic pH (12), and this leakage is enhanced in the presence of cells (12,18,19). Considerable uptake of free CF by cells occurs in some cases (Ho and Huang, unpublished observation). Furthermore, CF has been found to exit from cells once taken up intracellularly (14). All these drawbacks produce complications in data interpretation.

Fluorescent Lipids

A number of fluorophore-modified lipids have been used to label liposomes and to study the subcellular localization of liposomes after the uptake. N-modified phosphatidylethanolamine (PE) in the form of N-(4-nitrobenzo-2-oxa-1,3-diazole)-PE (N-NBD-PE) (20,21), N-(fluorescein isothiocyanyl)-PE (N-FITC-PE) (20,21), and N-rhodamine-PE (20) has been used. These lipids are not subject to the exchange reaction between liposomes and cells as shown by Struck and Pagano (20) and are therefore excellent markers for liposomes. Other acyl chain-modified lipids, such as NBD- phosphatidylcholine (PC) (20), NBD-PE, and NBD-PG (20), are subject to phospholipid exchange and/or enzymatic modification. Although these lipids are not suitable as liposome markers, they are excellent fluorescent analogs of PC in studies involving PC exchange and metabolism (20,22). Unlike the radioactive lipids, fluorescent lipids can be readily localized in the cells by fluorescence microscopy. Quantitation of the cell-associated fluorescent lipids is a problem since the sensitivity of the bulk fluorescence measurement is often much less than that of the radioactivity measurement. Furthermore, fluorescence is sensitive to quenching factors in the incubation medium and/or cells. Fluorescence quantitation of individual cells has been attempted by direct photometric measurement (21) or by photon counting (23), with the latter giving a greater sensitivity. These studies are quite tedious since measurements have to be done on many cells to obtain statistically significant data. Finally, the dequenching of concentrated fluorescent lipid in the liposome bilayer, similar to the dequenching of CF, has been used to study the transfer of phospholipids between liposomes (24) but has not yet been used in studies of liposome-cell interactions.

Resonance Energy Transfer

This technique also utilizes fluorescent lipids, but its sophistication deserves a separate discussion. Resonance energy transfer is a phenomenon in which the excitation energy of a fluorophore (donor) is transferred to a second fluorophore (acceptor), resulting in the fluorescence of the second fluorophore. The transfer efficiency depends on such factors as the orientations of the two fluorophores, overlap of the emission spectrum of the donor and the excitation spectrum of the acceptor, and the distance between the fluorophores (25). When two suitable types of fluorescent lipid are present in the same liposome membrane, the close proximity results in efficient energy transfer which can be readily measured by either the enhanced fluorescence of the acceptor or the quenched fluorescence of the donor (26). N-NBD-PE (donor) and N-rhodamine-PE (acceptor) have been used by Pagano and his colleagues to assess the fusion of liposomes with cells (23,27,28). Since the resonance energy transfer depends critically on the distance between the fluorophores, a dilution of phospholipid by mixing with cellular lipids as a result fo fusion would significantly decrease the energy transfer efficiency. If liposomes are taken up by adsorption or endocytosis, no lipid dilution occurs and hence there is no decrease in energy transfer. Since N-modified PE is not exchangeable with cellular lipids, the foregoing interpretation is not complicated by lipid exchange. This relatively simple technique is presumably very specific for liposome-cell fusion (23,27).

Electron Microscopy

Since liposomes prepared by most of the published methods are 0.5 μm or less in diameter, direct visualization of individual liposomes by optical microscopy is difficult if not impossible. Visualization with the electron microscope has been the method of choice, particularly in viewing liposome-treated cells. Intact or modified liposomes adsorbed to the lymphocyte surface have been seen in thin-section transmission electron microscopy (EM) (17). The adsorbed liposomes can also be observed by scanning EM, although the resolution is not as good as

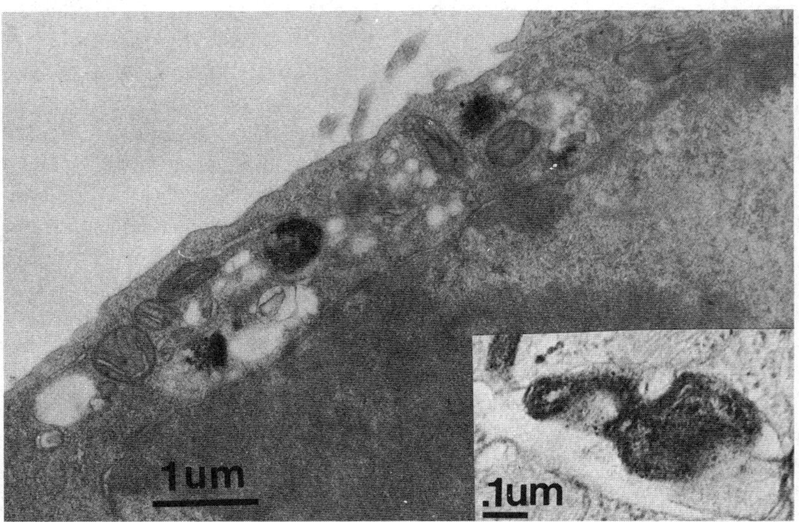

Figure 1 Thin-section micrograph of Chinese hamster V79 cells treated with
LUVs containing BCIP for 1 hr at 37°C. The thin section was not poststained.
An electron-dense deposit can be seen in the phagolysosomes. Insert is a high
magnification of the dense deposit. (From Ref. 34.)

with transmission EM (29,30). Unequivocal identification of liposomes fusing
with or being endocytosed by cells is difficult due to the numerous membranous
structures on the cell surfaces or inside the cells. However, in the case of MLVs,
which display a characteristic "onionskin" morphology (30-32,53), identification
becomes easier. Cytochemical markers are often needed to help in identifying
liposome components. Horseradish peroxidase (30,53) and ferritin (33) entrapped
in MLVs or LUVs demonstrated the endocytosis and/or adsorption of intact liposomes by cells. A new cytochemical marker, Br,Cl-indolyl phosphate (BCIP), has
recently been developed in the author's laboratory for specific demonstration of
liposome endocytosis (34). BCIP is a small, soluble molecule which can be
trapped even in the smallest SUV. It is a substrate for the lysosomal acid phosphatase. The dephosphorylated product spontaneously condenses into an indigo
blue which precipitates to form an electron-dense deposit inside the phagolysosomes. Figure 1 shows a thin-section TEM micrograph of Chinese hamster V79
cells treated with LUVs containing BCIP. The appearance of the electron-dense
deposit in phagolysosomes was eliminated by such treatments as cytochalasin B
and NaN_3 plus deoxyglucose and chloroquine, all known inhibitors of endocytosis. Since the formation of the indigo blue derivative takes place only when BCIP

is in physical contact with the enzyme, this marker is specific for liposomes that are phagocytosed and subsequently delivered to phagolysosomes as a result of phagosome-lysosome fusion. EM autoradiography has also been used to observe the subcellular distribution of liposomal lipids (17,29,35,49) and contents (17). The question of whether the distribution of the radioactive lipids in liposome-treated cells as seen in EM autoradiography represents the true location of the liposomal lipids rather than being an artifact of EM processing and handling has been investigated. In one study, cells treated with liposomes containing [³H]PC were added to untreated cells and the mixture was subsequently processed for EM autoradiography. A significant number of silver grains were found in the untreated cells, indicating a redistribution of the [³H]PC among cells (36). However, an identical study done by another laboratory using different cell types has resulted in no appearance of [³H]PC in the untreated cells (29). The issue remains unsettled.

Other Biophysical Methods

Fluorescence recovery after photobleaching (FRAP) has been used to measure the lateral diffusion coefficient of fluorescent label on cell surfaces. In this technique, a spot on the cell surface membrane is bleached by an intense laser beam. The recovery of fluorescence intensity of the spot is continuously monitored as the fluorophores in the neighboring area diffuse into the spot (37). The lateral diffusion coefficient (D) of the fluorophore is calculated from the rate of the fluorescence recovery. Cells treated with liposomes containing fluorescent lipids are expected to show a D value on the order of 10^{-7}-10^{-8} cm^2/sec if the fluorescent lipids are incorporated in the plasma membranes as a result of liposome-cell fusion and/or lipid exchange, whereas the fluorescent lipids should be essentially immobile if the intact liposomes adsorb on the cell surfaces (38,39). Another technique, using spin-labeled phospholipids, can also provide information on the mixing of liposomal lipids with cellular lipids (40). Liposomes containing high concentrations of spin-labeled phospholipids show strong exchange broadening in the electron paramagnetic resonance (EPR) spectrum due to the close proximity of the spin labels in liposome bilayer. Upon transfer into cell membranes by either fusion or lipid exchange, the exchange broadening disappears as a result of dilution. Another technique based on the dilution of the lipid probes into cellular lipids was described by Owen (106). A relatively high concentration of N-pyrenesulfonyl PE was incorporated into SUVs such that the fluorescence of the pyrene excimers in the lipid bilayer was prominent. Upon fusion of liposomes with cells, or transfer of the probe lipids to cell membranes, excimer fluorescence was significantly reduced due to the dilution of the probe lipids with cellular lipids. This method and the EPR method, although sensitive and potentially quantitative, cannot easily distinguish between the fusion and the lipid exchange mechanism. A fourth technique, γ-ray-perturbed angular correlation spectroscopy, monitors the release of the liposomal trapped content (33,41). In this method, the rotational correlation time of $^{111}In^{3+}$ encapsulated in liposomes is measured. Leakage of $^{111}In^{3+}$ from the liposomes results in the complex formation of this rare earth cation with proteins in the surrounding medium or cells and a decrease in the rotational correlation time. Since the measurement can be done in intact cells or animals, it offers a convenient assay method for the fate of liposome contents once taken up by the cells (42).

INTERACTION OF PROTEIN-FREE LIPOSOMES WITH CELLS

Historically, mechanistic studies of liposome-cell interactions started with liposomes made of lipids only. A vast amount of literature on the subject has appeared in the last few years. Unfortunately, the results of different investigators were not always in agreement. Substantial confusion still exists regarding the mechanisms of liposome-cell interaction. A great deal of the controversy arises from the facts that (a) liposomes can interact with cells by more than one mechanism; (b) the experimental techniques used to elucidate these mechanisms are not always faultless; (c) liposomes used by different investigators differ in lipid composition, method of formation, purity of lipids, and other experimental conditions; and (d) the cell systems used also differ from one study to another. We will present various mechanisms of interaction which have been proposed. Whenever possible, the evidence for and against each mechanism will be critically evaluated.

Mechanisms of Interaction

Fusion

Fusion of liposomes with cells is defined as the unification of the outermost lipid bilayer of the liposome with the plasma membrane of the cell. A necessary consequence of fusion is the mixing of the liposomal aqueous contents with the cytoplasm of the cell. Another important criterion for fusion is the mixing of the liposomal lipids with the plasma membrane lipids. Early reports of liposome-cell fusion were based largely on the apparent uptake of liposomal lipids and/or their contents by the cells (31,35,43-49) and the insensitivity of such uptake to metabolic inhibitors and/or glutaraldehyde fixation (35,44,45). Although it is important to show that the rate of uptake for liposomal lipids is the same as that for the aqueous content, this is not sufficient to prove liposome-cell fusion since liposome adsorption and endocytosis would give the same result (17,44). The injection of the aqueous contents of liposomes into the cytoplasm has best been shown with the fluorescence dequenching of carboxyfluorescein (13,15-17). The cell fluorescence usually appears diffused over the entire cell (13,17). This technique has been used to demonstrate fusion of SUVs prepared by sonication with the rod outer segments (13), lymphocytes (13,16,17), and fibroblasts (29). However, drawbacks of this technique (discussed earlier) have cast some doubts on the quantitation of the degree of fusion. Fusion of SUVs comprised of dioleyl PC with lymphocytes has also been demonstrated by the EM radioautography of cells treated with liposomes containing [^{3}H] inulin (17). Silver grains were found in the cytoplasm in this study.

There has been a considerable amount of controversy as to what types of liposomes can fuse with cells. Based largely on the uptake of radiolabeled liposomes and the sensitivity of this uptake to metabolic inhibitors, it was claimed that SUVs can fuse with cells if the liposomes are negatively charged, fluid at the incubation temperature, and Ca^{2+} ions are present in the incubation medium. These observations were not consistent with the results obtained by other laboratories in which SUVs comprised of neutral PC were demonstrated to fuse with cells even in the absence of Ca^{2+} (13,15-17,29,35). It is generally agreed that below the phase transition temperature of the liposomes, adsorption of intact liposomes to the cells instead of fusion is the major mechanism of interaction (17, 29,39). Ostro et al. have reported the optimal conditions for the cellular uptake of LUV-encapsulated RNA (50). They found that the uptake of RNA by cells in the late G_I phase of the cell cycle was not inhibitable by cytochalasin B. This uptake was assumed to be due to liposome-cell fusion. They also reported that the liposome uptake was enhanced when the cell membranes were fluidized by either pretreatment of cells with cholesterol-free liposomes or by treatment of cells with Piracetam. Inclusion in the liposomal membranes of a fusogenic lipid (e.g., lyso PC) is generally believed to increase the fusion efficiency (46,47,50-53), although a high concentration of lyso PC is likely to result in leaky liposomes (16,50). It is not clear whether SUVs are more likely to fuse than are larger liposomes (16,45). Theoretically, one would expect this to be true, because SUVs have higher surface tension due to the greater radius of curvature and because they are more approachable to the cell surfaces due to their smaller size.

Perhaps the most convincing evidence for liposome-cell fusion comes from the reports that biologically active substances encapsulated in liposomes can be introduced into the interior of the cells. Some of these substances are impermeable to cell membranes and can express their biological effects only when delivered to the cell interior. Examples include the introduction of adenosine 3′,5′-cyclic monophosphate (cyclic AMP) for the inhibition of cell growth (48) and activation of intracellular enzymes (54); initiation factor (eIF-3) for protein synthesis (55); ricin for the killing of ricin-resistant cells (56); actinomycin D for the killing of drug-resistant cells (57,58); Ca^{2+} ions for the activation of mast cells (59); monovalent cations for the excitation of neurons (60); amyloglucosidase for the removal of cytosolic glycogens in Pombe's fibroblasts (61); mRNA for the syntheses of foreign proteins (62,63); virus and viral genomes for the infection of resistant cells (64–66); and the metaphase chromosomes for the expression of a new phenotype in the treated cells (67). Although the results of these experiments are consistent with a fusion mechanism, it is not clear in most studies to what extent fusion of liposomes with cells has occurred. Introduction of minute amounts of biologically active substances is often enough to elicit the effects observed. As a matter of fact, several recent studies have suggested that liposome-cell fusion is *not* a major mechanism of interaction, regardless of the type of liposome used. Szoka et al. (39) have observed a punctate fluorescence distribution on mouse L929 cells incubated with SUVs containing a fluorescent lipid, indicating that most of the cell-associated liposomes are adsorbed on the cell surface. Furthermore, the adsorbed liposomes showed little lateral movement on the treated cells when measured using the FRAP technique. Hallett and Campbell used liposomes containing obelin, a Ca^{2+}-dependent photoprotein, to study the interaction of liposomes with isolated adipocytes (68). No cytoplasmic delivery of obelin was detected for negatively or positively charged liposomes under any conditions tested. They concluded that liposome fusion with cells was not a major mechanism of liposome uptake. Struck et al. (27) used SUVs containing N-NBD-PE and N-rhodamine-PE and the technique of resonance energy transfer to study the interaction of liposomes with Chinese hamster V79 cells. They concluded that under the experimental conditions used, the major mechanism of interaction was the adsorption of intact liposomes on cell surfaces. The same group later reported that liposome fusion with cells became increasingly important when the concentration of liposomes in the incubation medium decreased; at 0.05 μg of phospholipid per milliliter, virtually all cell-associated liposomes were taken up by fusion (23). At higher concentrations (e.g., 1–1000 μg/ml, the concentration range used in most studies), less than 10% of liposome uptake could be attributed to fusion. The same results were obtained when SUVs containing negatively charged lipids [e.g., phosphatidylserine (PS) and phosphatidic acid (PA)] were used in the presence or absence of Ca^{2+} in the incubation medium (23). Although these studies deny fusion as a major mechanism of interaction, they do not exclude the possibility that a small fraction of the total liposome uptake is due to fusion. The observed delivery of entrapped substances into cells could still be the result of fusion.

Adsorption

This term refers to the stable attachment of liposomes on the cell surface. The attached liposomes may be intact or modified by the medium or cell components. Significant adsorption of liposomes occurs at incubation temperatures near or below the phase-transition temperature (Tc) of the liposomes (17,29,39,49). Thus, the apparent uptake of dimyristoylphosphatidycholine (DMCP) liposomes by V79 cells at 2°C is about twice of that at 37°C (49). This peculiar temperature dependence of uptake was further investigated in mouse lymphocytes by Huang

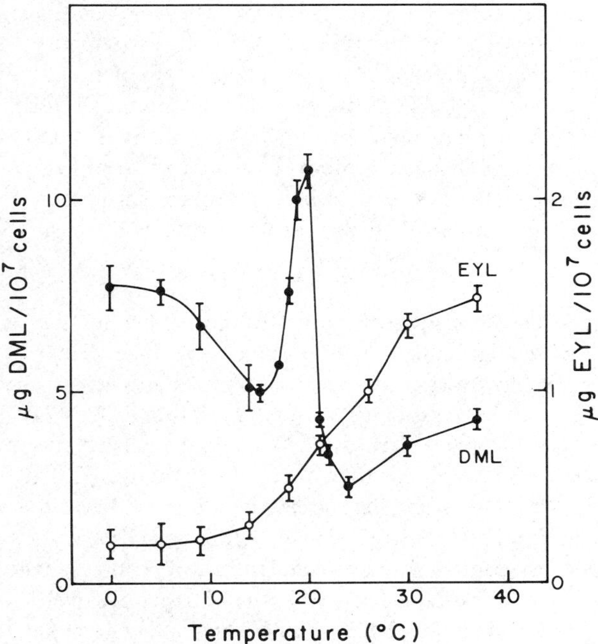

Figure 2 Temperature-dependent uptake of SUVs comprised of DMPC (DML) or egg PC (EYL) by mouse thymocytes. Incubation was for 1 hr. (From Ref. 17.)

et al. (17). As shown in Figure 2, a peak of uptake was observed near the Tc of DMPC-SUV; and the uptake at 2°C was considerably higher than that of 37°C. Thin-section EM and fluorescence microscopy of liposome-treated cells revealed numerous adsorbed liposomes on the surface of the cells treated at 2°C (17). The enhanced liposome adsorption for solid liposomes was also observed for SUVs composed of dipalmitoylphosphatidylcholine (DPPC) and sphingomyelin (29). In the case of mouse lymphocytes, the adsorption of liposomes was correlated with their ability to augment lectin binding to the liposome-treated cells and with lectin-induced mitogenesis (69). In Pagano's laboratory, two cell surface polypeptides of molecular weight 60,000 were inferred to be responsible for the enhanced adsorption of DPPC liposomes, as revealed by cell surface radioiodination and sodium dodecyl sulfate polyacrylamide gel electrophoresis (SDS-PAGE) autoradiography (29). Szoka and co-workers also found that solid SUVs containing neutral, positive, or negative charges adsorbed to the surface of mouse L929, ELD, HTC, and L1210 cells 3- to 10-fold more than did fluid liposomes (39). Rabbit platelet adhesion to glass surfaces coated

with phospholipid films was studied by Margolis et al. (70). They showed that
platelets adhered readily to solid phospholipid films of DPPC, distearoylphosphatidylcholine (DSPC), and sphingomyelin. However, platelet adhesion did not
occur when fluid films composed of egg PC, dioleophosphatidylcholine (DOPC),
and PE were used unless they were cross-linked by either glutaraldehyde or O_sO_4.
The molecular explanation for the enhanced liposome adsorption near or below
their Tc is not clear. It is also not clear if the adsorbed liposomes can subsequently undergo endocytosis, fusion, or other types of interaction with cells.

Lipid Exchange

Lipid exchange involves the transfer of liposomal lipids to the cell membrane
with concomitant transfer of cellular lipids to the liposome membrane. This
transfer is not accompanied by the transfer of liposome contents into cells.
Pagano and Huang first observed the appearance of cellular lipids in DOPC-SUV
which had been incubated with V79 cells (35). At 2°C the lipid exchange mechanism is the major pathway responsible for the uptake of radioactive PC. At
37°C other processes dominate. Sendra and Pagano had later observed that the
transfer of liposomal dioleophosphatidylethanolamine (DOPE) to cells was accompanied by the transfer of the same amount of cellular PE to liposomes; hence,
a bona fide "exchange" reaction was established (71). They further observed
that (a) exchange required physical contact between liposomes and cells, (b) exchange was specific with respect to the head group structure of the phospholipid, and (c) exchange was reduced by 50% after cells were treated with trypsin
(71). These observations strongly suggest that a cell surface exchange protein(s)
similar to the well-known soluble phospholipid exchange protein (PLEP) is responsible for this process (72), although such protein(s) has never been identified. Since most of the DOPE exchanged could be derivatized by trinitrophenyl
sulfonate (TNBS) under conditions in which TNBS was impermeable to cells,
and the cellular PE on the liposomes could be removed by using soluble PLEP,
they also concluded that lipid exchange took place between the outer monolayers of liposome bilayers and the plasma membrane bilayers of the cells (71).
Struck and Pagano (20) prepared a number of fluorescent phospholipids in which
the fluorophores were located either on the acyl chains of PE, PC, PG, and ceramide or on the head groups of PE. These fluorescent lipids were incorporated
into SUVs and their ability to undergo an exchange reaction with cells was
studied. The results showed that the acyl-modified PE and PC were exchanged
readily but not the N-modified PE or acyl-modified PG and ceramide, further
indicating the head group specificity of the exchange process. The exchanged
fluorescent phospholipids underwent rapid lateral diffusion at a rate similar to
that of the native cell surface lipids as measured by the FRAP technique (20).

   Another type of lipid transfer between liposomes and cells has been described for cholesterol (73,74). Apparently, cholesterol molecules are rather

free to move out of the liposome membranes and enter the cell membranes, and vice versa (75). Upon incubation with cholesterol containing liposomes, the cell membranes tend to equilibrate with liposomes with respect to the cholesterol content. The direction of the net flow of cholesterol depends on whether the cholesterol content of the liposomes is higher or lower than that of the cell membranes. Thus, the level of cholesterol, usually expressed as the cholesterol/ phospholipid ratio, of the cell membranes can be readily modulated. Many studies have used this approach to elucidate the role of plasma membrane cholesterol in cellular functions such as cell proliferation (e.g., lymphocyte mitogenesis) (69,76, 142) and the transport activities of the cells (77-81).

Endocytosis

Endocytosis involves the ingestion of intact liposomes by cells through a process resembling the phagocytosis of particles by cells. The endocytosed liposomes often appear in the cellular digestive apparatus, the phagolysosomes. The uptake of liposomes by endocytosis can be readily seen in thin-section TEM (30, 32-34,44,82), sometimes with the help of encapsulated cytochemical markers such as horseradish peroxidase (30,34,82), ferritin (33), or Br,Cl-indolyl phosphate (34). The extent of liposome endocytosis depends on the types of cells used. The highly phagocytic macrophage cells apparently endocytose liposomes with little selectivity (Hinckley, French, and Huang, unpublished observation), although the initial rate of endocytosis may depend on the lipid composition of liposomes (32). However, the endocytic uptake of liposomes is enhanced if specific glycolipids are present in the liposome membranes. Wu et al. reported that SUVs containing a 6-aminomannosyl derivative of cholesterol were endocytosed by mouse peritoneal macrophages at a much higher rate than were SUVs containing other types of glycosylcholesterol or cholesterol itself (33). This result suggests that the endocytosis of liposomes containing specific ligands may be mediated by a specific receptor system in the macrophage cells. They indicated further that the contents of liposomes stayed encapsulated even after they were ingested into phagolysosomes (33).

For cells that are less endocytic than macrophages, the degree of liposome endocytosis depends on the type of liposome and the cells used. It has been proposed that negatively charged solid liposomes and neutral fluid and solid liposomes are taken up predominantly by endocytosis (45). Other studies suggested that positively charged fluid liposomes could also be endocytosed (30,83). The evidence provided by these studies is based largely on the sensitivity of the liposome uptake to drugs known to inhibit endocytosis, such as cytochalasin B, NaN_3, and deoxyglucose (30,45). Endocytosis does not occur at an incubation temperature below 18-21°C (84). Sensitivity to low temperature is often used as evidence for the endocytic uptake of liposomes, although other uptake mechanisms may also be temperature sensitive.

Extraction of Cellular Proteins

Heustis and her colleagues have reported that cell membrane proteins were transferred to SUVs (but not to MLVs) composed of DMPC when they were incubated with human red cells (85). They identified seven polypeptide species transferred; none of them was a major polypeptide of the red cell membrane. The liposome-extracted polypeptides could subsequently be transferred back to the red cells. One peculiar consequence of this extraction was a massive influx of Na^+ ion into the red cells, which led to their eventual osmotic lysis (86). Since the treated cells were still impermeable to other cations, anions, and neutral small molecules, and since the Na^+ ion influx could be prevented by tetrodotoxin, they proposed that liposomes had selectively extracted a membrane protein system involved in the regulation of Na^+ ion permeability. Dunnick et al. have also reported the transfer of ^{125}I-labeled spleen cell surface proteins to the sonicated liposomes (87). The ectoenzyme, 5′-nucleotidase, of the rat adipocytes was also found to transfer from cells to liposomes (68). In all these reports it was unclear by which mechanism that cell membrane proteins were transferred to liposomes.

Cell-Induced Liposome Leakage

In a study using CF as a liposome-entrapped marker, Szoka et al. recognized that the presence of cells enhanced the leakage of CF from liposomes (12). Similar observations were reported by other investigators using hepatoma cells (18,19), hepatocytes (19), and adipocytes (68). All these authors suggested that the leakage was a result of direct liposome-cell contact. It is known that liposomes interact with serum albumin and lipoproteins, leading to massive leakage of the liposome contents (88,105). It is possible that the cell-induced liposome leakage is mediated by the serum lipoproteins adsorbed on the cell surface. This possibility was, however, ruled out in a recent study by Van Renswoude and Hoekstra (19). It was shown that the extent of CF leakage from SUVs correlated directly with the number of liposomes that became cell associated and that the leakage occurred only when liposomes contained more than 30 mol% of cholesterol or more than 20 mol% of cholesterol plus at least 10 mol% of negatively charged lipids. In contrast to this, in the case of serum lipoprotein-induced leakage, several authors have reported that increasing cholesterol content in liposomes actually decreases liposome leakage (89,90,105). Van Renswoude and Hoekstra also showed that the leakage-inducing activity of the cell could be removed by either trypsin or neuraminidase treatment of the cells (19). The activity could be solubilized in the medium by freezing and thawing of either intact cells or isolated plasma membranes. As discussed in the preceding section, it is not clear at present whether the liposome leakage is a result of the extraction of cellular proteins by liposomes.

Summary

Although liposomes can fuse with cells in vitro, recent studies have revealed that fusion is not a major mechanism of interaction, regardless the size, charge, and fluidity of the liposomes. However, even minor amounts of fusion could be enough to introduce encapsulated materials into cells which can be measured by the assay of their biological activity. The adsorption of liposomes is the predominant pathway for the uptake of liposomes at incubation temperatures near or below the Tc of the liposome. Endocytosis of intact liposomes occurs at temperatures above 18-21°C and is inhibited by a number of drugs. The extent of liposome endocytosis depends largely on the type of cells used. Cells of high endocytic activity (e.g., macrophages) apparently ingest liposomes of any type, although liposomes containing specific glycolipid (e.g., 6-aminomannosyl lipids) are taken up at an enhanced rate. Lipid exchange occurs between the outer monolayer of the liposome bilayer and the outer monolayer of the plasma membrane bilayer and is probably mediated by cell surface-bound exchange protein(s). Exchange is specific with respect to the head group structure of the phospholipid. It is the dominant interaction mechanism for fluid liposomes of low incubation temperatures. Cell membrane proteins can be transferred to liposomes upon incubation; it leads to the osmotic lysis of red cells. Liposomes become leaky when they come in contact with some cells. The leakage is probably mediated by cell membrane protein(s), perhaps glycoprotein(s).

INTERACTION OF PROTEOLIPOSOMES WITH CELLS

The development of techniques of membrane reconstitution has advanced rapidly in the last decade. Various membrane proteins, integral or peripheral, have been inserted into the membranes of liposomes of different size, charge, and fluidity. The interaction of proteoliposomes (liposomes with proteins or polypeptides either attached on or embedded in the bilayer membranes) with cells has received increasing attention because of the promising probability that specific cellular functions may be modified through treatment of the liposomes. For example, liposomes containing protein or glycoprotein antigens can stimulate lymphocyte proliferation and differentiation (see Chap. 8 for details). Recent attempts to use liposomes as drug carriers have prompted the development of methods with which target-specific antibody can be covalently attached to liposomes. The interaction of antibody-coated liposomes with target or nontarget cells is interesting for obvious reasons. In the following sections, demonstrated mechanisms of interaction will be discussed, although the available information is not as abundant as it is for protein-free liposomes.

Mechanisms of Interaction

Soon after the fusion potential of liposomes was realized, the use of proteoliposomes as vehicles to introduce exogenous membrane proteins to cell membranes
was attempted. The acetylcholine receptor of *Torpedo* was purified and reconstituted into PC liposomes which were subsequently incubated with cultured
fibroblasts. A small amount of acetylcholine receptor appeared on the surfaces
of treated cells (91). This low level of receptor association was later demonstrated as being primarily the result of liposome adsorption (34). A different
approach was to reconstitute into the same liposome preparation both the acetylcholine receptor and the fusion protein of Sendai virus. The latter has well-docu-
mented fusion-promoting activity (92). Proteoliposomes containing both proteins showed enhanced fusion activity compared to liposomes containing receptors
alone (34). Furthermore, cells treated with fusion-competent liposomes could
serve as targets for an antibody-dependent, complement-mediated cytolysis by
antireceptor antibody. This was not true for cells treated with liposomes containing receptor alone (see Fig. 3). These results indicate that the receptor molecules
are likely to be inserted into the plasma membranes of the treated cells, as would
be expected if fusion were occurring. In a similar study, mouse major histocompatability antigen, $H\text{-}2K^k$, was co-reconstituted with Sendai viral proteins on liposome membranes. Incubation of the proteoliposomes with cells of an alternate
histocompatability resulted in the transfer of $H\text{-}2K^k$ to the cell surface as assessed
by these cells' susceptibility to lysis by specifically primed cytotoxic T lymphocytes (93). Also, proteoliposomes containing reconstituted Ca^{2+}-ATPase have
been fused with erythrocytes, resulting in the transfer of adenosine triphosphate
(ATP)-dependent Ca^{2+} uptake activity in these cells (94). It has also been shown
that beef heart cytochrome oxidase can be transferred to erythrocyte membranes
by liposome-cell fusion (95).

Fusion has also been induced by the use of polyethylene glycol. Szoka et al.
showed that liposomes containing glycolipids could be effectively bound to fibroblasts by the use of lectins such as concanavalin A and *Ricinus communis* agglutinin, presumably due to the bridging effect of the lectins. Subsequent addition of
polyethylene glycol to the liposome-cell complex resulted in a high efficiency of
fusion of the bound liposomes to the cells. This was evaluated by the injection
of a fluorescent marker into the cells (96). Similarly, proteoliposomes containing
the hepatic asialoglycoprotein receptor were fused with fibroblasts in the presence of polyethylene glycol. The turnover of the exogeneous receptors on the
fibroblasts was then studied in detail (97).

Liposomes containing Sendai viral glycoproteins have also been used to introduce biologically active substances into cells. Fragment A of diphtheria toxin is
an adenosine diphosphate (ADP)-ribosyl transferase which catalyzes a reaction
that inactivates elongation factor 2 and hence the entire cellular protein synthesis

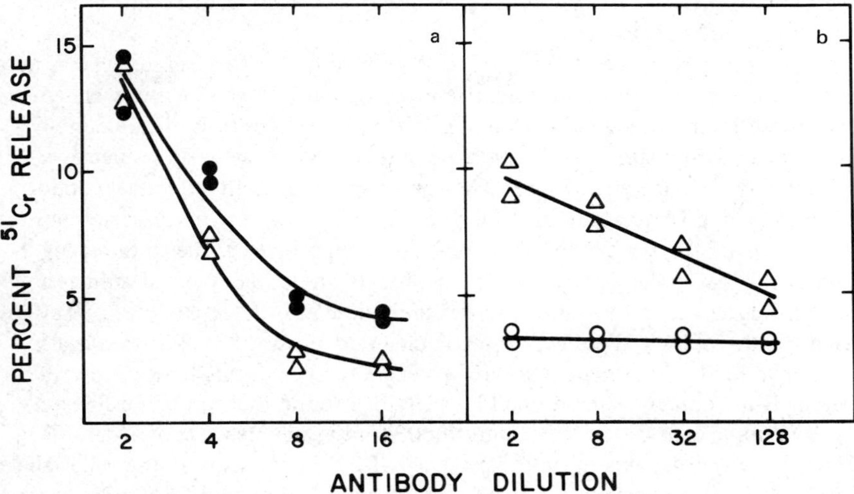

Figure 3 Antibody-dependent, complement-mediated lysis of mouse L929 cells pretreated 1 hr at $37°C$ with liposomes containing acetylcholine receptor and Sendai virus envelop proteins ($\triangle$), with liposomes containing receptor alone ($\bigcirc$), or with 2×10^4 hemagglutinating units of live Sendai virus ($\bullet$). Cell lysis expressed as specific ^{51}Cr release was tested on anti-Sendai antiserum (a) or anti-acetylcholine receptor antiserum (b) at different serum dilutions. (From Ref. 34.)

machinery, leading to cell death (98,99). Since cells are normally impermeable to fragment A, it is not cytotoxic when applied as a free protein. Okada and colleagues have shown that the cytotoxicity of fragment A is significantly enhanced when it is encapsulated in proteoliposomes containing Sendai glycoproteins (100,101). The enhanced cytotoxicity is presumably the result of an efficient delivery of fragment A into cells by liposome-cell fusion. This approach appears very promising with respect to its potential applications in cancer chemotherapy since one molecule of fragment A introduced into a cell is enough to kill it (101). Recently, an interesting report from the same laboratory indicated that liposomes with encapsulated fragment A, in the absence of viral proteins, can selectively kill human embryonic lung cells infected with subacute sclerosing panencephalitis virus but not the uninfected cells (103). Although the mechanism of the selective killing is still unknown, it is possible that the liposomes can selectively fuse with the infected cells since there are fusion proteins expressed on the surface of these cells but not on the uninfected cells (103). This result has brought up an intriguing possibility of using liposomes as carriers for antiviral agents since many enveloped viruses induce the expression of fusion proteins on the infected cell surfaces.

Liposomes have been used as vehicles to introduce enzymes into cell. A number of enzymes, mainly those defective in lysosomal storage diseases, were encapsulated in liposomes and delivered into cells by endocytosis. For example, horseradish peroxidase was encapsulated in MLVs which were subsequently coated with heat-aggregated IgM. The proteoliposomes were effectively endocytosed by dogfish phagocytes, which are normally low in peroxidase activity (82). Apparently, the enhanced endocytosis of liposomes is due to the recognition of the IgM molecules by the Fc receptors of the phagocytes. A similar approach was also used to introduce hexosaminidase into the leukocytes of Tay-Sachs patients (104). Fc-receptor-mediated endocytosis of liposomes was also demonstrated by Leserman et al. (107). These workers, using liposomes containing N-nitrophenyl-substituted PE (DNP-PE), found that uptake of liposomes by lymphoma cells was enhanced up to 600-fold in the presence of anti-DNP IgG. The uptake was dependent on the presence of the Fc-receptor on the cell surface. Interestingly, liposome-encapsulated methotrexate only killed those cells containing Fc receptors having a high degree of receptor-mediated endocytosis (107). These works have demonstrated the possibility of using Fc receptors as a liposome targeting strategy.

Liposome Targeting in Vitro

As discussed in Chapter 9, the use of liposomes as drug carriers in vivo has faced serious problems, not the least of which is the lack of target-cell specificity. Various attempts to coat the liposome surfaces with some "recognition molecules" for specific delivery of the encapsulated drugs have been reported. Most of the investigators used antibody as a targeting molecule. We shall review the various methods with which antibody molecules can be attached to the liposome surfaces. The work on the mechanisms of interaction between the targeted liposomes and cells has just begun. A summary of the literature is presented.

Antibody as a Targeting Molecule

Gregoriadis and Neerunjun prepared liposomes by brief sonication in the presence of antibody and ^{111}In-labeled bleomycin. After the removal of unbound antibody and untrapped bleomycin, the liposomes were incubated with cells. When antifibroblast antibody was used, selective binding of liposomes with fibroblast cells, but not with HeLa or AKR-A cells, was observed (108). Similar specific binding of liposomes with HeLa or AKR-A cells was observed when anti-HeLa or anti-AKR-A antibody was used, respectively. This was the first successful attempt at antibody-mediated liposome targeting. Weinstein et al. (109) took a different approach. In their studies, liposomes containing a haptenated lipid,

N-DNP-PE, were found to bind with trinitrophenylated cells in the presence of anti-DNP antibody. This binding was about 20-fold higher than that in the absence of antibody. Since anti-DNP cross-reacted with the trinitrophenyl (TNP) hapten, the enhanced binding was probably due to antibody bridging between liposomes and cells. However, the authors had also shown that the antibody-induced binding was not accompanied by the intracellular delivery of the liposome encapsulated marker, CF. The latter result indicated that the bound liposomes did not fuse with cells, probably because the antibody-antigen complexes prevented a close juxtaposition between the liposomes and cell membranes.

The antibody molecules used in much of the early targeting work were not covalently attached to liposomes. To provide greater stability of the antibody-coated liposomes, various methods to covalently couple antibody to the liposomes were developed. There are basically two different strategies used in coupling. The first one uses preformed liposomes and the antibody is coupled directly or indirectly by the use of coupling reagents. The other approach is to derivatize antibody covalently with lipophilic molecules and subsequently incorporate these amphipathic antibodies into liposomes under conditions similar to those used in membrane reconstitution. Although significant amounts of antibody molecules can be attached to liposomes using either method, the first approach cannot easily avoid potential deleterious effects on the liposome lipids and/or cont^nts due to the relatively harsh coupling conditions. Furthermore, some of these methods are prone to homocoupling between liposomes or between antibodies, which is completely avoided in the second approach. The second type of coupling method has another advantage, the ease of incorporation of other membranous proteins into the antibody-coated liposomes. Tables 2 and 3 present a summary of various coupling methods published to date.

Monoclonal antibody was used in a few of these reports (115,118,120). Whether monoclonal antibody is superior to the conventional polyclonal antibody is not clear at present, but the ease of obtaining a large quantity of antibody and the homogeneity of these molecules are likely to provide better experimental systems. Furthermore, some laboratories purposely used Fab' or F(ab')$_2$ instead of the whole IgG as the targeting molecules to avoid the possible Fc receptor-mediated endocytosis of liposomes by macrophages (113,114,117,119). However, in preliminary experiments, we observed that antibody-targeted liposomes did not show enhanced uptake by the mouse peritoneal macrophages compared to liposomes without antibodies (Hinkley and Huang, unpublished results). The coupling of antibody to liposomes probably occurs mainly in the Fc region of the IgG, resulting in a relatively "buried" configuration of the Fc region, making it unavailable to the macrophage receptors.

In most cases, antibody-targeted liposomes showed enhanced specificity in binding to antigen-bearing cells (113-119). Whether such selective binding leads to enhanced internalization of the encapsulated contents has not been vigorously

Table 2 Methods for Covalent Coupling of Antibody to Preformed Liposomes

Liposome type	Lipid coupled	Antibody	Coupling method	Direct or indirect coupling	Remarks (references)
SUV	PE	IgG	Glutaraldehyde	Direct	Low-level coupling with possible homocoupling (110,111)
SUV	PE	IgG	Carbodiimide	Direct	Homocoupling likely (112)
LUV	IO_4^--oxidized glycolipids	F(ab')$_2$ and IgG	Reduced Schiff's base	Direct	High-level coupling, oxidation of liposome components likely (113,114)
SUV	Dithiopyridine-PE	Thiolated monoclonal IgG and protein A	Disulfide bonds	Direct	Arm between IgG and liposomes (115)
SUV	Biotinyl-PE	Biotinyl IgG	Through biotin-avidin binding	Indirect	Antibody specific to tumor-specific glycolipids (116)
LUV	Dithiopyridine-PE	Fab'	Disulfide bonds	Direct	High-level coupling (117)

Table 3 Incorporation of Lipid-Conjugated Antibody into Liposomes

Lipid moiety	Antibody	Method of conjugation	Incorporation method	Remarks (reference)
Fatty acid	Monoclonal IgG	Activated ester	Detergent dialysis	Antibody-attached liposomes specifically bind to target cells (118)
PE	$F(ab')_2$	Carbodiimide	Direct adsorption	Liposomes were leaky (119)
N-(N^α-iodoacetyl, N^ϵ-dansyllysyl)-PE	Bence Jones dimer	Alkylation of sulfhydryl groups	Direct adsorption	Antibody can also be attached to red cell ghost (120)

tested. As a matter of fact, one report indicated that actinomycin D encapsulated in antibody-targeted liposomes did not show enhanced potency in inhibiting the RNA synthesis of the target cells compared to free drug (116). The results of Weinstein et al. (109), as discussed above, suggested that simple targeting with antibody may not enhance content delivery by liposome-cell fusion. Although the bound liposomes could still be endocytosed by the target cells, the efficiency of internalization would depend largely on the ability of the cell to do so. Furthermore, the encapsulated drug would be delivered to the phagolysosomes of the cells and may not reach the proper target compartment. These considerations have prompted further engineering efforts on the antibody-targeted liposomes. For instance, incorporation of a fusogen such as the Sendai F protein in the antibody targeted liposomes may enhance fusion of the bound liposomes with cells. This type of approach is likely to be the direction of future research in this area.

Other Targeting Molecules

Gregoriadis and Neerunjun have noncovalently coated liposomes with asialofetuin and have shown enhanced uptake of liposomes by hepatic tissue (108). This route of targeting is presumably mediated by the asialoglycoprotein receptors on the hepatocytes. Other carbohydrate targeting strategies [e.g., utilizing phospho-mannosyl receptors on fibroblasts (121)] has not been used. Potentially feasible strategies such as attaching hormone molecules to liposomes and targeting them to receptor-containing cells have not yet been explored.

USES OF LIPOSOMES IN CELL BIOLOGY

Although much research effort has been directed toward the use of liposomes as carriers of pharmaceuticals in chemotherapy applications, liposomes have also been used in various in vitro studies aimed at modifications of cellular morphology, organization, metabolism, growth, and other specialized functions. The potential uses of liposomes in cell biological studies are illustrated by the following examples.

Lipid Dynamics in Cellular Membranes

In Pagano's laboratory, SUVs of defined phospholipid composition have been used to introduce specific phospholipids to the plasma membranes of cultured fibroblasts. NBD-labeled PC and PE (labeled at the acyl chains) were introduced at 2°C to the outer monolayers of plasma membrane bilayers by phospholipid exchange. A bright peripheral ring of fluorescence was observed on all treated cells (20). The peripherally located fluorescent PC and PE showed a lateral diffusion coefficient of $2\text{-}4 \times 10^{-9}$ cm^2/sec as measured by the FRAP technique, indicating that they are properly mixed with cellular lipids in membranes. Upon raising the incubation temperature to 37°C, rapid internalization of fluorescent lipids occurred, probably as a result of cellular lipid turnover. Acyl chain-labeled NBD-PA was also taken up by cells when cells were incubated with liposomes at 2°C. However, the fluorescence of the treated cells appeared on intracellular membranes in this case, indicating rapid internalization even at 2°C (22). Extraction of cellular lipids from the treated cells indicated that more than 50% of the NBD-PA had been hydrolyzed to NBD-diglycerides. These results suggest a new approach to the study of cellular phospholipid metabolism at a particular subcellular location. Schroit and Pagano have used a different approach to study the lateral movement of exogenous phospholipids in the cell membranes (122). They treated V79 fibroblasts with SUV composed of DOPC or DPPC containing N-trinitrophenyl PE (TNP-PE) and fused the treated cells with untreated cells using Sendai virus. The polykaryons were then stained with fluorescent anti-TNP antibody. The percent of polykaryons that showed uniform distribution of fluorescence over the entire hybrid cell surface was measured at different time periods after fusion. A lateral diffusion coefficient for TNP-PE was calculated from the half-time of antigen mixing. TNP-PE introduced to cells by DOPC liposomes (fluid at the incubation temperature) could diffuse freely in the membrane with $D \geqslant 0.6 \times 10^{-8}$ cm^2/sec, whereas TNP-PE introduced to cells by DPPC liposomes (solid at the incubation temperature) failed to diffuse over the hybrid cell surface. These results are consistent with the notion that fluid liposomes can fuse with cells and solid liposomes prefer to adsorb on the cell surface. SUVs made of N-trinitrophenyl-1-acyl-2-(NBD-caproyl)-PE, a fluorescent and haptenated phospholipid analog, were used by the same authors to study the redistribution of phospholipids in EL-4 lymphoma cells (123). After incubation at 2°C, cells showed peripheral ring fluorescence and a lateral diffusion coefficient for the lipid analog of 8.5×10^{-9} cm^2/sec, indicating proper integration of this phospholipid into the plasma membranes of the treated cells. Upon warming up to 37°C, rapid internalization of fluorescence was observed. However, warming the cells in the presence of bound anti-TNP antibodies resulted in a rapid capping of fluorescence instead of internalization. Capping was inhibited by azide and cytochalasin B. A similar type of antibody-induced capping was observed in fibroblasts using another haptenated phospholipid, TNP-DOPE (122). These observations are somewhat surprising, because capping generally requires divalent antibody and multivalent antigen. It is possible that the monovalent phospholipid analog behaves as a multivalent entity in the cell membranes as a result of association with certain membrane proteins or formation of lipid aggregates due to incomplete intermixing with cell lipids. Alternatively, the capping may be explained by a membrane-flow model (124). Although no definitive answer to these different possibilities is yet available, these observations have demonstrated the usefulness of liposomes in studying the dynamics of membrane lipids.

Cell Labeling

Conventional techniques for specific immunofluorescence labeling of cells include direct and indirect methods. The sensitivity of direct immunofluorescence labeling is quite limited because only a few fluorophores can be conjugated to an antibody molecule. The indirect methods, although having enhanced sensitivity, require multiple steps and are thus prone to nonspecific labeling. The recent development in covalent coupling of antibody to liposomes has offered a new possibility: the use of liposomes as a fluorescence labeling agent. If the fluorophores are located in liposomes, either in the lipid phase or the aqueous phase, a single binding event of antibody to a cell surface antigen would bring a large number of fluorophores to the cell surface. This "immunoliposome" labeling has recently been demonstrated using liposomes coated with monoclonal anti-H-2K^k antibody (21). N-NBD-PE and N-FITC-PE were used as fluorophores in liposome bilayers. The fluorescent immunoliposomes labeled mouse L-929 cells, which express anti-H-2K^k antigens, but not the A-31 cells, which express H-2^d antigens (Fig. 4). The fluorescence intensity of the labeled cells was quantitated by microscopic photometry. It was shown that cells labeled with immunoliposomes exhibited four- to sixfold higher fluorescent intensities than did cells labeled by either the direct or indirect immunofluorescence method. Futhere development of the method, such as using more fluorescent lipids in liposomes, could provide even higher labeling intensity. Since this method involves minimal manipulation of cells and does not interfere with cell viability, its application to cell labeling in conjunction with cell sorting is a promising possibility. Specific fluorescence labeling of cells by liposomes was also studied by Leserman et al. (125). It was shown that SUVs containing DNP-PE and entrapped CF could specifically bind to myeloma MOPC 315 cells which expressed surface Ig molecules which bind specifically to the DNP hapten.

Transfer of Exogenous Antigen to Cells

Martin and MacDonald prepared positively charged MLV containing 10% dinitrophenylaminocaproylphosphatidylethanolamine (DNP-Cap-PE) and 20% lyso-PC (52). Human erythrocytes treated with these liposomes could be lysed with anti-DNP serum in the presence of complement. The appearance of the new antigenic determinant on the cell surface was also demonstrated by indirect immunofluorescence labeling and immunoferritin electron microscopy. Hale et al. have used reconstituted liposomes containing Sendai fusion protein and the H-2K^k antigen to insert H-2 antigens into cells (93). The treated cells could then be lysed by cytotoxic T lymphocytes. Acetylcholine receptors were similarly introduced to mouse L-cells, rendering them susceptible to lysis by antireceptor serum in the presence of complement (34; and Fig. 3). Since it has been shown that an antigen molecule cannot fix complement and mediate an immune cytolysis unless it is inserted into the bilayers of cell membranes (126), these studies offer evidence that the

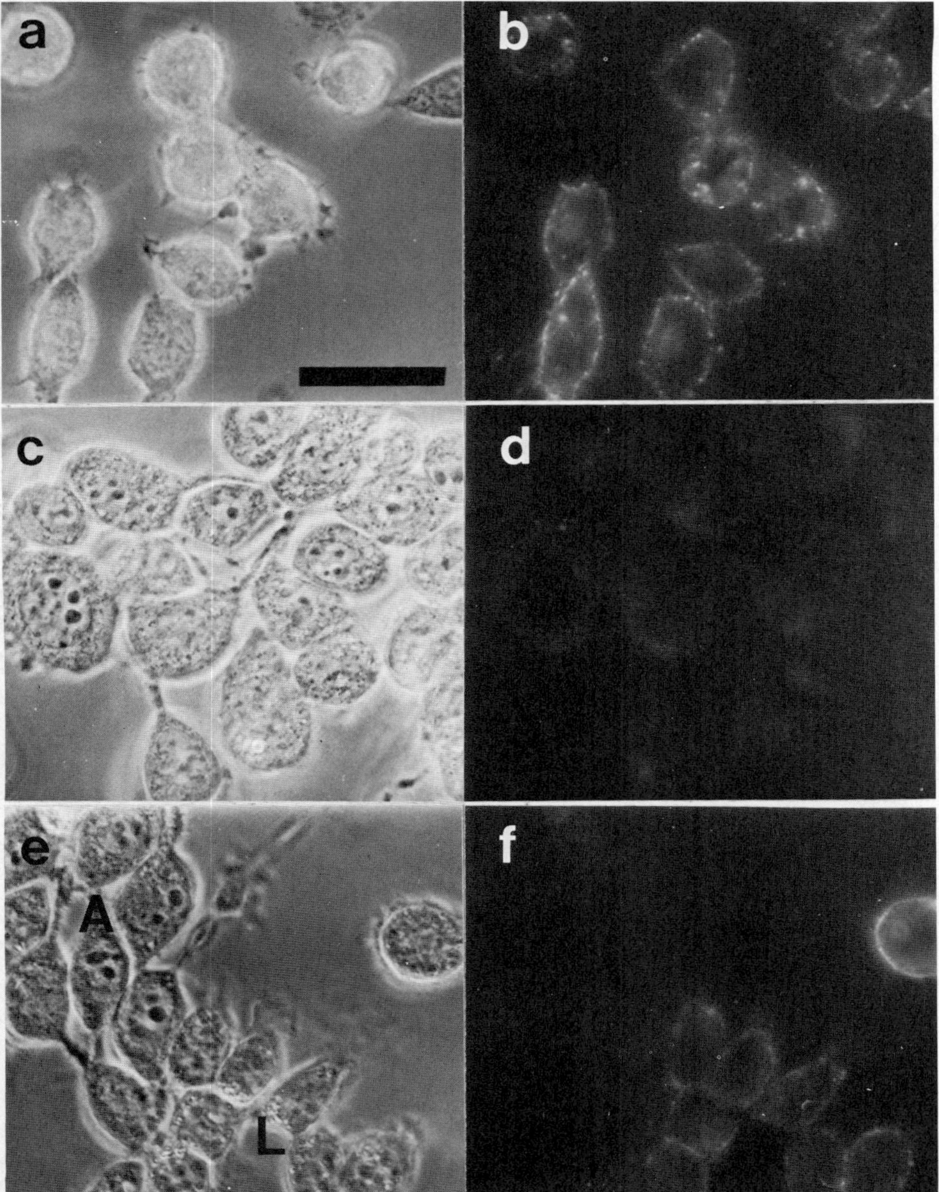

Figure 4 Target-specific binding of immunoliposomes. Monoclonal anti-H-2^k antibody was covalently attached to liposomes containing N-NBD-PE, a fluorescent phospholipid. Cultures of mouse L929 cells [H-2^k type, parts (a) and (b)], or A31 cells [H-2^d type, parts (c) and (d)], or a mixed culture of both cell types [parts (e) and (f)] were incubated with immunoliposomes at 4°C for 3.5 hr. Photographs on the left are phase contrast and those on the right are fluorescence. Cell colonies of either cell type are marked with L (L929 cells) or A (A31 cells) in part (e). (From Ref. 118.)

transfer of antigen by liposomes represents a proper integration of the foreign antigens into the cell membranes. These observations provide possibilities for mechanistic studies of how different antigen molecules interact with each other in cell membranes, a current concern in cellular immunology (see Chap. 6).

Modification of Cell Morphology, Transport, Metabolism, and Growth

When an (ethylenedinitrilo)tetraacetic acid (EDTA)-treated Chinese hamster V79 cell suspension was incubated with SUVs composed of DPPC, adsorption of intact liposomes on the cell surface was seen with scanning electron microscopy (SEM) (29). In addition, the microvilli of the treated cells appeared flattened and held close to the cell body. Few microvilli were found to extend outward. A similar SEM study was carried out with mouse embryo fibroblasts and epithelial cells (127). After incubation of both cell types with SUVs composed of egg PC, one could observe a more than twofold increase in the number of microvilli on the cell surface. Cells also became less flattened. Such dramatic morphological effects of liposomes were found only for SUVs composed of egg PC; SUVs composed of DPPC or MLVs composed of egg PC were not effective. Whether these morphological changes are the result of liposome-cell fusion or lipid exchange was not explored further.

Incubation of cells with liposomes, particularly those containing cholesterol, produces marked modifications in various transport activities of the cells. Bruckdorfer et al. (77) and Grunze and Denticke (78) showed that erythrocytes became osmotical more fragile and more permeable to nonelectrolytes after a partial depletion of membrane cholesterol by incubation with SUVs composed of egg PC. The D-glucose transport activity of erythrocytes was also decreased after substantial removal of membrane cholesterol by liposome incubation (79). A similar observation was made by Baldassare et al. using mouse LM cells (80). They found that the transport of 3-O-methylglucose was significantly reduced after incubation of the cells with PC liposomes and that the transport activity could be restored by subsequent incubation with liposomes containing cholesterol. Rat thymocytes depleted of cholesterol showed an increase in ouabain-sensitive Na^+ influx and a concomitant decrease in K^+ influx (81). These studies demonstrated the usefulness of liposomes in elucidating the role of membrane cholesterol in the modulation of transport activities. Other transport studies have shown that the bidirectional transport of methotrexate, a potent cytotoxic agent used in cancer therapy, was significantly increased when Ehrlich ascites tumor cells were incubated with positively charged liposomes (128). Such effects appeared to be specific since the transport of folic acid, a structural analog of methotrexate, was unaffected. Whether this observation is related to the cholesterol depletion discussed above is not clear. Enrichment of membrane cholesterol also affects the activities of membrane-bound enzymes. After incubation with liposomes containing DPPC and cholesterol but not with liposomes composed of DPPC or egg PC alone, normal rat kidney fibroblasts exhibited strikingly decreased activities of adenylate cyclase and (Na^+,K^+)-ATPase (129). This and a similar observation made using human platelet cells (130) may be related to the studies of Inbar and Shinitzky (131), who reported that the tumorigenicity of YAC lymphoma cells in mouse was significantly reduced after the cells were treated in vitro with liposomes containing PC and cholesterol, but not PC alone. The observed effect was correlated with the change of the membrane cholesterol/phospholipid ratio and the microviscosity measured by the fluorescence depolarization of diphenylhexatriene. Their general thesis was that the turmorigenicity of the lymphoma cell was directly controlled by the physical state of membrane lipids as modulated by the cholesterol content (132).

The ability of liposomes to modify the proliferative potential of cells was reported by Alderson and Green (76). They showed that the concanavalin A-stimulated cell proliferation was suppressed when the cholesterol/phospholipid ratio of normal lymphocytes was increased by liposome treatment (76). However, this observation is not consistent with that of Ozato et al., who pretreated mouse lymphocytes for a short time (about 30 min) with liposomes and found that solid SUVs (DPPC, DMPC, or egg PC with cholesterol) could significantly increase the mitogen-stimulated cell proliferation, whereas fluid liposomes (egg PC alone) had no effect (69). They attributed this differential effect to the ability of solid liposomes to adsorb to cell surfaces much more readily than the fluid vesicles (17). While these conflicting results remain unresolved, it is clear that liposomes are valuable tools for the modification of cell surface structure, composition, and even specific metabolic functions and growth behaviors.

Muscle and Neurobiology

Interaction of liposomes with cultured myoblast cells was studied by Sandra (133). Adhesion of liposomes to the cell surface at temperatures below the Tc was accompanied by a marked inhibition of cell fusion activity. This inhibition could be partially reversed by culturing the treated cells above the Tc of the liposomes. The glycocalyx of a cultured muscle cell line was modified by treatment of cells with liposomes containing foreign gangliosides or glycoproteins, resulting in complex changes in cell adhesion, morphology, and viability (134). Adsorption of DPPC liposomes containing 10% stearylamine to frog skeletal muscle fibers at 4°C was observed by Leung (135). Incubation at higher temperatures resulted in an effective transfer of the liposome-encapsulated fluorescent dyes into the muscle fibers without affecting the time course of muscle twitch tension (135) or the excitability of muscle membranes (136). These studies suggested that liposomes could be used as effective vehicles to inject impermeable substances into muscle cells. A recent report by O'Loughlin et al. has confirmed this possibility by demonstrating the ability of liposomes to introduce eukaryotic initiation factor 3 (eIF-3), regulatory components of eIF 3, or mRNA into myogenic cells. Introduction of these compounds resulted in the specific regulation of muscle protein synthesis (55). SUVs composed of DOPC have been used to inject CF into the rod outer segment of frog retina (13) and have also been used to deliver retinol congeners into retina cells for the regeneration of visual pigments after bleaching (137). The value of liposome-mediated intracellular delivery was further substantiated by the studies of Rahamimoff et al., who used SUVs composed of egg PC containing different internal ions to treat frog sartorius nerve-muscle preparations (60). The results showed that both the evoked and spontaneous transmitter release at the neuromuscular junction was significantly increased when Ca^{2+} or Na^+, but not K^+ ions, were encapsulated in liposomes. The neurite outgrowth of cultured neuroblastoma cells was stimulated by liposomes composed of PS and PC (138). This effect of liposomes, shown by Sandra et al. (139), was a phenomenon elicited by a variety of phospholipids, including DOPC, DPPC, PS, and PE. It was also shown that fatty acids such as oleic and palmitic acids were equally, if not more effective in inducing neurite outgrowth (139). They concluded that it was the metabolites of the exogenous phospholipid, rather than the initial liposome-cell interaction, that gave rise to this effect.

Induction of Cell Fusion

Fluid, negatively charged SUVs can induce cell-to-cell fusion in the presence of Ca^{2+} (140). Cells tested in this study were BHK hamster cells, 3T3 and L929 mouse cells. Both monolayer and suspension cells could be fused by this method to form polykaryons. Since the same laboratory reported that only fluid and negatively charged liposomes could fuse with cells in the presence of Ca^{2+} (45), they concluded that cell-to-cell fusion may be related to the liposome-to-cell fusion. Martin and MacDonald found that SUVs containing DSPC, lyso-PC, and stearylamine could effectively fuse Ehrlich ascites tumor cells, RK, VERO, and KB cells, and chicken erythrocytes without appreciable cell lysis (51). Little or no fusion activity was found for liposomes containing PC alone, PC and lyso-PC, or PC and stearylamine. Although the fusion-promoting activity of liposomes is comparable to that of Sendai virus, it is generally considered to be lower than that of polyethylene glycol.

Other Applications in Cell and Molecular Biology

Liposomes have been extensively used in studies of both humoral and cellular immunology. Since these studies are reviewed in detail later in this book and in other recent publications (141), they will not be discussed here. For studies using liposomes as carriers for viruses, nucleic acids, cloned genes, or chromosome preparations, the readers are referred to Chapter 5.

CONCLUDING REMARKS

Both protein-free liposomes and proteoliposomes have proven useful in various applications in cell biology. Many of these applications rely on a thorough understanding of the mechanisms by which liposomes interact with cells. From the information discussed in this chapter it is clear that the actual mechanism of interaction is very complex. The information summarized in this chapter should provide a useful starting point for these studies. Recent advances in the formation and utilization of proteoliposomes have provided new opportunities in liposome research. The knowledge gained in studying the interactions of these liposomes with cells in vitro should help in understanding their behavior in vivo for such important applications as cancer chemotherapy and enzyme-replacement therapy.

ACKNOWLEDGMENTS

The research in the author's laboratory has been supported by NIH Grants GM23473 and CA24553 and by the Muscular Dystrophy Association of America. I thank Anthony Huang for a critical reading of the manuscript.

REFERENCES

1. Pagano, R. E. and J. N. Weinstein. 1978. Interactions of liposomes with mammalian cells. Annu. Rev. Biophys. Bioeng. 7:435.
2. Kimelberg, H. K. and E. Mayhew. 1978. In *Critical Reviews in Toxicology*, L. Golberg (Ed.). CRC Press, Cleveland, Ohio, pp. 25-79.
3. Poste, G. 1980. In *Liposomes in Biological Systems,* G. Gregoriadis and A. C. Allison (Eds.). Wiley, New York, Chap. 4.
4. Papahadjopoulos, D., (Ed.). 1978. *Liposomes and Their Uses in Biology and Medicine.* Ann. N.Y. Acad. Sci., Vol. 308.
5. Juliano, R. L. and D. Layton. 1980. In *Drug Delivery Systems,* R. L. Juliano (Ed.). Oxford University Press, New York, Chap. 6.
6. Chajek, T., L. Aron, and C. J. Fielding. 1980. Interaction of lecithin: cholesterol acyltransferase and cholesteryl ester transfer protein in the transport of cholesteryl ester into sphingomyelin liposomes. Biochemistry 19:3673.
7. Young, P. M. and P. Brecher. 1981. Cholesteryl ester transfer from phospholipid vesicles to human high density lipoproteins. J. Lipid Res. 22:944.
8. Adrian, G. and L. Huang. 1979. Entrapment of proteins in phosphatidylcholine vesicles. Biochemistry 18:5610.
9. Stein, Y., G. Halperin, and O. Stein. 1980. Biological stability of ^{3}H-cholesteryl oleyl ether in cultured fibroblasts and intact rat, FEBS Lett. 111:104.
10. Paltauf, F. 1968. Die Synthese von radioaktiv markierten langkettigen Alkyl-Choleseterinathern. Monatsh. Chem. 99:1277.

11. Klausner, R. D., N. Kumar, J. N. Weinstein, R. Blumenthal, and M. Flavin. 1981. Interaction of tublin with phospholipid vesicles. J. Biol. Chem. 256:5579.

12. Szoka, F. C., K. Jacobson, and D. Paphajopoulos. 1979. The use of aqueous space markers to determine the mechanism of interaction between phospholipid vesicles and cells. Biochim. Biophys. Acta 551:195.

13. Weinstein, J. N., S. Yoshikami, P. Henkart, R. Blumenthal, and W. A. Hagins. 1977. Liposome-cell interaction: transfer and intracellular release of a trapped fluorescent marker. Science 195:489.

14. Blumenthal, R., J. N. Weinstein, S. O. Sharrow, and P. Henkart. 1977. Liposome-lymphocyte interaction: saturable sites for transfer and intracellular release of liposome contents. Proc. Natl. Acad. Sci. USA 74:5603.

15. Hagins, W. A. and S. Yoshikami. 1977. In *International Symposium on Photoreception,* H. B. Barlow and P. Fatt (Eds.). Academic Press, New York, p. 97.

16. Ralston, E., R. Blumenthal, J. N. Weinstein, S. O. Sharrow, and P. Henkart. 1980. Lysophosphatidylcholine in liposomal membranes: enhanced permeability but little effect on transfer of a water-soluble fluorescent marker into human lymphocytes. Biochim. Biophys. Acta 597:543.

17. Huang, L., K. Ozato, and R. E. Pagano. 1978. Interaction of phospholipid vesicles with murine lymphocytes. I. Vesicle-cell adsorption and fusion as alternate pathways of uptake. Membr. Biochem. 1:1.

18. Van Renswoude, A. J. B. M., P. Westenberg, and G. L. Scherphof. 1979. In vitro interaction of zajdela ascites hepatoma cells with lipid vesicles. Biochim. Biophys. Acta 558:22.

19. Van Renswoude, A. J. B. M. and D. Hoekstra. 1981. Cell-induced leakage of liposome contents. Biochemistry 20:540.

20. Struck, D. K. and R. E. Pagano. 1980. Insertion of fluorescent phospholipids into the plasma membrane of a mammalian cell. J. Biol. Chem. 255:5404.

21. Huang, A., S. J. Kennel, and L. Huang. 1982. Immunoliposome labelling: a sensitive and specific method for cell labelling. J. Immunol. Methods (in press).

22. Pagano, R. E., A. J. Schroit, and D. K. Struck. 1981. In *Liposomes: From Physical Structure to Therapeutic Applications,* C. G. Knight (Ed.). Elsevier/North-Holland Biomedical Press, Amsterdam.

23. Pagano, R. E., O. C. Martin, and D. K. Struck. 1981. Personal communication.

24. Nichols, J. N. and R. E. L. Pagano. 1981. Kinetics of soluble lipid monomer diffusion between vesicles. Biochemistry 20:2783.

25. Stryer, L. 1978. Fluorescence energy transfer as a spectroscopic ruler. Annu. Rev. Biochem. 47:819.

26. Fung, B. K.-K. and L. L. Stryer. 1978. Surface density determination in membranes by fluorescence energy transfer. Biochemistry 17:5341.

27. Struck, D. K., D. Hoekstra, and R. E. Pagano. 1981. Use of resonance energy transfer to monitor membrane fusion. Biochemistry 20:4903.

28. Pagano, R. E., O. Martin, A. J. Schroit, and D. K. Struck. 1982. Formation of asymmetric phospholipid membranes via spontaneous transfer of fluorescent lipid analogues between vesicle populations. Biochemistry (in press).

29. Pagano, R. E. and M. Takeichi. 1977. Adhesion of phospholipid vesicles to Chinese hamster fibroblasts: role of cell surface proteins. J. Cell Biol., 74:531.

30. Magee, W. E., C. W. Goff, J. Schoknecht, M. D. Smith, and K. Cherian. 1974. The interaction of cationic liposomes containing entrapped horseradish peroxidase with cells in culture. J. Cell Biol. 63:492.

31. Pagano, R. E., L. Huang, and C. Wey. 1974. Interaction of phospholipid vesicles with cultured mammalian cells. Nature 252:166.

32. Raz, A., C. Bucana, W. E. Fogler, G. Poste, and I. J. Fidler. 1981. Biochemical, morphological, and ultrastructural studies on the uptake of liposomes by murine macrophages. Cancer Res. 41:487.

33. Wu, P., G. W. Tin, and J. D. Baldeschwieler. 1981. Phagocytosis of carbohydrate-modified phospholipid vesicles by macrophage. Proc. Natl. Acad. Sci. USA 78:2033.

34. Ho, S. C. 1980. Phospholipid vesicles as carriers to transfer biological molecules into cells. Ph.D. dissertation, Univ. of Tennessee, Knoxville.

35. Pagano, R. E. and L. Huang. 1975. Interaction of phospholipid vesicles with cultured mammalian cells. II. Studies of mechanism. J. Cell Biol. 67:49.

36. Poste, G., C. W. Porter, and D. Paphadjopoulos. 1978. Identification of a potential artifact in the use of electron microscope autoradiography to localize saturated phospholipids in cells. Biochim. Biophys. Acta 510:256.

37. Axelrod, D., D. E. Koppel, J. Schlessinger, E. Elson, and W. W. Webb. 1976. Mobility measurement by analysis of fluorescence photobleaching recovery kinetics. Biophys. J. 16:1055.

38. Edidin, M. 1981. In *Comprehensive Biochemistry, Membrane Structure, and Function,* J. B. Finean and R. H. Michell (Eds.). Elsevier/North-Holland, Amsterdam.

39. Szoka, F., K. Jacobson, Z. Dezko, and D. Papahadjopoulos. 1980. Fluorescence studies on the mechanism of liposome-cell interactions in vitro. Biochim. Biophys. Acta 600:1.

40. Maeda, T. and H. M. McConnell. 1979. Specificity of memory cells raised against trinitrophenyl-conjugated syngeneic cells. Proc. Natl. Acad. Sci. USA 76:1537.

41. Hwang, K. J. and M. R. Mauk. 1977. Fate of lipid vesicles in vivo: a gamma-ray perturbed angular correlation study. Proc. Natl. Acad. Sci. USA 74:4991.

42. Mauk, M. R., R. C. Gamble, and J. D. Baldeschwieler. 1980. Targeting of lipid vesicles: specificity of carbohydrate receptor analogues for leukocytes in mice. Proc. Natl. Acad. Sci. USA 77:4430.

43. Papahadjopoulos, D., E. Mayhew, G. Poste, S. Smith, and W. J. Vail. 1974. Incorporation of lipid vesicles by mammalian cells provides a potential method for modifying cell behavior. Nature 252:163.

44. Batzri, S. and E. D. Korn. 1975. Interaction of phospholipid vesicles with cells: endocytosis and fusion as alternative mechanisms for the uptake of lipid-soluble and water-soluble molecules. J. Cell Biol. 66:621.

45. Poste, G. and D. Papahadjopoulos. 1976. Lipid vesicles as carriers for introducing materials into cultured cells: influence of vesicle lipid composition on mechanism(s) of vesicle incorporation into cells. Proc. Natl. Acad. Sci. USA 73:1603.

46. Martin, F. and R. MacDonald. 1974. Liposomes can mimic virus membranes. Nature 252:161.

47. Martin, F. and R. MacDonald. 1976. Lipid vesicle-cell interactions. I. Hemagglutination and hemolysis. J. Cell Biol. 70:494.

48. Papahadjopoulos, D., G. Poste, and E. Mayhew. 1974. Cellular uptake of cyclic AMP captured within phospholipid vesicles and effect on cell-growth behavior. Biochim. Biophys. Acta 363:404.

49. Huang, L. and R. E. Pagano. 1975. Interactions of phospholipid vesicles with cultured mammalian cells. I. Characterization of uptake. J. Cell Biol. 67:38.

50. Ostro, M. J., D. Lavelle, W. Paxton, B. Matthews, and D. Giacomoni. 1980. Parameters affecting the liposome-mediated insertion of RNA into eucaryotic cells in vitro. Arch. Biochem. Biophys. 201:392.

51. Martin, F. and R. MacDonald. 1976. Lipid vesicle-cell interactions. II. Induction of cell fusion. J. Cell Biol. 70:506.

52. Martin, F. and R. MacDonald. 1976. Lipid vesicle-cell interactions. III. Introduction of a new antigenic determinant into erythrocyte membranes. J. Cell Biol. 70:515.

53. Weissmann, G., C. Cohen, and S. Hoffstein. 1977. Introduction of enzymes, by means of liposomes, into non-phagocytic human cells in vitro. Biochim. Biophys. Acta 498:375.

54. Culpepper, J. A. and A. Y.-C. Liu. 1981. Induction of tyrosine aminotransferase in H-35 hepatoma cells by cAMP captured in phospholipid vesicles. J. Cell Biol. 88:89.

55. O'Loughlin, J., L. Lehr, A. Havaranis, and S. M. Heywood. 1981. Encapsulation of "core" eIF3, regulatory components of eIF3 and mRNA into liposomes, and their subsequent uptake into myogenic cells in culture. J. Cell Biol. 90:160.

56. Dimitriadis, G. J. and T. D. Butters. 1979. Liposome-mediated ricin toxicity in ricin-resistant cells. FEBS Lett. 98:33.

57. Papahadjopoulos, D. and G. Poste. 1976. Drug-containing lipid vesicles render drug-resistant tumor cells sensitive to adinomycin D. Nature 261:699.

58. Papahadjopoulos, D., G. Poste, V. J. Vail, and J. L. Biedler. 1976. Use of lipid vesicles as carriers to introduce actinomycin D into resistant tumor cells. Cancer Res. 36:2988.

59. Theoharides, T. C. and W. W. Douglas. 1978. Secretion in mast cells induced by calcium entrapped within phospholipid vesicles. Science 201:1143.

60. Rahamimoff, R., H. Meiri, S. D. Erulkar, and Y. Barenholz. 1978. Changes in transmitter release induced by ion-containing liposomes. Proc. Natl. Acad. Sci. USA 75:5214.

61. Roerdink, F. H. 1978. Liposomes as enzyme carriers: uptake by rat liver and by cultured fibroblasts from a patient with Pompe's disease (glycogenosis type II). Ph.D. dissertation, University of Groningen.

62. Ostro, M. J., D. Giacomoni, D. Lavelle, W. Paxton, and S. Dray. 1978. Evidence for translation of rabbit globin mRNA after liposome-mediated insertion into a human cell line. Nature 274:921.

63. Dimitriadis, G. J. 1978. Translation of rabbit globin mRNA introduced by liposomes into mouse lymphocytes. Nature 274:923.

64. Wilson, T., D. Paphadjopoulos, and R. Taber. 1977. Biological properties of poliovirus encapsulated in lipid vesicles: antibody resistance and infectivity in virus-resistant cells. Proc. Natl. Acad. Sci. USA 74:3740.

65. Wilson, T., D. Papahadjopoulos, and R. Taber. 1979. The introduction of poliovirus RNA into cells via lipid vesicles (liposomes). Cell 17:77.

66. Fraley, R., S. Subramani, P. Berg, and D. Papahadjopoulos. 1980. Introduction of liposome-encapsulated SV40 DNA into cells. J. Biol. Chem. 255: 10431.

67. Mukherjee, A. B., S. Orloff, J. D. Butler, T. Triche, P. Lalley, and J. D. Schulman. 1978. Entrapment of metaphase chromosomes into phospholipid vesicles (lipochromosomes): carrier potential in gene transfer. Proc. Natl. Acad. Sci. USA 75:1361.

68. Hallett, M. B. and A. K. Campbell. 1980. Uptake of liposomes containing the photoprotein obelin by rat isolated adipocytes: adhesion, endocytosis or fusion? Biochem. J. 192:587.

69. Ozato, K., L. Huang, and R. E. Pagano. 1978. Interactions of phospholipid vesicles with murine lymphocytes. II. Correlation between altered surface properties and enhanced proliferative response. Membr. Biochem. 1:27.

70. Margolis, L. B., A. N. Tikhonov, and E. Yu. Vasilieva. 1980. Platelet adhesion to fluid and solid phospholipid membranes. Cell 19:189.

71. Sendra, A. and R. E. Pagano. 1979. Liposome-cell interactions: studies of lipid transfer using isotopically asymmetric vesicles. J. Biol. Chem. 254: 2244.

72. Wirtz, K. W. A. 1974. Transfer of phospholipids between membranes. Biochim, Biophys. Acta 344:95.

73. Bruckdorfer, K. R., P. A. Edwards, and C. Green. 1968. Properties of aqueous dispersions of phospholipid and cholesterol. Eur. J. Biochem. 4:506.

74. Bruckdorfer, K. R., J. M. Graham, and C. Green. 1968. Incorporation of steroid molecules into lecithin sols, β-lipoproteins and cellular membranes. Eur. J. Biochem. 4:512.

75. Shinitzky, M. and M. Inbar. 1974. Difference in microviscosity induced by different cholesterol levels in the surface membranes lipid layer of normal lymphocytes and malignant lymphoma cells. J. Mol. Biol. 85:603.

76.   Alderson, J. C. E. and C. Green. 1975. Enrichment of lymphocytes with cholesterol and its effect on lymphocyte activation. FEBS Lett. 52:208.

77.   Bruckdorfer, K. R., R. A. Demel, J. de Gier, and L. L. M. van Deenen. 1969. The effect of partial replacements of membrane cholesterol by other steroids on the osmotic fragility and glycerol permeability of erythrocytes. Biochim. Biophys. Acta 183:334.

78.   Grunze, M. and B. Denticke. 1974. Changes of membrane permeability due to extensive cholesterol depletion in mammalian erythrocytes. Biochim. Biophys. Acta 356:125.

79.   Masiak, S. J. and P. G. LeFevre. 1974. Effects of membrane steroid modification on human erythrocyte glucose transport. Arch. Biochem. Biophys. 162:442.

80.   Baldassare, J. J., Y. Saito, and D. F. Silbert. 1979. Effect of sterol depletion on LM cell sterol mutants: changes in the lipid composition of the plasma membrane and their effects on 3-O-methylglucose transport. J. Biol. Chem. 254:1108.

81.   Kramers, M. T. C., J. Patrick, J. M. Bottomley, P. J. Quinn, and D. Chapman. 1980. Studies of liposome interactions with rat thymocytes. Eur. J. Biochem. 110:579.

82.   Weismann, G., D. Bloomgarden, R. Kaplan, C. Cohen, S. Hoffstein, T. Collins, A. Gotlieb, and D. Nagle. 1975. A general method for the introduction of enzymes, by means of immunoglobulin-coated liposomes, into lysosomes of deficient cells. Proc. Natl. Acad. Sci. USA 72:88.

83.   Jansons, V. K., P. Weis, T. Chen, and W. R. Redwood. 1978. In vitro interaction of L1210 cells with phospholipid vesicles. Cancer Res. 38:531.

84.   Silverstein, S., R. Steinman, and Z. A. Cohn. 1977. Endocytosis. Annu. Rev. Biochem. 46:669.

85.   Bouma, S. R., F. W. Drislane, and W. H. Heustis. 1977. Selective extraction of membrane-bound proteins by phospholipid vesicles. J. Biol. Chem. 252:6759.

86.   Heustis, W. H. 1977. A sodium-specific membrane permeability defect induced by phospholipid vesicles treatment of erythrocytes. J. Biol. Chem. 252:6764.

87.   Dunnick, J. K., J. D. Rooke, S. Aragon, and J. P. Kriss. 1976. Alteration of mammalian cells by interaction with artificial lipid vesicles. Cancer Res. 36:2385.

88.   Zborowski, J., F. Roerdink, and G. Scherphof. 1977. Leakage of sucrose from phosphatidylcholine liposomes induced by interaction with serum albumin. Biochim. Biophys. Acta 497:183.

89.   Gregoriadis, G. and C. Davis. 1979. Stability of liposomes in vivo and in vitro is promoted by their cholesterol content and the presence of blood cells. Biochem. Biophys. Res. Commun. 89:1287.

90.   Kirby, C., J. Clarke, and G. Gregoriadis. 1980. Effect of the cholesterol content of small unilamellar liposomes on their stability in vivo and in vitro. Biochem. J. 186:591.

91. Huang, L. and A. K. Ritchie. 1978. Phospholipid vesicles as vehicles for transplanting acetylcholine receptors to the surfaces of cultured Chinese hamster cells. Ann. N.Y. Acad. Sci. 308:439.

92. Scheid, A. and P. W. Choppin. 1974. Identification of biological activities of paramyxovirus glycoproteins: activation of cell fusion, hemolysis and infectivity by proteolytic cleavage of an inactive precursor protein of Sendai virus. Virology 57:475.

93. Hale, A. H., M. J. Ruebush, D. S. Lyles, and D. T. Harris. 1980. Antigen-liposome modification of target cells as a method to alter their susceptibility to lysis by cytotoxic T lymphocytes. Proc. Natl. Acad. Sci. USA 77:6105.

94. Eytan, G. D. and E. Eytan. 1980. Fusion of proteoliposomes and cells: ATP-dependent Ca^{+2} uptake into erythrocytes catalysed by Ca^{+2}-ATPase from skeletal muscle. J. Biol. Chem. 255:4492.

95. Gad, A. E., R. Broza, and G. D. Eytan. 1979. Fusion of cells and proteoliposomes: incorporation of beef heart cytochrome oxidase into rabbit erythrocytes. FEBS Lett. 102:230.

96. Szoka, F., K.-E. Magnusson, J. Wojcieszyn, Y. Hou, Z. Derzko, and K.. Jacobson. 1981. Use of lectins and polyethylene glycol for fusion of glycolipid-containing liposomes with eukaryotic cell. Proc. Natl. Acad. Sci. USA 78:1685.

97. Baumann, H., E. Hou, and D. Doyle. 1980. Insertion of biologically active membrane proteins from rat liver into the plasma membrane of mouse fibroblasts. J. Biol. Chem. 255:10001.

98. Honjo, T., Y. Nishizuka, O. Hayaishi, and I. Kato. 1968. Diphtheria toxin-dependent adenosine diphosphate ribosylation of aminoacyl transferase II by diphtheria toxin and inhibition of protein synthesis. J. Biol. Chem. 243:3553.

99. Gill, D. M., A. M. Pappenheimer, R. Brown, and J. J. Kurnick. 1969. Studies on the mode of action of diphtheria toxin. VII. Toxin-stimulated hydrolysis of nicotinamide adenine dinucleotide in mammalian cell extracts. J. Exp. Med. 129:1.

100. Uchida, T., J. Kim, M. Yamaizumi, Y. Miyake, and Y. Okada. 1979. Reconstitution of lipid vesicles associated with HVJ (Sendai virus) spikes: purification and some properties of vesicles containing nontoxic fragment A of diphtheria toxin. J. Cell Biol. 80:10.

101. Uchida, T., M. Yamaizumi, and Y. Okada. 1977. Reassembled HVJ (Sendai virus) envelopes containing non-toxic mutant proteins of diphtheria toxin show toxicity to mouse L cells. Nature 266:839.

102. Yamaizumi, M., E. Mekada, T. Uchida, and Y. Okada. 1978. One molecule of diphtheria toxin fragment A introduced into a cell can kill the cell. Cell 15:245.

103. Ueda, S., T. Uchida, and Y. Okada. 1981. Selective killing of subacute sclerosing panencephalitis virus-infected cells by liposomes containing fragment A of diphtheria toxin. Exp. Cell Res., 132:259.

104.   Cohen, C. M., G. Weissmann, S. Hoffstein, Y. Awasthi, and S. K. Srirastava. 1976. Introduction of purified hexosaminidase A into Tay-Sachs leukocytes by means of immunoglobulin-coated liposomes. Biochem. 15:452.

105.   Allen, T. M. and L. G. Cleland. 1980. Serum-induced leakage of liposome contents. Biochim. Biophys. Acta 597:418.

106.   Owen, C. S. (1980). A membrane-bound fluorescent probe to detect phospholipid vesicle-cell fusion. J. Membr. Biol. 54:13.

107.   Leserman, L. D., J. N. Weinstein, R. Blumenthal, and W. D. Terry. 1980. Receptor-mediated endocytosis of antibody-opsonized liposomes by tumor cells. Proc. Natl. Acad. Sci. USA 77:4089.

108.   Gregoriadis, G. and D. E. Neerunjun. 1975. Homing of liposomes to target cells. Biochem. Biophys. Res. Commun. 65:537.

109.   Weinstein, J. N., R. Blumenthal, S. O. Sharrow, and P. A. Henkart. 1978. Antibody-mediated targeting of liposomes: binding to lymphocytes does not ensure incorporation of vesicle contents into the cells. Biochim. Biophys. Acta 509:272.

110.   Torchilin, V. P., V. S. Goldmacher, and V. N. Smirnov. 1978. Comparative studies on covalent and noncovalent immobilization or protein molecules on the surface of liposomes. Biochem. Biophys. Res. Commun. 85: 983.

111.   Torchilin, V. P., B. A. Khaw, V. N. Smirnov, and E. Haber. 1979. Preservation of antimyosin antibody activity after covalent coupling to liposomes. Biochem. Biophys. Res. Commun. 89:1114.

112.   Dunnick, J. K., I. R. McDougall, S. Aragon, M. L. Goris, and J. P. Kriss. 1975. Vesicle interactions with polyamino acids and antibody: in vitro and in vivo studies. J. Nucl. Med. 16:483.

113.   Heath, T. D., B. A. Macher, and D. Papahadjopoulos. 1981. Covalent attachment of immunoglobulins to liposomes via glycosphingolipids. Biochim. Biophys. Acta 640:66.

114.   Heath, T. D., R. T. Fraley, and D. Papahadjopoulos. 1980. Antibody targeting of liposomes: cell specificity obtained by conjugation of F(ab')$_2$ to vesicle surface. Science 210:539.

115.   Leserman, L. D., J. Barbet, F. Kourilsky, and J. N. Weinstein. 1981. Targeting to cells of fluorescent liposomes covalently coupled with monoclonal antibody or protein A. Nature 288:602.

116.   Urdal, D. L. and S. Hakomori. 1980. Tumor-associated ganglio-N-triosylceramide: target for antibody-dependent, avidin-mediated drug killing of tumor cells. J. Biol. Chem. 255:10509.

117.   Martin, F. J., W. L. Hubbell, and D. Papahadjopoulos. 1981. Immunospecific targeting of liposomes to cells: a novel and efficient method for covalent attachment of Fab' fragments via disulfide bonds. Biochemistry 20:4229.

118.   Huang, A., L. Huang, and S. J. Kennel. 1980. Monoclonal antibody covalently coupled with fatty acid: a reagent for in vitro liposome targeting. J. Biol. Chem. 255:8015.

119. Jansons, V. K. and P. L. Mallett. 1981. Targeting liposome: a method for preparation and analysis. Anal. Biochem. 111:54.

120. Sinha, D. and F. Karush. 1979. Attachment ot membranes of exogenous immunoglobulin conjugated to a hydrophobic anchor. Biochem. Biophys. Res. Commun., 90:554.

121. Fischer, H. D., A. Gonzalez-Noriega, W. S. Sly, and D. J. Morre. 1980. Phosphomannosyl-enzyme receptor in rat liver: subcellular distribution and role in intracellular transport of lysosomal enzymes. J. Biol. Chem. 255:9608.

122. Schroit, A. J. and R. E. Pagano. 1978. Introduction of antigenic phospholipids into the plasma membrane of mammalian cells: organization and antibody-induced lipid redistribution. Proc. Natl. Acad. Sci. USA 75:5529.

123. Schroit, A. J. and R. E. Pagano. 1981. Capping of a phospholipid analog in the plasma membrane of lymphocytes. Cell 23:105.

124. Bretscher, M. S. 1976. Directed lipid flow in cell membranes. Nature 206:21.

125. Leserman, L. D., J. N. Weinstein, R. Blumenthal, S. O. Sharrow, and W. D. Terry. 1979. Binding of antigen-bearing fluorescent liposomes to the murine myeloma tumor MOPC 315. J. Immunol. 122:585.

126. Morein, B., D. Barz, U. Kojzinowski, and V. Schirramacher. 1979. Integration of a virus membrane protein into the lipid bilayer of target cells as a prerequisite for immune cytolysis: specific cytolysis after virosome-target cell fusion. J. Exp. Med. 150:1383.

127. Margolis, L. B. and L. D. Bergelson. 1979. Lipid-cell interactions: induction of microvilli on the cell surface by liposomes. Exp. Cell Res. 119: 145.

128. Fry. D. W., J. C. White, and I. D. Goldman. 1979. Alterations of the carrier-mediated transport of an anionic solute, methotrexate, by charged liposomes in Ehrlich ascites tumor cells. J. Membr. Biol. 50:123.

129. Klein, I., L. Moore, and I. Pastan. 1978. Effect of liposomes containing cholesterol on adenylate cyclase activity of cultured mammalian fibroblasts. Biochim. Biophys. Acta 506:42.

130. Sinha, A. K., S. J. Shattil, and R. W. Colman. 1976. Cyclic AMP metabolism in cholesterol-rich platelets. Fed. Proc. 35: 1714.

131. Inbar, M. and M. Shinitzky. 1974. Increase of cholesterol level in the surface membrane of lymphoma cells and its inhibitory effect on ascites tumor development. Proc. Natl. Acad. Sci. USA 71:2128.

132. Inbar, M. and M. Shinitzky. 1974. Cholesterol as a bioregulator in the development of inhibition of leukemia. Proc. Natl. Acad. Sci. USA 71:4229.

133. Sandra, A. 1980. Interaction of phospholipid vesicles with cultured myogenic cells: effects on cell fusion. Exp. Cell Res. 125:411.

134. Barratt, D. G., J. D. Rogers, F. J. Sharom, and C. W. M. Grant. 1978. Direct modification of the glycocalyx of a cultured muscle cell line by incorporation of foreign gangliosides and an integral membrane glycoprotein. J. Supramol. Struct. 8:119.

135. Leung, J. G. M. 1980. Liposomes: effect of temperature on their mode of action on single frog skeletal muscle fibers. Biochim. Biophys. Acta 597: 427.
136. Venkatakrishnan, R., J. Leung, D. T. Mason, and J. Wikman-Coffelt. 1979. Liposomes for the transport of an impermeable fluorescent dye into muscle fibers. Biochem. Med. 21:209.
137. Yoshikami, S. and G. N. Noll. 1978. Isolated retinas synthesize visual pigments from retinol congeners delivered by liposomes. Science 200: 1393.
138. Chen, J. S., A. Del Fa, A. Di Luzio, and P. Calissano. 1976. Liposome-induced morphological differentiation of murine neuroblastoma. Nature 263:604.
139. Sandra, A., W. B. Paltzer, and M. J. Thomas. 1981. Morphological differentiation of murine neuroblastoma induced by liposomes. Exp. Cell Res. 132:473.
140. Papahadjopoulos, D., G. Poste, and B. E. Schaeffer. 1973. Fusion of mammalian cells by unilamellar lipid vesicles: influence of lipid surface charge, fluidity, and cholesterol. Biochim. Biophys. Acta 323:23.
141. Tom, B. H. and H. R. Six, (Eds.). 1980. *Liposomes and Immunobiology.* Elsevier/North-Holland, New York.
142. Ip, S. H. C., J. Abraham, and R. A. Cooper. 1980. Enhancement of blastogenesis in cholesterol-enriched lymphocytes. J. Immunol. 124:87.

4

Liposome Reconstitution: Applications in Cell Physiology

P. Malathi / College of Medicine and Dentistry of New Jersey, Rutgers Medical School, Piscataway, New Jersey

INTRODUCTION

In recent years it has been increasingly clear that biological membranes are the final arbiters of cell integrity and function. These membranes are in general composed of a bilayer of lipids into which are also intercalated proteins by virtue of specific groups of hydrophobic amino acid residues interacting with the hydrocarbon chains of the lipids (1). Several functions vital to the cell reside in these membranes. These functions are attributable mainly to protein components, although the lipids may also contribute in a specific way. Such functions include vectorial enzymes, receptors, and mediated transport systems.

Although much is known about the characteristics of these various membrane functions, progress in the isolation and characterization of the molecular species responsible for such functions has perforce been slow. This is largely due to the characteristics of the membrane components themselves; they are amphiphiles (2), having hydrophilic and hydrophobic regions not readily rending themselves to available techniques for separation and characterization of water-soluble proteins. The problem of characterizing transport proteins is further compounded by technical difficulties in assaying the activity of any such isolated species. Two general procedures have been in vogue as reviewed by Tanner (3). One is specifically to label the carrier system in the membrane before attempting to isolate it and the other is to reassemble the isolated components into an artificial membrane by reconstituting the protein into phospholipid bilayers which may then be used to measure specific translocation fucntions. Two types of phospholipid bilayer model membranes can be found described in the literature; one consists of a planar membrane and the other of numerous vesicles in an aqueous milieu, the liposomes. The planar model membranes separating two infinitely variable aqueous compartments would be ideally suitable for such studies. However, reconstitutions into planar bilayers have not been as productive as those into liposomes. Although liposomes have the limitation of having a small internal space, much information has accumulated in recent years from reconstitution into liposomes of isolated membrane proteins. Combining the two techniques, that is, reconstituting into liposomes and then fusing these with planar membranes, is now gaining favor, expecially in transmembrane voltage measurements.

Reconstitution into liposomes of membrane transport activity for identification and characterization, although the most productive approach to date, has its limitations as succinctly pointed out by Racker (4). As he states: "The primary purpose of resolutions and reconstitutions is indeed the simplification of the system to the minimal number of components required for functions." What may be sacrificed in the process is the possible interaction of other membrane components of the native membrane with the system in question. In spite of such reservations we have come to understand the behavior of many membrane components using the liposome-reconstitution approach.

METHODS AND LIMITATIONS OF LIPOSOME RECONSTITUTION

The various techniques used to intercalate proteins into liposomes have been reviewed in detail by Racker (5). The procedures involve (a) solubilization and isolation of membrane components with their catalytic membrane functions intact, (b) reconstitution into liposomes, and (c) assay of the function in question. Kinetically, the reassembled system must, of course, behave like the native membrane system if valid conclusions are to be drawn. For this, the artificially constructed system must have the intercalated functions oriented in the same way as in the native membranes and a lipid milieu that does not alter natural function. The choice of lipids and method of reconstitution to achieve optimal activity rests with the investigator. The isolation of integral hydrophobic membrane proteins is fraught with problems and extensive reviews on the subject are available (6,7). Once solubilized and isolated from the membrane, the catalytic function can be assembled into liposomes in several ways (5). In general, the methods involve either a self-assembly of reconstituted liposomes from a micellar mixture of lipids, hydrophobic proteins, and a suitable detergent, followed by slow removal of detergent by prolonged dialysis (8), or the insertion of protein into preformed liposomes by controlled sonication (9). The detergent-dialysis procedure which was the first demonstration of reconstitution into liposomes (8) is relatively slow and cannot be used with membrane proteins that are sensitive to prolonged exposure to detergents. A modification of the method avoids dialysis by diluting the detergent, which may also spare the protein from any deleterious effects of the detergent (5). The sonication procedure is very rapid and avoids contact with detergents. The isolated proteins are mixed with preformed liposomes and subjected to sonication in a bath sonicator (5,9). The disadvantage lies in the use of the sonicator, whose power output is difficult to reproduce rigidly. Here again the investigator has to optimize the conditions. The sonication method has been modified by including a freeze-thaw step prior to sonication (10). The time of sonication required is shorter, and apparently larger liposomes result from the procedure. Whatever the procedure employed, the aim is to preserve the catalytic membrane function of the component in question.

The choice of phospholipids, detergents, and the particular procedure for reconstitution have therefore to be decided upon by the investigator. It is not the intent of this chapter to elaborate on these methods. Rather, we will attempt to summarize how such reconstitution methods have clarified our understanding of in vivo systems.

RECONSTITUTED SYSTEMS

Ion Transport Systems

Ion transport systems abound in biological systems, such fluxes being crucial to the maintenance of normal homeostasis and cell function. Of the many such systems documented and investigated by reconstitution (11), the adenosine triphosphate (ATP)-driven ion pumps have been investigated most extensively. Reconstitution of these pumps has facilitated the study of a wide range of properties, including kinetics, stoichiometry, specific lipid requirements, and the effects of transmembrane potential. It has also revealed the structural requirements of particular systems. Advances made with reconstitution of ATP-dependent proton, Ca^{2+}-Mg^{2+}, and Na^+-K^+ pumps are discussed here.

Proton Pumps

Proton translocation pumps are present in a variety of cells and the subject has been thoroughly reviewed (12). These pumps generate a large transmembrane electrochemical potential for H^+ using energy derived from oxidation, light, or ATP hydrolysis. The energy in the electrochemical potential is used to drive cellular functions, including ATP synthesis, in keeping with Mitchell's chemiosmotic hypothesis of energy coupling in oxidative phosphorylation (13).

Biological membranes that can synthesize ATP possess a reversible H^+-translocating ATPase complex called $F_0 \cdot F_1$ as an energy-transducing coupling mechanism (14-25). The precise mechanisms involved in the coupling of H^+ translocation to phosphorylation by the ATP synthetase are not yet clear, but purification and reconstitution of these systems has opened the eay to an understanding of mechanisms. The $F_0 \cdot F_1$ complex from various biomembranes ranging from bacteria to mammalian mitochondria (14-25) have striking comparable structural features. The F_1 fragment consists of the ATPase, which is easily dissociated from $F_0 \cdot F_1$ into a water-soluble component. The membrane integral hydrophobic moiety F_0 renders F_1 sensitive to oligomycin or N,N'-dicyclohexylcarbodiimide (DCCD) and this property has been used as a criterion of $F_0 \cdot F_1$ association during purification. F_0 is concerned with proton translocation. The $F_0 \cdot F_1$ complex has been widely investigated because the two functional fragments of the complex can be separated from one another and studied for their individual functions.

The first successful attempt to isolate and characterize the properties of the components of the mitochondrial oxidative phosphorylation chain by reconstitution into liposomes was reported by Kagawa snd Racker (8). Oligomycin-sensitive ATPase ($F_0 \cdot F_1$) was reconstituted into liposomes and shown to catalyze oligomycin-sensitive ATP-driven proton translocation. The reconstitution technique reported in that paper has since been widely employed in a number of studies. The mitochondrial $F_0 \cdot F_1$ was dispersed into a micellar solution with exogenous phospholipids derived from soybeans and sodium cholate as detergent. Removal of cholate by slow dialysis resulted in closed reconstituted liposomes that catalyzed oligomycin-sensitive, ATP-driven proton translocation similar to those of submitochondrial particles. Since that observations with mitochondria, more information about $F_0 \cdot F_1$ has come from the thermophilic bacterium PS3 (23). $F_0 \cdot F_1$ isolated from this bacterium and reconstituted into liposomes translocated protons when the system was coupled to ATP hydrolysis (26,27). The electrochemical proton gradient (intra- to extravesicular) formed by ATP hydrolysis was about 300 mV (26) and ATP was synthesized when a $\Delta\bar{\mu}H^+$ value greater than 200 mV was established across the liposomes (17,28). The F_1-ATPase isolated from various sources has at least five subunits, with an additional subunit that inhibits ATPase found in mammalian mitochondria (20). The five subunits of F_1 from PS3 have been purified into reconstitutable active forms and their roles delineated (29,30). The F_0 fragment from PS3 and *Escherichia coli* $F_0 \cdot F_1$ has three subunits characterizing it (31,32). An $F_0 \cdot F_1$ of nine subunits has been purified from chloroplasts (33). All of these $F_0 \cdot F_1$ preparations show $^{32}P_i$-ATP exchange (31-33), $\Delta\bar{\mu}H^+$-dependent ATP synthesis (17,33) and ATP-driven proton pumping (26, 28,31). The purified PS3 F_0 fragment composed of three polypeptides has been further dissociated into a fragment of only two polypeptides by removal of one of the constituent polypeptides on carboxymethyl cellulose (34). The dipeptide fragment mediates all the activities of F_0, such as H^+ translocation, DCCD sensitivity, F_1 binding, and energy-transducing reactions with $F_1 \cdot F_0$ has been assayed for its proton translocating function after incorporation into liposomes and imposition of a membrane potential by a diffusion gradient of K^+ (35,36). The initial rate of H^+ translocated was found to be linearly related to the amount of F_0 incorporated into the membrane and to the magnitude of the membrane potential. Proton translocation was inhibited by oligomycin when preparations were derived from mitochondria (37) or by DCCD (35,36) when preparations were obtained from bacteria. The DCCD was shown to react with the smallest subunit fragment of F_0; a proteolipid component (36,38). Reconstitution studies further revealed that proton translocation is abolished by nitration of tyrosine. The loss of activity parallels the derivatization of the only tyrosyl residue of the proteolipid subunit, with complete loss occurring when a third of the proteolipid is modified (39). From their various studies with reconstituted F_0, Kagawa and co-workers concluded that F_0 is composed essentially of two subunit fragments, one a proteolipid mediating H^+ conductance and the other mediating binding of the proton channel to F_1 (35). From these studies with F_0, F_1, and $F_0 \cdot F_1$ it is clear that F_1 binds to F_0, blocking the proton channel. In the presence of ADP and P_i, a gating mechanism in F_1 allows H^+ conductance through the channel together with ATP synthesis. On the other hand, during ATP hydrolysis, the gating mechanism permits the flow of protons in the opposite direction. The gating mechanism has also been characterized by Kagawa and co-workers (39) using reconstitution, where it has been found that the γ subunit of F_1 is the gate in the ATPase. It is to be expected that more insight into the mechanism of $F_0 \cdot F_1$ function will be forthcoming by pursuing the reconstitution approach.

It is worthwhile to emphasize here the manipulations that reconstituted liposomes allow in the studies described above. Electrochemical proton gradients were readily established by incubating liposomes in a low-pH buffer and valinomycin and then quickly adding a higher-pH buffer containing KCl. Both the pH gradient and electrogenic K^+ gradient were required for optimal ATP synthesis by $F_0 \cdot F_1$. Proton fluxes were measured with a pH electrode or estimated by quenching of 9-aminoacridine fluorescence.

The liposome-reconstitution appraoch has also been successfully used with purified components of the mitochondrial respiratory chain. The three coupling sites of the chain have each been reconstituted and shown to catalyze proton translocation (40-44). When functionally linked to the ATP synthetase complex, such preparations exhibit oxidative phosphorylation (45).

Bacteriorhodopsin (BRh) is a membrane protein synthesized by halophilic bacteria when grown under low O_2 tension (46). Under normal aerobic conditions the bacteria synthesize ATP by using O_2 and do not use BRh. At low O_2 tension the organisms can synthesize ATP in the presence of BRh and light. It was proposed (46) that BRh generates an electrochemical proton gradient across the cell membrane by functioning as a light-driven proton pump, with the energy in the gradient being utilized for ATP synthesis. BRh is a simple protein of molecular weight 25,000 and makes up over 75% of the membrane of the bacteria that synthesize it (47-49). Retinal is covalently attached 1:1 with BRh protein (48,50). Reconstitution of BRh into liposomes has confirmed its function as a light-driven proton pump. Racker (9) first showed BRh reconstitution using sonication in the absence of detergent. BRh liposomes showed light-dependent uptake of protons (9,51). The process was reversed in the dark and inhibited by proton carriers. The orientation of the protein in these liposomes was obviously reversed since in intact bacteria protons are released in light. Electron microscopic evidence supported this assumption (52). Proton transport was electrogenic (51) and illumination of BRh liposomes resulted in the generation of a membrane potential (53,54). Indeed, it has been demonstrated that the electrochemical potential generated by BRh can drive ATP synthesis by mitochondrial or bacterial F_1 incorporated into liposomes (17,51,55). Electrical measurements made by fusing reconstituted liposomes to one side of a planar thick lipid membrane have confirmed these observations (56-59).

Sarcoplasmic Reticulum Ca^{2+} Pump

The sarcoplasmic reticulum (SR) is composed of closed membrane tubules found between myofibrils in muscle cells and regulates the calcium ion content in the cytoplasm upon which the activity of the contractile proteins is dependent. When the muscle is stimulated, calcium stored within the SR is released into the cytoplasm. During relaxation, Ca^{2+} is retaken and stored once again in the SR (60). This reuptake of Ca^{2+} from the cytoplasm is mediated by a pump located in the SR and it was suggested that the pump was a Ca^{2+}-Mg^{2+}-activated ATPase (61). Reconstitution of isolated Ca^{2+}-ATPase has confirmed this. In early investigations (62) it was reported that a detergent-solubilized microsomal fraction from muscle exhibited Ca^{2+}-ATPase and Ca^{2+} uptake, both of these functions were susceptible to phospholipase C, but the loss of activities could be restored by the addition of lecithin. Active Ca^{2+} transport mediated by the Ca^{2+}-Mg^{2+} pump was later demonstrated when a purified ATPase was reconstituted into liposomes and shown to accumulate Ca^{2+} (63). The transport of Ca^{2+} by the reconstituted ATPase was ATP and Mg^{2+} dependent and inhibited by (ethylenedinitrilo)tetraacetic acid (EDTA) (63-66). The reconstituted ATPase also exhibited Ca^{2+}-Mg^{2+}-dependent ATP hydrolysis (63-66), Ca^{2+}-dependent enzyme phosphorylation (66), P_i-ATP exchange (63,66), and synthesis of ATP from ADP and P_i (66) under appropriate conditions.

    Ca^{2+}-ATPase has a critical requirement for a very specific lipid environment in order to exhibit enzymatic activity. Under optimal conditions the enzyme is embedded in bilayers composed of zwitterionic lipids with fluid fatty acid chains (67). Apparently, about 30 lipid molecules, of the 90 associated with each native enzyme molecule are equally distributed between the two sides of the bilayer,

forming an annulus that maintains and modulates enzyme activity (67). Ca^{2+}
accumulation in reconstituted liposomes was demonstrable provided that enzyme
activity was protected within the liposome bilayers. A wide range of phospholipids were shown to provide membranes within which the Ca^{2+}-ATPase was
functional (67). The electrogenic nature of the pump was also estab''shed by
the use of reconstituted liposomes (68). Ca^{2+} uptake was found to be regulated
by the electrical voltage difference imposed across the liposomal membrane by
K^+ and valinocycin. The structural requirements of the enzyme for transport
function have also been studied by reconstitution into liposomes and into planar
membranes. The enzyme is a simple protein not having any subunit fragments
(69,70). Tryptic cleavage of the 100,000-dalton Ca^{2+}-ATPase yields fragments
of 55,000, 45,000, 30,000, and 20,000 daltons (71,72). When the liposome-
incorporated enzyme was exposed to trypsin for 1 min, the 100,000-dalton enzyme was cleaved into two fragments of 45,000 and 55,000 daltons (72,73). The
ionophoric site was found associated with the 55,000-dalton fragment when
assayed for divalent ion conductance in black lipid membranes or liposomes.
Following further cleavage, the activity was found associated with the 20,000-
dalton fragment (73-75). Ionophoric activity has later been ascribed to a 13,000-
dalton fragment (76). However, these studies have all been based on ionophoric
ability using divalent metal ions. The SR transports only Ca^{2+} and Sr^{2+}, the two
ions that activate the ATPase. Other ions, such as Ba^{2+}, Mg^{2+}, and Mn^{2+}, have no
such property but are conducted by the "ionophores" studied. The specificity
of the ionophoric fragments for Ca^{2+} transport are therefore open to question
since their selectivity for Ca^{2+} is rather low.

The reconstitution studies with the holoenzyme exhibited all the properties
of the native SR membrane, but quantitatively the preparations were inferior.
Ca^{2+} transport was vastly improved, almost to that observed in SR (77), by the
addition to liposomes of a proteolipid fraction of 12,000 daltons purified from
SR (78). This proteolipid has been ascribed a "coupling" function in ion translocation. Without this peptide Ca^{2+} translocation was found to be uncoupled
from ATP hydrolysis (79). However, no stoichiometric relationship between
this proteolipid and the ATPase could be found (67,80), although some indirect
role for it in Ca^{2+}-ATPase function has not been ruled out (67). No ionophoric
properties were found associated with the proteolipid (81), but it did reduce the
permeability of the SR membrane to water and nonspecific ions.

The requirement of a specific lipid environment for Ca^{2+} ATPase activity was
referred to earlier (67). In the native SR, the phospholipid:protein ratio is 80:1,
and when this ratio is reduced to below 30:1, enzymatic activity decreases linearly until it is irreversibly lose at a ratio of 15:1 (64,82). It was suggested that
ATPase was functionally stable only when surrounded by an annulus of phospholipid (67,83). The lipid requirement for Ca^{2+} translocation was also investigated
but the results have not been as clear cut. Ca^{2+}-ATPase liposomes transported

Ca^{2+} if the liposomes were made from soybean lipids or from a mixture of purified phosphatidylcholine and phosphatidylethanolamine (66,84,85). Blocking
of the amino group of phosphatidylethanolamine by acetylation resulted in a loss
of Ca^{2+} transport unless long-chain alkylamines were added to the liposomes (84).
It was concluded that a free amino group in the membrane is a prerequisite for
functional reconstitution. Ca^{2+} transport therefore seems to be defined by the
nature of the phospholipid polar head group, whereas the ATPase function depends only on the fatty acid residues (64,82,83). The two functions of this pump
are therefore regulated by different parts of the lipid milieu. Bennett et al. (86)
tested a series of synthetic perospholipids and concluded that different phospholipids can support Ca^{2+} uptake into liposomes and that native SR lipid is not essential for Ca^{2+} accumulation and ATPase activity. In their opinion the inability of
certain lipids to support Ca^{2+} accumulation is a consequence of their inability to
form intact liposomes. In further studies on the lipid requirement for proper
Ca^{2+}-ATPase activity, Hidalgo et al. (87), using lipid substitution, examined phosphoenzyme formation and its subsequent dephosphorylation. Their studies revealed that the phosphoenzyme was stabilized only by an ordered lipid environment.

Na$^+$-K$^+$ Pump

(Na$^+$,K$^+$)-ATPase is a ubiquitous enzyme found in cellular plasma membranes and
it has been estimated that about a third of the ATP produced by basal respiration
it consumed by (Na$^+$,K$^+$)-ATPase of mammalian membranes (88). The enzyme
regulates intracellular Na$^+$ and K$^+$ concentrations and proof that the enzyme is in
fact the coupled Na$^+$-K$^+$ pump came from reconstitution studies incorporating
purified enzyme from various sources into liposomes by the detergent dialysis
method (89-93). Na$^+$ active transport into liposomes in the presence of added
ATP was demonstrated. Na$^+$ transport was inhibited by ouabain if the inhibitor
was included in the intravesicular space, showing that the same asymmetry for
ouabain sensitivity was preserved in the reconstituted preparations as in the
natural membranes. Furthermore, since in this pump system, transport of Na$^+$
and K$^+$ are coupled, it was also possible to measure K$^+$ transport. When the
ATPase was reconstituted into liposomes, including ^{42}K or ^{86}Rb in the vesicles,
active extrusion of K$^+$ or Rb$^+$ was seen, with Na$^+$ being simultaneously and actively taken up in the presence of ATP. Stoichiometric analysis revealed the
amount of Na$^+$ and K$^+$ transported to be in a ratio of 1.43. Ouabain inhibited
K$^+$ efflux if present intravesicularly (91). The reconstituted systems exhibited
all the properties of the native pump. However, the enzyme activities were low
because of the susceptibility of the enzyme to the detergent-cholate used in all
these studies. In later studies using the freeze-thaw-sonication method (which ex-
cludes detergents), reconstituted liposomes with higher transport activities were
obtained (94). In all these studies the enzyme was reconstituted randomly; that

is, only about 50% of the sites were oriented in the native manner and the measurement of Na^+ ingress and K^+ egress in the presence of externally added ATP was the reverse of what one obtains in the whole cell.

The enzyme consists of two types of polypeptide chains, a phosphorylated catalytic subunit of 100,000 daltons and a glycopeptide of 55,000 daltons (95, 96). The ratio of the catalytic subunit to the glycoprotein varies (96). All the functions of the enzyme seem to reside in the larger subunit and the role of the glycoprotein is not clear. Antibodies raised against the glycoprotein inhibit ATPase activity but not ouabain binding (97,98). The electrogenicity of the pump from some sources ($3Na^+:2K^+$) has been demonstrated in reconstituted liposomes using $^{14}CSCN^-$ (99,100). The movement of Cl^- with Na^+ to maintain charge equilibrium has also been demonstrated in liposomes (100).

Glucose Transporters

Erythrocytes

D-Glucose is specifically taken up by erythrocytes by a facilitated diffusion mechanism (101). The protein mediating the process has been investigated by reconstitution (10,102,103). Kasahara and Hinkle (102) purified a 55,000-dalton glycoprotein from erythrocyte membranes and reconstituted it into liposomes by the freeze-thaw-sonication method. The assay of the protein was based on its ability to transport D-glucose. Transport was inhibited by agents such as cytochalasin B, 1-fluoro-2,4-dinitrobenzene, p-chloromercuribenzene sulfonic acid, and mercuric ions, all of which inhibit transport in erythrocytes. The uptake of D-glucose was stereospecific, that is, L-glucose was not transported and D-glucose uptake was inhibited by structural analogs (10,103). Cytocholasin B, a potent inhibitor of erythrocyte D-glucose transport (104-107), was shown to bind to the reconstituted liposomes, and this binding was competitively inhibited by D-glucose and other molecules that competitively interact with the carrier (108).

As determined by polyacrylamide gel electrophoresis (102), the transporter has been claimed to have a molecular weight of 55,000. This has been confirmed by immunological tagging (109). There has been some question as to whether the 55,000-dalton fraction arises as a result of proteolysis while still retaining transport activity (110,111). The immunological studies of Sogin and Hinkle do not support this since fresh erythrocyte ghosts showed cross-reaction with antibody only at the 55,000-dalton band. Whether the transporter in the native membrane is an aggregate of the monomeric 55,000-dalton protein remains to be established.

Adipocytes

The adipocyte plasma membrane also exhibits facilitated, stereospecific uptake of D-glucose which has been reconstituted into liposomes (112). As in intact fat cells, the reconstituted membranes showed inhibition by cytochalasin B, phloretin, phlorizin, and dipyridamole. A major protein of 94,000 daltons was thought to be the mediator. However, more recently (113), using reconstitution to monitor purification, the same laboratory has ruled out this protein as the glucose transporter. The number of glucose transport sites in adipocytes is very low compared to erythrocytes (113) and to date, highly purified transporting fractions have shown no identifiable protein.

Epithelia

The transport of D-glucose across the luminal brush border of kidney and small intestine is an active process coupled to a gradient of Na^+ across the membrane (114). Both Na^+ and D-glucose bind to the carrier to form a ternary complex which transports both species into the cell. The carrier is stereospecific and has an absolute requirement for Na^+ for glucose transport. A greater understanding of the mechanisms involved has come from studies with isolated membrane vesicles (115-117). In attempts to characterize the protein involved, we have used liposome reconstitution of solubilized membrane components (118-122). The reconstitution method that gave the best results was brief controlled sonication of a mixture of membrane proteins and preformed liposomes made from soybean phospholipids. The reconstituted preparation exhibited all the characteristics of D-glucose transport observed in membrane vesicles; transport was Na^+-dependent, stereospecific, inhibited by phlorizin and other sugar analogs that share the same entry mode, and the rate of uptake depended on the magnitude of the membrane potential. Isolation and purification of the transport moiety has proved to be an arduous task. The brush border membranes have numerous intrinsic proteins, the glucose transporter constituting only about 0.4% of the total protein (123). Initial attempts using immobilized lectin chromatography (120) yielded a five-fold purification of transport activity. The preparation contained four major protein bands. Exposure of native membranes to the controlled proteolytic action of subtilisin and papain yielded a preparation with glucose transport function with a major band on polyacrylamide gel electrophoresis at 160,000 daltons. This band seems to be a dimer of an 80,000-dalton polypeptide (122). Lin et al. (124) have reported using a phlorizin polymer affinity chromatography resin to purify the glucose transporter. Although a 20- to 30-fold purification was reported, the protein has not been identified. Further studies should establish the identity of this protein as the glucose transporter.

The reconstitution approach has also been used in attempts to identify amino acid transport systems in various cell types (125-128). Here again, while the reconstituted preparations have shown transport characteristics similar to the native systems, purification and identification of the carriers have not been successful.

Mitochondrial Adenosine Diphosphate (ADP)-ATP Countertransport

ATP synthesized in the mitochondria is transferred to the cytoplasm by a transport protein residing in the inner mitochondrial membrane, which exchanges ATP for ADP. The enzyme has been purified to homogeneity (129-130), and reconstituted in liposomes (131-134). The reconstituted system shows the same properties seen in the native mitochondrial milieu and is regulated by the membrane potential. In beef heart a mitochondrial transporter of molecular weight 60,000 seems to be the sole mediator of ADP-ATP exchange. Flux constants for ADP and ATP were also measured in the reconstituted systems with an asymmetric exchange observed.

CONCLUSIONS

In the past few years a plethora of reports on reconstitution have been documented. Not all of these have been referred to here. What emerges is that the technique has proved immensely useful in some instances and has at least proved to be a useful tool in others. In the case of carriers, the method is invaluable in following purification where no enzymatic function is exhibited by the transporting moiety. The technique itself is still in its infancy and although many systems have veen reconstructed to mimic the in situ membrane systems, few data on the molecular mechanisms involved have come from these studies. However, a great deal of optimism has been generated, and coupled to finer physiochemical techniques, much information should be forthcoming.

REFERENCES

1.   Singer, S. J. and G. L. Nicolson. 1972. The fluid mosaic model of the structure of cell membranes. Science 175:720.
2.   Singer, S. J. 1971. The molecular organization of biological membranes. In *Structure and Function of Biological Membranes*, L. I. Rothfield (Ed.). Academic Press, New York, pp. 145-222.
3.   Tanner, M. J. A. 1979. Isolation of integral membrane proteins and criteria for identifying carrier proteins. In *Current Topics in Membranes and Transport*, F. Bonner and A. Kleinzeller (Eds.). Academic Press, New York, pp. 1-51.
4.   Racker, E. 1977. Perspectives and limitations of resolutions-reconstitution experiments. J. Supramol. Struct. 6:215.
5.   Racker, E. 1979. Reconstitution of membrane processes. In *Methods of Enzymology*, S. Fleischer and L. Packer (Eds.). Academic Press, New York, pp. 699-711.

6. Helenius, A. and K. Simons. 1975. Solubilization of membranes by detergents. Biochim. Biophys. Acta 415:29.
7. Maddy, A. H. and M. J. Dunn. 1976. The solubilization of membranes. In *Biochemical Analysis of Membranes*, A. H. Maddy (Ed.). Chapman & Hall, London, pp. 177-197.
8. Kagawa, Y. and E. Racker. 1971. Partial resolution of the enzymes catalyzing oxidative phosphorylation. J. Biol. Chem. 246:5477.
9. Racker, E. 1973. A new procedure for the reconstitution of biologically active phospholipid vesicles. Biochim. Biophys. Res. Commun. 55:224.
10. Kasahara, M. and P. C. Hinkle. 1976. Reconstitution of D-glucose transport catalyzed by a protein fraction from human erythrocytes in sonicated liposomes. Proc. Natl. Acad. Sci. USA 73:396.
11. Montal, M. 1976. Experimental membranes and the mechanisms of bioenergy transductions. Annu. Rev. Biophys. Bioeng. 5:119.
12. Fillingame, R. H. 1980. The proton-translocation pumps of oxidative phosphorylation. Annu. Rev. Biochem. 49:1079.
13. Mitchell, P. 1972. Chemiosomotic coupling in energy transduction: a logical development of biochemical knowledge. J. Bioenerg. 3:5.
14. Mitchell, P. 1966. Chemiosomotic coupling in oxidative and photosynthetic phosphorylation. Biol. Rev. Camb. Philos. Soc. 41:445.
15. Serrano, R., B. I. Kanner, and E. Racker. 1976. Purification and properties of the proton-translocating adenosine triphosphatase complex of bovine heart mitochondria. J. Biol. Chem. 251:2453.
16. Ryrie, I. J. and P. F. Blackmore. 1976. Energy-linked activities in reconstituted yeast adenosine triphosphatase proteoliposomes: adesine triphosphate formation coupled with electron flow between ascorbic and ferricyamide. Arch. Biochem. Biophys. 176:127.
17. Sone, N., M. Yoshida, H. Hirata, and Y. Kagawa. 1977. Adenosine triphosphate synthesis by electrochemical proton gradient in vesicles reconstituted from purified adenosine triphosphatase and phospholipids of thermophilic bacterium. J. Biol. Chem. 252:2956.
18. Winget, G. D., N. Kanner, and E. Racker. 1977. Formation of ATP by the adenosine triphosphatase complex from spinach chloroplasts reconstituted together with bacteriorhodopsin. Biochim. Biophys. Acta 460:490.
19. Kozlov, I. A. and V. P. Skulachev. 1977. H^+-adenosine triphosphatase and membrane energy coupling. Biochim. Biophys. Acta 463:29.
20. Senior, A. E. 1979. The mitochondrial ATPase. In *Membrane Proteins in Energy Transduction*, R. A. Capaldi (Ed.). Marcel Dekker, New York, pp. 233-278.
21. Panet, R. and D. R. Sanadi. 1976. Soluble and membrane ATPase of mitochondria chloroplasts and bacteria: molecular structure enzymatic properties and functions. Curr. Top. Membr. Transp. 8:99.
22. Nelson, N. 1976. Structure and function of chloroplast ATPase. Biochim. Biophys. Acta 456:314.
23. Kagawa, Y. 1978. Reconstitution of the energy transformer gate and

channel subunit reassembly crystalline ATPase and ATP synthesis. Biochim. Biophys. Acta 505:45.

24. Haddock, B. A. and C. W. Jones. 1977. Bacterial respiration. Bacteriol Rev. 41:47.

25. Downie, J. A., F. Gibson, and G. B. Cox. 1979. Membrane adenosine triphosphatase of prokaryotic cells. Annu. Rev. Biochem. 48:103.

26. Sone, N., M. Yoshida, H. Hirata, H. Okamoto, and Y. Kagawa. 1976. Electrochemical potential of protons in vesicles reconstituted from purified, proton-translocating adenosine triphosphatase. J. Membr. Biol. 30:121.

27. Sone, N., M. Yoshida, H. Hirata, and Y. Kagawa. 1977. Reconstitution of vesicles capable of energy transformation from phospholipids and adenosine triphosphatase of a thermophilic bacterium. J. Biochem (Tokyo) 8:519.

28. Sone, N., Y. Takeuchi, M. Yoshida, and K. Ohno. 1977. Formation of electrochemical proton gradient and adenosine triphosphatase in proteoliposomes containing purified adenosine triphosphatase and bacteriorhodopsin. J. Biochem. (Tokyo) 82:1751.

29. Yoshida, M., H. Okamoto, N. Sone, H. Hirata, and Y. Kagawa. 1977. Reconstitution of thermostable ATPase capable of energy coupling from its purified subunits. Proc. Natl. Acad. Sci. USA 73:936.

30. Yoshida, M., N. Sone, H. Hirata, and Y. Kagawa. 1977. Reconstitution of adenosine triphosphatase of thermophilic bacterium from purified individual subunits. J. Biol. Chem. 252:3480.

31. Sone, N., M. Yoshida, H. Hirata, and Y. Kagawa. 1975. Purification and properties of a DCCD-sensitive adenosine triphosphatase from a thermophilic bacterium. J. Biol. Chem. 250:7917.

32. Foster, D. L. and R. H. Fillingame. 1979. Energy-transducing H^+-ATPase of *Escherichia coli:* purification, reconstitution and subunit composition. J. Biol. Chem. 254:8230.

33. Pick, U. and E. Racker. 1979. Purification and reconstitution of the N,N'-dicyclohexylcarbodiimide-sensitive ATPase complex from spinach chloroplasts. J. Biol. Chem. 254:2793.

34. Sone, N., M. Yoshida, H. Hirata, and Y. Kagawa. 1978. Resolution of the membrane moiety of the H^+-ATPase complex into two kinds of subunits. Proc. Natl. Acad. Sci. USA 75:4219.

35. Okomoto, H., N. Sone, H. Hirata, M. Yoshida, and Y. Kagawa. 1977. Purified proton conduction in proton translocating adenosine triphosphatase of a thermophilic bacterium. J. Biol. Chem. 252:6125.

36. Negrin, R. S., D. L. Foster, and R. H. Fillingame. 1980. Energy transducing H^+-ATPase of *Escherichia coli.* Reconstitution of proton translocation activity of the intrinsic membrane sector. J. Biol. Chem. 255:5643.

37. Shchipakin, V., E. Chuchlova, and Y. Evtodienko. 1976. Reconstitution of mitochondrial H^+-transporting system proteoliposomes. Biochem. Ciophys. Res. Commun. 69:123.

38. Sone, N., M. Yoshida, H. Hirata, and Y. Kagawa. 1979. Carbodiimide-binding protein of H^+-translocating ATPase and inhibition of H^+ conduction by dicyclohexylcarbodiimide. J. Biochem. (Tokyo) 85:503.

39. Sone, N., K. Ikeha, and Y. Kagawa. 1979. Inhibition of proton conduction by chemical modification of the membrane moiety of proton translocation ATPase. FEBS Lett. 97:61.

41. Hinkle, P. C., J. A. Kim, and E. Racker. 1972. Ion Transport and respiratory control in vesicles formed from cytochrome oxidase and phospholipids. J. Biol. Chem. 247:1338.

41. Hinkle, P. C. 1973. Electron transfer across membranes and energy coupling. Fed. Proc. 32:1988.

42. Leung, K. H. and P. C. Hinkle. 1975. Reconstitution of ion transport and respiratory control in vesicles formed from reduced coenzyme Q-cutochrome and reductase and phospholipids. J. Biol. Chem. 250:8647.

43. Ragan, C. I. and P. C. Hinkle. 1975. Ion transport and respiratory control in vesicles formed from reduced nicotinamide adenine dinucleotide coenzyme Q reductase and phospholipids. J. Biol. Chem. 250:8472.

44. Racker, E. and A. Kandrach. 1971. Reconstitution of the third site of oxidative phosphorylation. J. Biol. Chem. 246:7069.

45. Ragan, C. I. 1979. Reconstitution of energy conservation. In *Methods in Enzymology,* S. Fleischer and L. Packer (Eds.). Academic Press, New York, pp. 715-736.

46. Oesterhelt, D. and W. Stoeckenius. 1973. Functions of a new photoreceptor membrane. Proc. Natl. Acad. Sci. USA 70:2853.

47. Kushwaha, S. C., M. Kates, and W. G. Martin. 1975. Characterization and composition of the purple and red membrane from *Halobacterium cutirubrum.* Can. J. Biochem. 53:284.

48. Oesherhelt, D. and W. Stoeckenius. 1971. Rhodopsin-like protein from the purple membrane of *Halobacterium halobium.* Nature New Biol. 233:149.

49. Bridgen, J. A. and I. D. Walker. 1976. Photoreceptor protein from the purple membrane of *Halobacterium halobium:* molecular weight and retinal binding site. Biochemistry 15:792.

50. Oesterhelt, D., M. Meetzen, and L. Schulmann. 1973. Reversible dissociation of the purple complex in bacteriorhodopsin and identification of 13-cis and all-trans-retinal as its chromophores. Eur. J. Biochem. 40:453.

51. Racker, E. and W. Stoeckenius. 1974. Reconstitution of purple membrane vesicles catalyzing light-driven protons uptake and adenosine triphosphate formation. J. Biol. Chem. 249:662.

52. Korenbrat, J. I. 1977. Ion transport in membrane: incorporation of biological ion-translocating proteins and model membrane systems. Annu. Rev. Physiol. 39:19.

53. Kayushin, L. P. and V. P. Skulachev. 1974. Bacteriorhodopsin as an electrogenic proton pump: reconstitution of bacteriorhodopsin proteoliposomes generating $\Delta\psi$ and ΔpH. FEBS Lett. 39:39.

54. Racker, E. and C. Hinkle. 1974. Effect of temperature on the function of a proton pump. J. Membr. Biol. 17:181.

55. Yoshida, M., N. Sone, H. Hirata, Y. Kagawa, Y. Takeuchi, and K. Ohno. 1975. ATP synthesis by purified DCCD sensitive ATPase incorporated

into reconstituted purple membrane vesicles. Biochem. Biophys. Res. Commun. 67:1295.

56. Barsky, E. L., Z. Daneshazy, L. A. Drachev, M. D. Il'ina, A. A. Jasaitis, A. A. Kondrashin, V. D. Samuilov, and V. P. Skulachev. 1976. Reconstitution of biological molecular generators of electric current: bacteriochlorophyll and plant chlorophyll complexes. J. Biol. Chem. 251:7066.

57. Drachev, L. A., V. N. Frolov, A. D. Kaulen, E. A. Lieberman, S. A. Ostroumov, V. G. Plakunova, A. Y. Semenov, and V. P. Skulachev. 1976. Reconstitution of biological/molecular generators of electric current: bacteriorhodopsin. J. Biol. Chem. 251:7059.

58. Drachev, L. A., A. A. Jasaitis, A. D. Kaulen, A. A. Kandrashin, V. C. La, A. Y. Semenov, I. I. Severina, and V. P. Skulachev. 1976. Reconstitution of biological molecular generators of electric current cytochrome oxidase. J. Biol. Chem. 251:7072.

59. Drachev, L. A., A. A. Jasaitis, H. Mikelsaar, I. B. Nemecek, A. Y. Semenov, E. G. Semenova, I. I. Severina, and V. P. Skulachev. 1976. Reconstitution of biological molecular generators of electric current. H^+-ATPase. J. Biol. Chem. 251:7077.

60. Inesi, G. 1972. Active transport of calcium ions in sarcoplasmic membranes. Annu. Rev. Biophys. Bioeng. 1:191.

61. Hasselbach, W. 1963. Relaxing factor and the relaxation of muscle. Prog. Biophys. Biophys. Chem. 14:167.

62. Martonosi, A. 1968. Sarcoplasmic reticulum. Solubilization of microsomal adenosine triphosphatase. J. Biol. Chem. 243:71.

63. Racker, E. 1972. Reconstitution of a calcium pump with phospholipids and a purified Ca^{++}-adenosine triphosphatase from sarcoplasmic reticulum. J. Biol. Chem. 247:8198.

64. Warren, G. B., P. A. Toon, N. J. M. Birdsall, A. G. Lee, and J. C. Metcalfe. 1974. Reconstitution of a calcium pump using defined membranes components. Proc. Natl. Acad. Sci. USA 71:622.

65. McLennan, D. H., T. J. Ostwald, and P. S. Stewart. 1974. Structural components of the sarcoplasmic reticulum membrane. Ann. N.Y. Acad. Sci. 227:527.

66. Knowles, A. F. and E. Racker. 1975. Properties of a reconstituted calcium pump. J. Biol. Chem. 250:3538.

67. Bennett, J. P., K. A. McGill, and G. B. Warren. 1980. The role of lipids in the functioning of a membrane protein: the sarcoplasmic reticulum calcium pump. In *Current Topics in Membranes and Transport,* F. Bronner and A. Kleinzeller (Eds.). Academic Press, New York, pp. 127-164.

68. Zimniak, P. and E. Racker. 1978. Electrogenicity of Ca^{2+}-transport catalyzed by the Ca^{2+}-ATPase from sarcoplasmic reticulum. J. Biol. Chem. 253:4631.

69. McLennan, D. H. 1970. Purification and properties of an adesine triphosphatase from sarcoplasmic reticulum. J. Biol. Chem. 245:4508.

70. Meissner, G., G. E. Connor, and S. Fleischer. 1973. Isolation of sarco-

plasmic reticulum by zonal centrifugation and purification of Ca^{2+}-pump and Ca^{2+}-binding proteins. Biochim. Biophys. Acta 298:246.

71. Inesi, G. and D. Scales. 1974. Tryptic cleavage of sarcoplasmic reticulum protein. Biochemistry 13:3298.

72. Stewart, P. S. and D. H. Maclennan. 1974. Surface particles of sarcoplasmic reticulum membrane. Structural features of the adenosine triphosphtase. J. Biol. Chem. 249:985.

73. Shamoo, A. E., T. E. Ryan, P. S. Stewart, and D. H. Maclennan. 1976. Localization of ionophore activity in a 20,000 Dalton fragment of the adenosine tryphosphatase of sarcoplasmic reticulum. J. Biol. Chem. 251: 4147.

74. Shamoo, A. E., T. L. Scott, and T. E. Ryan. 1977. Active calcium transport via coupling between the enzymatic and ionphoric sites of Ca^{++} + Mg^{++}-ATPase. J. Supramol. Struct. 6:345.

75. Shamoo, A. E. 1978. Ionophorus properties of the 20,000 Dalton fragment of $(Ca^{2+} + Mg^{2+})$-ATPase in phosphatidylcholine: cholesterol membranes. J. Membr. Biol. 43:227.

76. Maclennan, D. H., A. Klip, R. Reithmeier, M. Michalak, and K. P. Campbell. 1979. Possible sites of ion flow in the sarcoplasmic reticulum membrane. In *Membrane Bioenergetics,* C. P. Lee, G. Schatz, and L. Ernster (Eds.). Addison-Wesley, Reading, Mass., pp. 255-266.

77. Hasselbach, W. and M. Makinose. 1961. Die Calciumpumpe der Erschlaffungsgrane des Muskels und ihre Abhangigkeit von der ATP-Spaltung. Biochem. Z. 333:518.

78. Maclennan, D. H., C. C. Yip, G. H. Iles, and P. Seeman. 1972. Isolation of sarcoplasmic reticulum proteins. Cold Spring Harbor Symp. Quant. Biol. 37:469.

79. Racker, E. and E. Eytan. 1975. A coupling factor from sarcoplasmic reticulum required for the translocation of Ca^{2+} ions in a reconstituted and Ca^{+2} ATPase pump. J. Biol. Chem. 250:7533.

80. Sigrist, H., K. Sigrist-Nelson, and C. Gritler. 1977. Single phase butanol extraction: a new tool for proteolipid isolation. Biochem. Biophys. Res. Commu. 74:178.

81. Laggner, P. and P. E. Graham. 1976. The effect of proteolipid from sarcoplasmic reticulum on the physical properties of artificial phospholipid membranes. Biochim. Biophys. Acta 433:311.

82. Warren, G. B., P. A. Toon, N. J. M. Birdsall, A. G. Lee, and J. C. Metcalfe. 1974. Reversible lipid titrations of the activity of pure adenosine triphosphatase-lipid complexes. Biochemistry 13:5501.

83. Warren, G. B., N. J. M. Birdsall, A. G. Lee, and J. C. Metcalfe. 1974. Lipid substitution: the investigation of functional complexes of single species of phospholipid and purified calcium transport protein. In *Membrane Proteins in Transport and Phosphorylation,* G. Azzone, M. Klingenberg, E. Quagliariello, and N. Siliprandi (Eds.). North-Holland, Amsterdam, pp. 1-120.

84. Knowles, A. F., A. Kandrach, E. Racker, and H. G. Khorana. 1975. Acetyl

phosphatidylethanolamine in the reconstitution of ion pumps. J. Biol. Chem. 250:1809.

85. Knowles, A. F., E. Eytan, and E. Racker. 1976. Phospholipid-protein interactions in the Ca^{2+} adenosine triphosphatase of sarcoplasmic reticulum. J. Biol. Chem. 251:5161.

86. Bennett, J. P., G. A. Smith, M. D. Honslay, T. R. Hesketa, J. C. Metcalfe, and J. B. Warren. 1978. The phospholipid headgroup specificity of an ATP-dependent calcium pump. Biochim. Biophys. Acta 513:310.

87. Hidalgo, C., N. Ikemoto, and J. Gergely. 1976. Role of phospholipids in the calcium-dependent ATPase of the sarcoplasmic reticulum. J. Biol. Chem. 251:4224.

88. Whittam, R. and M. E. Ager. 1965. The connection between active cation transport and metabolism in erythrocytes. Biochem. J. 97:214.

89. Goldin, S. M. and S. W. Tong. 1974. Reconstitution of active transport catalyzed by the purified sodium and potassium ion-stimulated adenosine adenosine triphosphatase from canine renal medulla. J. Biol. Chem. 249: 5907.

90. Gilden, S., H. M. Rhee, and L. E. Hokin. 1974. Sodium transport by phospholipid vesicles containing purified sodium and potassium ion activated adenosine triphosphatase. J. Biol. Chem. 249:7432.

91. Hilden, S. and L. E. Hokin. 1975. Active potassium transport coupled to active sodium transport in vesicles reconstituted from purified sodium and potassium ion-activated adenosine triphosphatase from the rectal gland of Squalus acanthias. J. Biol. Chem. 250:6296.

92. Goldin, S. M. 1977. Active transport of sodium and potassium ions by the sodium and potassium ion-activated adenosine triphosphatase from renal medulla. J. Biol. Chem. 252:5630.

93. Anner, B. M., L. K. Lane, A. Schwartz, and B. J. R. Pitts. 1977. A reconstitution Na^{+}-K^{+} pump in liposomes containing purified (Na^{+}-K^{+}) ATPase from kidney medulla. Biochim. Biophys. Acta 467:340.

94. Hokin, L. E. and J. F. Dixon. 1979. In Na, K-ATPase Structure and Kinetics, J. C. Skou and J. G. Novlry (Eds.). Academic Press, London, p. 47.

95. Perrone, J. R., J. F. Hackney, J. F. Dixon, and L. E. Hokin. 1975. Molecular properties of purified (sodium + potassium)-activated adenosine triphosphatases and their subunits from the rectal gland of Squalus acanthias and the electric organ of Electrophorus electricus. J. Biol. Chem. 250:4178.

96. Hokin, L. E. 1975. Purification and molecular properties of the (sodium + potassium)-adenosine triphosphatase and reconstitution of coupled + potassium transport in phospholipid vesicles containing purified enzyme. J. Exp. Zool. 194:197.

97. Rhee, H. M. and L. E. Hokin. 1975. Inhibition of the purified sodium-potassium activated adenosinetriphosphatase from the rectal gland of Squalus acanthias by antibody against the glycoprotein subunit. Biochem. Biophys. Res. Commun. 63:1139.

98. Rhee, H. M. and L. E. Hokin. 1979. Inhibition of ouabain-binding to (Na^{+}

+ K$^+$)ATPase by antibody against the catalytic subunit but not by antibody against the glycoprotein subunit. Biochim. Biophys. Acta 558:108.

99.   Hokin, L. E. 1979. Reconstitution of the sodium-potassium pump. In *Membrane Bioenergetics,* C. P. Lee, G. Schatz, and L. Ernster (Eds.). Addison-Wesley, Reading, Mass., p. 281.

100.   Dixon, J. F. and L. E. Hokin. 1980. The reconstituted (Na,K)-ATPase is electrogenic. J. Biol. Chem. 255:10681.

101.   Widdas, W. F. 1980. The asymmetry of the hexose transfer system in the human red cell membrane. In *Current Topics of Membranes and Transport,* F. Bronner and A. Kleinzeller (Eds.). Academic Press, New York, pp. 165–223.

102.   Kasahara, M. and P. C. Hinkle. 1977. Reconstitution and purification of the D-glucose transporter from human erythrocytes. J. Biol. Chem. 252:7384.

103.   Kahlenberg, A. and C. A. Zala. 1977. Reconstitution of D-glucose transport in vesicles composed of lipids and intrinsic protein (zone 4,5) of the human erythrocyte membrane. J. Supramol. Struct. 7:287.

104.   Taverna, R. D. and R. G. Langdon. 1973. Reversible association of cytochalasin B with the human erythrocyte membrane. Inhibition of glucose transport and the stoichometry of cytochalasin binding. Biochim. Biophys. Acta 323:207.

105.   Lin, S., D. V. Santi, and J. A. Spudich. 1974. Biochemical studies on the mode of action of cytochalasin B. J. Biol. Chem. 249:2268.

106.   Taylor, N. F. and G. L. Gagneja. 1975. A model for the mode of action of cytochalasin B inhibition of D-glucose transport in the human erythrocyte. Can. J. Biochem. 53:1078.

107.   Jung, C. Y. and A. L. Rampal. 1977. Cytochalasin B binding sites and glucose transport carrier in human erythrocyte ghosts. J. Biol. Chem. 252:5456.

108.   Sogin, D. C. and P. C. Hinkle. 1980. Binding of cytochalasin B to human erythrocyte glucose transporter. Biochemistry 19:5417.

109.   Sogin, D. C. and P. C. Hinkle. 1980. Immunological identification of the human erythrocyte glucose transporter. Proc. Natl. Acad. Sci. USA 77:5725.

110.   Phutrakul, S. and M. N. Jones. 1979. The permeability of bilayer lipid membranes on the incorporation of erythrocyte membrane extracts and the identification of the monosaccharide transport properties. Biochim. Biophys. Acta 550:188.

111.   Mullins, R. E. and R. G. Langdon. 1980. Maltosyl isothiocyanate: an affinity label for the glucose transporter of the human erythrocyte membrane. I. Inhibition of glucose transport. Biochemistry 19:1199.

112.   Shanahan, M. F. and M. P. Czech. 1977. Purification and reconstitution of the adipocyte plasma membrane D-glucose transport system. J. Biol. Chem. 252:8341.

113.   Carter-Su, C., D. J. Pillion, and M. P. Czech. 1980. Reconstituted D-

glucose transport from the adipocyte plasma membrane: chromatographic resolution of transport activity from membrane glycoproteins using immobilized cancanavalin A^+. Biochemistry 19:2374.

114. Crane, R. K. 1977. The gradient hypothesis and other models of carrier-mediated active transport. Rev. Physiol. Biochem. Pharmacol. 78:98.

115. Crane, R. K. and F. C. Dorando. 1980. On the mechanism of Na^+-dependent glucose. Ann. N.Y. Acad. Sci. 339:46.

116. Dorando, F. C. and R. K. Crane. 1980. The effect of membrane potential on the apparent mechanism of Na^+-dependent glucose transport. Fed. Proc. 39:2159.

117. Dorando, F. C. and R. K. Crane. 1981. The kinetics and mechanism of Na^+-gradient-coupled glucose transport. Fed. Proc. 40:1893.

118. Crane, R. K., P. Malathi, and H. Preiser. 1976. Reconstitution of specific Na^+-dependent D-glucose transport in liposomes by Triton X-100 extracted proteins from purified brush border membranes of hamster small intestine. Biochem. Biophys. Res. Commun. 71:1010.

119. Crane, R. K., P. Malathi, and H. Preiser. 1976. Reconstitution of Na^+-dependent glucose transport in liposome vesicles with Triton-100 extract of rabbit kidney tubular brush border membranes. FEBS Lett. 67:214.

120. Crane, R. K., P. Malathi, H. Preiser, and P. Fairclough. 1978. Some characteristics of kidney Na^+-dependent glucose carrier reconstituted into sonicated liposomes. Am. J. Physiol. 234:E1.

121. Fairclough, P., Malathi, H. Preiser, and R. K. Crane. 1979. Reconstitution into liposomes of glucose active transport from the rabbit renal proximal tubule: characteristics of the system. Biochim. Biophys. Acta 553: 295.

122. Malathi, P., H. Preiser, and R. K. Crane. 1980. Protease-resistant integral brush border membrane proteins and their relationship to sodium-dependent transport of D-glucose and L-alanine. Ann. N.Y. Acad. Sci. 358:253.

123. Tannenbaum, C., G. Toggenburger, M. Kessler, A. Rothstein, and G. Semenza. 1977. High-affinity phlorizin binding to brush border membranes from small intestine: identity with (a part of) the glucose transport system, dependence on the Na^+-gradient, partial purification. J. Supramolec. Struct. 6:519.

124. Lin, J. T., M. E. M. Da Cruz, S. Riedel, and R. Kinne. 1981. Partial purification of hog kidney sodium-D-glucose cotransport system by affinity chromatography on a phlorizin polymer. Biochim. Biophys. Acta 640:43.

125. Johnstone, R. M. and C. Bardin. 1976. Amino acids in reconstituted vesicles derived from plasma membranes of Ehrlich ascites cells. J. Cell. Physiol. 89:801.

126. Cecchini, G., G. S. Payne, and D. L. Oxender. 1978. Reconstitution of neutral amino acid and transport systems from Ehrlich ascites tumor cells. Membr. Biochem. 1:269.

127. Nishino, H., L. G. Tillotson, Robert M. Schiller, Ken-Ichi Inui, and Kurt J. Isselbacher. 1978. Sodium-stimulated active transport of aminoisobu-

tyric acid by reconstituted vesicles from partially purified plasma membranes of mouse fibroblasts transformed by simian virus 40. Proc. Natl. Acad. Sci. USA 75:3856.

128.   Hirata, H. 1979. Solubilization and purification of alanine carrier from thermophilic bacteria and reconstitution into vesicles capable of transport. In *Methods in Enzymology*, S. Fleischer and L. Packer (Eds.). Academic Press, New York, pp. 430–435.

129.   Klingenberg, M., H. Aquila, and P. Riccio. 1979. Isoaltion of functional membrane proteins related to or identical with the ADP, ATP carrier of mitochondria. In *Methods in Enzymology*, S. Fleischer and L. Packer (Eds.). Academic Press, New York, pp. 407–414.

130.   Lauquin, G. J. M., G. Brandolin, F. Boulay, and P. V. Vignais. 1979. Puritication of an atractyloside-binding protein related to the ADP/ATP transport system in yeast mitochondria. In *Methods in Enzymology*, S. Fleischer and L. Packer (Eds.). Academic Press, New York, pp. 414–418.

131.   Kramer, R. and M. Klingenberg. 1977. Reconstitution of inhibitor binding properties of the isolated adenosine $5'$-diphosphate, adenosine $5'$-triphosphate carrier-linked binding protein. Biochemistry 16:4954.

132.   Kramer, R. and M. Klingenberg. 1979. Reconstitution of adenine nucleotide transport from beef heart mitochondria. Biochemistry 18:4209.

133.   Kramer, R. and M. Klingenberg. 1979. Reconstitution of adenine nucleotide transport from beef hear mitochondria. Biochemistry 18:4209.

134.   Kramer, R. and M. Klingenberg. 1980. Modulation of the reconstituted adenine nucleotide exchange by membrane potential. Biochemistry 19: 556.

Liposomes as a Tool in Molecular Biology: A Comparison to Other Methodologies

Marc J. Ostro / The Liposome Company, Princeton, New Jersey

D. Giacomoni / College of Medicine, University of Illinois, Chicago, Illinois

INTRODUCTION

The discovery in the late 1940s and early 1950s that polynucleotides (DNA and RNA) constituted the "stuff" of which genes are made led to a revolution in biology since it allowed for a synthesis between the sciences of genetics and bio-chemistry. It became possible to define genotype and phenotype in chemical terms: the genotype being a reflection of the sequences of nucleotides that compose DNA (or RNA) and the phenotype being the expression of these sequences in the form of proteins. From this synthesis arose the new science of molecular biology, which investigates genetic phenomena by biochemical techniques.

One important aspect of these studies has been the elucidation of the biological activity of purified nucleic acids and proteins within a xenogeneic cellular environment. Originally, when viral nucleic acids were used, the experimental rationale was to confirm the fact that nucleic acids carried genetic information. At the present time the insertion of macromolecules into viable cells is viewed as a means of investigating a variety of biosynthetic, catabolic, and regulatory events in an intact cell. However, the technology involved with the delivery of a macromolecule into a target cell is not without its difficulties unless, of course, the cellular insertion of the molecule in question is mediated by a normal physiological phenomenon. When the barrier presented by the cell membrane cannot be traversed by a macromolecule, one of two methodologies must be chosen: (a) the cell can be broken and the molecule can be added directly to the lysate, or (b) the molecule, by mechanical means, can be microinjected into the intact cell. The former methodology includes all the cell-free translation and coupled transcription-translation systems, while the latter includes either direct "syringe-type" microinjection into the cell or the sequestration of the macromolecule within discrete vesicles (e.g., red cell ghosts or liposomes) followed by vesicle-mediated insertion of the sequestered molecule into the target cell by fusion or endocytosis.

Although a considerable volume of information has been obtained using cell-free systems, these data need not reflect processes that occur in intact cells. For example, Purchio et al. (1) have shown that fragments obtained from the 3' end of the avian sarcoma virus 35S RNA can be translated in a reticulocyte lysate cell-free system yielding a 60,000-dalton glycoprotein which is very similar to the native sarcoma envelope protein. When similar fragments are injected into chick embryo fibroblasts, no envelope protein is synthesized (2). Since in avian sarcoma virus-infected cells, splicing of 5' to 3' sequences must precede the production of envelope glycoprotein, critical regulatory events must be absent from the cell-free system (3).

If one accepts the premise that the regulatory mechanisms defined by the constituents of a cell may be altered when these constituents are released from their plasma membrane-imposed juxtaposition, the need for a technique that can insert macromolecules into an intact viable cell becomes evident. The purpose of this chapter is to discuss the use of liposomes as a vehicle by which such cellular intervention can be achieved and to convey to the reader the excitement (if this word can be properly applied to the patient process that characterizes much scientific progress) generated by research in this field.

DELIVERY OF MACROMOLECULES INTO CELLS BY TECHNIQUES THAT DO NOT EMPLOY LIPOSOMES

Although liposome technology constitutes this chapter's raison d'etre, we feel that to put the liposome-mediated delivery of macromolecules (RNA, DNA, protein, viruses, and subcellular components) into cells in a proper perspective, a review of the methods utilized to date to accomplish macromolecular transfer is essential.

Incorporation of Macromolecules by Cocultivation

RNA

Viral RNA Gierer and Schramm (4) in 1956 first demonstrated that a purified viral RNA could be infectious by using RNA extracted from tobacco mosaic virus, by the then new technique of phenol extraction (5), to infect tobacco leaves. The efficiency of infection was only 0.1-0.5% of the efficiency obtained when intact virus was used, although 10-fold higher than that obtained by Fraenkel-Conrat et al. (6). Following these reports the infectivity of other viral RNAs was investigated. The isolation of infectious RNA from Mengo virus, West Nile virus, and poliovirus was first described by Colter and co-workers (7). The infectivity of the RNA preparations was measured as median lethal dose (LD_{50}) and was calculated to be 0.1% of that of the intact virus. Alexander et al. (8) demonstrated the in vitro infectivity of poliovirus RNA by assessing its ability to form lytic plaques on HeLa cell monolayers. Soon, the infectivity of many viral RNAs was reported, so that in reviewing the field in 1961, Colter and Ellem reported 7 plant and 16 animal viruses that yielded infectious "naked" RNA (9).

The RNA preparations used always contained a small amount of protein (0.1-0.3%) that might have accounted for their infectivity. However, the high susceptibility of the preparations to enzymatic degradation by RNAse, their resistance to antiviral sera, and the fact that the sedimentation velocity of the infectious molecules was different from that of intact virus ruled out viral contamination in the infectious preparations.

Following an observation by Alexander et al. (10), it was established by Holland et al. (11,12) that the infectivity of poliovirus RNA could be increased 20-fold if the cell-RNA interaction was carried out in a hypertonic solutions (1 M NaCl). Later, Pagano and Vaheri succeeded in increasing by 100-fold the infectivity of poliovirus RNA by incubating it with cell monolayers in the presence of diethylaminoethyl (DEAE)-dextran (13), a technique that had been previously shown to enhance the infectivity of influenza viruses (14). The effectiveness of the DEAE-dextran treatment was confirmed by Bachrach (15) using hoof-and-mouth viral RNA. In this work a 10^4-fold increase in the RNA infectivity was established. Also, dimethyl sulfoxide (DMSO) was shown to increase the infectivity of Mengo virus RNA (16) (four- to sixfold increase) and of poliovirus RNA (17).

One of the most exciting properties of isolated viral RNA is that its infectivity is not restricted by membrane-imposed specificities. For example, Holland and co-workers (11,12) showed that poliovirus RNA infected HeLa cells (cells normally infected by the intact virus) and mouse L cells (cells naturally resistant to the virus) with equal efficiency. The viruses that developed from the RNA-infected cells were identical to the parent virus. These investigations confirmed that genetic information is encoded in the nucleic acid of the virus and that the protein's primary function is to provide a protective coat for the polynucleotide. Furthermore, these same workers (12) demonstrated that the inactivation of RNA by RNAse follows first-order kinetics, indicating that one break in the RNA is sufficient for its inactivation, thus demonstrating that the integrity of the genome is needed for infectivity.

Nonviral RNA   The demonstrated susceptibility of eukaryotic cells to infection by viral RNAs encouraged numerous attempts to induce phenotypic changes in eukaryotic cells by cocultivation of the cells with a variety of eukaryotic and prokaryotic RNA preparations. These experiments presented several major difficulties. First, unlike viral RNA, most eukaryotic mRNA is difficult to obtain in pure form and in sufficient quantities to permit proper experimentation, and second, few (if any) biochemical techniques utilized to detect mRNA translation have the sensitivity of the plaque assay used to detect viral production. In spite of these formidable problems, many laboratories have worked on the fate and the biological activity of nonviral RNA added directly to eukaryotic cells. By 1971 an extensive review of the subject by Bhargava and Shanmugan (18) contained a list of no fewer than 200 papers dealing with this subject.

The possibility that a measurable amount of RNA could be incorporated into cultured eukaryotic cells was established by experiments which demonstrated that treatment of cells with RNAse following cocultivation with radiolabeled RNA removed only 30-50% of the cell-associated radioactivity (19,20). Furthermore, actinomycin D did not inhibit radiolabeled RNA incorporation into the cells, an observation which eliminated the possibility that the cell-associated radioactivity observed was due to an extracellular degradation of the RNA followed by transport of the labeled nucleotides into the cell and their utilization in de novo RNA synthesis (20,21). Shanmugan and Bhargava (21) observed that liver cells could incorporate both rat liver RNA and *Escherichia coli* RNA but that *E. coli* RNA was degraded more rapidly. These data were confirmed and extended by Wang et al. (20), who demonstrated that rabbit spleen cells could incorporate small amounts of homologous RNA. It was shown that 30% of the incorporated homologous RNA could be reextracted as a high molecular weight RNA, while bacterial RNA was rapidly degraded inside the cells (20). The amount of RNA that cells can incorporate varies depending on the incubation conditions. Two groups reported that spleen cells could incorporate $4\text{-}10 \times 10^{10}$ daltons of RNA per cell (20,22). Higher values (9×10^{11} daltons per cell) were reported for mouse cells (23). One must be aware that in this type of experiment much of the RNA is degraded even before it is taken up (20), possibly due to the RNAse in the medium.

A very elegant demonstration that exogenous nonviral RNA molecules can be incorporated by mammalian cells was provided by Herrera et al. (24), who incorporated *E. coli* tRNA into a murine leukemia cell line. It was found that reextraction of the tRNA from the leukemia cells did not affect the ability of these RNA molecules to act as a substrate for *E. coli* leucine-activating enzyme, an enzyme which is specific for *E. coli* tRNA (25).

The first attempts to stimulate the synthesis of xenogeneic proteins in cells by treating them with exogenous mRNA were done by Niu and co-workers, who reported increased synthesis of serum albumin in Nelson mouse ascites cells treated in culture with liver RNA (26) and synthesis of a modest amount of tryptophan pyrrolidase in ascites cells treated with RNA from liver induced to produce that enzyme (27). Induction of albumin synthesis by liver RNA was also reported by Zimmerman et al. (28). In later years Tuohimaa et al. (29) succeeded in inducing the synthesis of a modest amount of avidin (1 μg/g tissue) in chick oviduct injected with RNA from the oviduct of progesterone-treated chicks. Similarly, Villee (30) observed a small (37%) increase in glucose-6-phosphate dehydrogenase activity in the uterine horn of a 21-day-old female rat injected with RNA from the uteri of an ovariectomized estradiol-treated rat. More recently, Mroczkowski et al. (31) reported uptake and translation of poly(A)-containing exogenous RNA by chick myoblast cultures. Production of creatine kinase was increased two- to threefold when poly(A) RNA from embryonic chick leg muscle was used, a synthesis that was not inhibited by actinomycin D. The modesty of the results obtained is probably due to the extensive breakdown of RNA during the RNA-cell interaction (20).

A special mention must be made of those researchers who attempted to induce predetermined immunological properties in normal lymphoid cells by means of RNA derived from immunocompetent cells (immune RNA). The interest for these researchers arose because it was felt that by studying the mode of action of "immune RNA" it would be possible to learn how immunocompetent cells develop and also because it was perceived that this field of research could lead to the development of immunotherapies.

The first indication that immunocompetence could be transferred by RNA-containing preparations came from the observations of Sterzl and Hrubesova (32), who reported that crude nucleoprotein extracts from spleens of rabbits immunized against *Salmonella paratyphi* elicited specific antibodies in 5-day-old rabbits that do not normally respond to this antigen. In 1961, Fishman (33) reported that RNA preparations from peritoneal exudate cells exposed to T2 bacteriophage induced the synthesis of anti-T2 antibodies when incubated with lymph node fragments from a normal animal. The presence of antibodies was assessed by a T2 neutralizing test. Later, and independently, Friedman (34) and Cohen and Parks (35) reported that spleen cells from normal mice could acquire the ability to form antibodies against sheep red blood cells when incubated with RNA preparations from spleens of mice immunized against that antigen. The active RNA had a sedimentation value of 8-12S, characteristic of many mRNAs (36). In these and in other studies it was assumed that the biological activity was due to the RNA molecules because of the ability of RNAse (but not DNAse or proteases) to destroy the biological activity of the preparations. These experiments, however, elicited much criticism. In fact, it was felt that even minute amounts of antigen in the RNA preparation could explain the results described. The basis for the criticisms was the finding by Campbell and Garvey (37,38) that antigens can be retained in liver, be associated with RNA, and (when in such form) be very active as immunogens (superantigen). The failure of Chin and Silverman (39) to induce myeloma immunoglobulin synthesis in lymphoid cells treated with RNA from myeloma cells seems to support this possibility since myeloma cells produce antibody-like molecules without exposure to an antigen. Two types of experiments, however, seem to prove that in at least a few cases transfer of immune competence was achieved via RNA. Adler et al. (40) reported the ability of RNA from T2-immunized rabbits to elicit the synthesis of anti- bodies against T2 with the allotype of the RNA donor. These experiments were continued by Bell and Dray (see, e.g., Ref. 41). More interesting is a paper by Schafer et al. (42), who reported the isolation of two RNA preparations from rabbits immunized against T2. One preparation could elicit anti-T2 antibodies in a spleen cell population deprived of cells with receptors for T2. The other preparation, which contained RNA-antigen complexes, could elicit anti-T2 anti- bodies only if cells with T2 receptors were present. It seems, therefore, that in this case, biologically active RNA (possibly mRNA) has been separated from a RNA-antigen complex (superantigen), thus lending support to the contention that RNA can transfer immunocompetence. The exact mechanism governing this putative transfer is not known.

DNA

Evidence that eukaryotic cells can incorporate biologically active DNA was ob- tained initially with viral DNA since it can be easily isolated in purified form and its infectivity can be tested easily and efficiently. The first report that infectious DNA had been isolated appeared in 1959. DiMayorca et al. (43) extracted poly- oma DNA by the phenol method and obtained a preparation (destroyed by DNAse but not by RNAse) that elicited the production of viruses when incubated with natural host cells. The viruses that developed from this infection were identical to the viruses from which the DNA was obtained. The infectivity of polyoma virus DNA was soon confirmed by Weil (44), who increased the efficiency of infection by using hypertonic buffers or EDTA (0.02%). Infectivity of SV40 DNA and its ability to induce T-antigen synthesis was reported by Gerber (45) and Black and Rowe (46), respectively. These authors also used hypertonic solutions during the DNA-cell incubation step. The infectivity of viral DNA could be further enhanced by the DEAE-dextran technique of Pagano and Vaheri (13), originally developed for poliovirus RNA. Using this method, McCutchan and Pagano (47) achieved an increase by a factor of 10^4 in the infectivity of SV40 DNA when 100-1000 $\mu g/\mu l$ of DEAE-dextran was used. Warden and Thorne (48) confirmed these observa- tions for polyoma DNA but when adenovirus I DNA was used only a low infec- tivity could be observed (49) in the presence of DEAE-dextran. However, higher infectivity for DNA from adenovirus was obtained when DNA was coprecipitated with calcium phosphate and then added to the cells. This technique, developed by Graham and Van der Eb (50), enhanced the infectivity of human adenovirus DNA from 0.1-0.5 PFU/μg DNA obtained with DEAE-dextran to 3-6 PFU/μg DNA with calcium phosphate. In the past 10 years these two techniques have been used extensively in DNA transfection experiments which have demonstrated the infectivity of a variety of other viral DNAs (51,52).

However, it is important to recall that not all early DNA transfections were accomplished using viral DNA. In 1962, Szybalska and Szybalski (53) incubated DNA from a human cell line positive for inosinic acid pyrophosphorylase with mutant cells lacking this enzyme. When the treated cells were cultured in a selective medium in which enzyme-negative cells could not grow, cell clones expressing the enzyme grew with a frequency of transformation of 4 $\times$ 10^{-4}, well above the reversion rate, indicating that a successful genetic transfer had occurred.

In contrast with viral RNA, viral DNA, even when it is not intact, can confer new genetic characteristics on recipient cells. This fact and the availability of restriction enzymes that cut DNA molecules in precisely defined fragments has given impetus to a new line of research involving the genetic transformation of eukaryotic cells. Very important are the studies of Graham et al. (54), who showed that fragmented DNA from adenovirus and SV40 could cause neoplastic transformation in rat cells in vitro. When defined restriction fragments were used, it was possible to size and map the DNA sequences with transforming ability. Maitland and McDougall (55) and Wigler et al. (56) isolated a restriction fragment that alone could change cells lacking thymidine kinase (TK^-) into TK^+ cells.

At first glance it would appear that transfer of genetic markers is of very limited scope. In fact, one might feel that only when markers are available that are easily identifiable (e.g., neoplastic transformation) or for which a selective medium can be easily devised [e.g., medium containing hypoxanthine, aminopterine, and thymidine (HAT), which selects for TK^+ cells] can these experiments be carried out. Fortunately, and thanks to the efforts of Wigler et al. (57), a general method to introduce genes into mammalian cells for which no selective medium exists has been developed. These authors performed cotransfection experiments by treating TK^- cells simultaneously with viral TK genes and other DNAs. It was found that many of the cells that had been transformed to express TK had also integrated the other DNA sequences into their genome at a relatively high frequency. Recently, this technique was used to insert mouse mammary tumor virus-specific DNA into mouse cells (58), where it was biologically active and retained its normal regulation. Limits imposed by evolutionary diversity on the possibility of functionally inserting the TK gene into eukaryotic cells have been described by Wigler et al. (59). Total DNA from a variety of different organisms was used to convert TK^- into TK^+ cells. The TK^+ clones that developed always contained the enzyme of the donor type. Interestingly, mouse, hamster, chicken, calf, and human DNA could convert mouse cells, whereas salmon, *Drosophila,* and slime mold DNAs could not. Thus, a genetic barrier to gene transfer may exist.

Proteins

The interest of the researchers who investigated the biological effects on eukaryotic cells of exogenously added RNA and DNA was focused on the macromolecules themselves since the purpose of many experiments was to study the processing and expression of the DNA and RNA molecules. Conversely, the main purpose of the experiments investigating protein-cell interactions was, and is, a practical one: to find an "enzyme replacement therapy" for diseases whose etiology has been shown to be rooted in the lack of a specific enzyme. This practical and clinical preoccupation has limited the scope of these investigations. Since enzymes, like other molecules, enter the cells, at least in part, by endocytosis and are therefore generally inserted into the lysosomes, most studies were performed using lysozomal enzymes for which the final destination in the cell represents a definite advantage. Several investigators have shown conclusively that cultured fibroblasts lacking lysozomal enzymes can take up the missing enzyme in culture. For example, Porter et al. (60) in 1971 and Wiesman et al. (61) in 1972 reported the uptake of arylsulfatase A by fibroblasts obtained from patients with leucodystrophy (a disease thought to be due to lack of arylsulfatase). Upon incubation of the cells with the enzyme for 3 days, the intracellular level of the enzyme increased 10-fold (up to 25% that of normal cells) (61) and remained at a high level for several days (60,61). Very low turnover after incorporation into fibroblasts was also reported for α-acetylglucosamidase (62) (half-life 4-7 days) and β-glucuronidase (63). In contrast with these data, α-galactosidase appeared to be rapidly inactivated within the cells (64), possibly because a plant enzyme was used in this replacement experiment.

The amount of exogenous enzyme that was taken up by fibroblasts was reported to be as high as 40% for α-iduronidase (65) or as low as 0.5% when α-acetylglucosamidase was used (62). Interestingly, albumin appears to be taken up at much lower levels (1/10 or less) than the lysozomal enzymes investigated (63,65). These in vitro experiments appear, as a whole, very successful in that in most cases a satisfactory uptake and a good persistence of the enzyme was observed. However, enzyme replacement therapy attempted by injection of the missing enzyme into the plasma of patients suffering from inborn errors of metabolism have not yielded the hoped for results. In 1964, Baudhuin et al. (66) first administered α-glucosidase from *Aspergillus niger* to a patient with type II glycogenosis. The patient died but postmortem analysis showed a decrease of glycogen. No decrease of glycogen was observed in a similar clinical trial by Lauer et al. (67) in 1968, although an increase of enzyme activity was observed. Desnick et al. (68), reviewing the field in 1976, reported no fewer than 20 clinical attempts to correct enzymic deficiencies in humans by direct administration of the enzyme. Many difficulties were encountered, such as (a) short half-lives of the administered enzymes, (b) difficulty in reaching the specific tissues, and most important, (c) immunological complications.

From the experiments reviewed above it appears that introduction of macromolecules into eukaryotic cells by cocultivation often yielded unsatisfactory results. There is no doubt that eukaryotic cells can incorporate macromolecules, in small amounts when RNA and DNA are added exogenously, but in larger amounts when some types of protein are used. The turnover of the macromolecules taken up also seems to vary, being very fast for nucleic acids and longer for proteins. Yet the two problems (low incorporation and fast turnover) do not seem to affect the studies involving DNA since in these studies genetic changes are investigated and very inefficient uptake can still lead to interesting results if the transformed cells are selected for properly. In experiments in which RNA-mediated phenotypic changes were investigated, the double difficulty of low incorporation and high turnover caused severe problems. In fact, in these experiments, even when successful, only very modest RNA-mediated effects were observed. Moreover, the difficulties involved in purifying specific mRNAs contributed to the paucity of meaningful data obtained. Certainly, if the function and fate of foreign RNA incorporated into eukaryotic cells is to be investigated, the techniques described above will not do. Finally, the relatively high level of enzymes incorporated into fibroblasts suggests that this methodology could provide an excellent model system to study the fate of exogenous proteins in human cells. This type of experimentation was not pursued, however, because clinical considerations were given priority. The clinical trials were unsuccessful, due primarily to immunological problems caused by the injected enzymes.

Microinjection of Macromolecules into *Xenopus* Oocytes

The concept of selectively injecting molecules and subcellular constituents into cells has attracted the attention of cell biologists since Barber devised an apparatus for the micromanipulation of single cells in 1904 (69). However, the application of this technology was not exploited until 1952, when Briggs and King (70) first demonstrated that nuclei from somatic cells could be transplanted into the eggs of *Rana pipiens* by a microinjection procedure. This methodology was improved by Elsdale et al. (71), who, by replacing these eggs with the oocytes of *Xenopus laevis,* were able to demonstrate that a higher percentage of cells remained viable following microinjection. The amphibian oocytes were found to be a convenient target for microinjection due to their large size (1.2 mm in diameter), their ability to accommodate injections of up to 5% of their total volume without deleterious side effects, and the fact that they can be maintained in culture for several weeks.

Although several groups have utilized this system to study the effect of the undifferentiated oocyte environment on the expression of genes carried within intact nuclei of fully differentiated cells, it was not until 1971 that the *Xenopus* oocyte was used to study the interaction of a highly purified xenogeneic macromolecule with the metabolic machinery of the cell (72). In this study, Lane and colleagues injected rabbit globin mRNA into oocytes and observed, in the presence of hemin, the production of rabbit hemoglobin. This constituted the first evidence that factors required for the translation of a foreign message existed in oocytes. Based on these data it was felt that the undifferentiated *Xenopus* oocyte might represent a "cellular test tube" which could help answer questions relating to translational and even transcriptional regulatory mechanisms.

In the 10 years following the original translation experiment by Lane et al. (72), the *Xenopus* oocyte system has been used to translate numerous messenger RNAs, including, among others, those coding for mouse and duck globin (73), αA2 cyrstallin (74), mouse K-light chain (75), honeybee promellitin (76), mouse immunoglobulin heavy chain (77), mouse β-glucuronidase (78), mouse interferon (79), human interferon (80), human histone proteins (81), and avian myeloblastosis virus proteins (82). It should be noted, however, that all messenger RNAs have not been successfully translated in the oocyte system, as evidenced by the fact that injection of Rauscher virus RNA failed to induce the production of any nonindigenous proteins (83).

The question as to whether the oocyte is able to perform proper epigenetic modifications of newly synthesized foreign proteins has also been addressed. Data indicate that many translation products seem to be correctly processed by the oocyte. (a) Injection of 14S crystallin mRNA into oocytes resulted in the faithful translation of the message as well as the correct acylation of the N-terminal methionine (74). (b) Messenger RNA isolated from the K-light-chain-producing myloma, when injected into *Xenopus* oocytes, yielded a peptide which was glycosylated in a manner analogous to the light chain glycosylation in the myeloma cell (75). Furthermore, when mRNA isolated from MOPC 321 (a tumor that produces nonglycosylated K chains) was microinjected into the oocyte, no glycosylation of the newly synthesized K chain was apparent (75). (c) Translation of protamine-specific mRNA in oocytes resulted in a correctly phosphorylated protein (84).

In contrast to the experimental results described above, some translation products were incorrectly processed by the oocyte, indicating that, at least in those instances, the posttranslational regulation within the oocyte was aberrant. For example, when honeybee promellitin mRNA was translated by the oocyte, the resulting precursor protein was not cleaved to mellitin. In addition, while the C-terminal glutamine in naturally occurring promellitin exists in the form of glutamine amide, oocyte translation of promellitin mRNA yielded a peptide whose C-terminal glutamine contained a free α-carboxylic acid (76). Peptide cleavage was also not apparent when mouse β-glucuronidase mRNA was translated in the oocyte system. Even though catalytically active tetrameric enzyme was produced, the individual peptide constituents appeared to be longer than the corresponding enzyme subunits isolated from the mouse (78). The reason for the improper modification of the precursor proteins is not known.

The *Xenopus* oocyte system has also been used to study problems concerning transcriptional regulation. However, microinjection experiments utilizing DNA present the added complication that the DNA must be inserted directly into the nucleus (germinal vesicle). Any DNA injected into the cytoplasm of the oocyte will not be transcribed (85). Initial experiments designed to determine the fidelity of transcription of microinjected DNAs were done using *Xenopus* DNA containing the genes for 5S ribosomal RNA. When DNA fragments containing the 5S rRNA genes (85) or similar DNA fragments cloned in a pMB9 plasmid (86) were injected into the nucleus of the oocyte, accurate transcription of the DNA to 5S rRNA could be demonstrated. Similar microinjection experiments using either *Xenopus* DNA coding for tRNA$_{met}$ (87) or DNA specific for yeast tRNA$_{tyr}$ (88) have been performed and the appropriate tRNA was routinely produced. The fact that faithful production of yeast tRNA$_{tyr}$ occurred implies that the oocyte has the ability to correctly splice out a 14-base insert that immediately follows the anticodon region of the tRNA, showing that the tRNA splice signals are evolutionarily conserved. In addition to rRNA and tRNA genes, DNA derived from SV40, adenovirus 5, plasmids containing cloned *Drosophila melanogaster* histone genes, ϕX174, ϕ80 plac, and col El plasmid have been microinjected into oocytes. In each experiment the appropriate RNA was produced (89).

Microinjection of Macromolecules into Differentiated Eukaryotic Cells

Studying the processes of transcription and translation by microinjecting DNA or RNA into *Xenopus* oocytes saddles the investigator with the dilemma of interpreting the results of negative experiment—those experiments which indicate that specific genetic or epigenetic phenomena are aberrantly regulated. The question of whether the observed faulty regulation is a function of the undifferentiated state of the target cell must always be asked. Concurrent with the development of the *Xenopus* oocyte system, Diacumakos and co-workers demonstrated that the technique of microinjection could also be used to insert molecules and subcellular constituents into fully differentiated cells (90). The ability to use many types of target cells as well as the elimination of the problem posed by the undifferentiated state of the oocyte made this procedure attractive.

In 1974, Graessmann and his colleagues recognized that the ability to microinject macromolecules into the cytoplasm and nuclei of a variety of cultured cells provided a mechanism for directly confirming the function of purified viral components (91). Initial experiments by this group (92) provided early evidence that SV40 T antigen was coded for by the viral genome. In this work, RNA transcribed from SV40 DNA in vitro was microinjected into the cytoplasm of either African green monkey kidney cells or primary mouse kidney cells and shown by immunofluorescence to produce T antigen. Further experiments by these and other workers using the microinjection system have confirmed that (a) T antigen stimulates host cell DNA synthesis (93), (b) the SV40 DNA sequence located at 0.67 map units is required for the transcription of early viral genes (94), and (c) SV40 DNA synthesis is required for production of late region V antigens (95).

Viral systems other than SV40 have also been studied using the microinjection technique. For example, the elucidation of the early activity of *src* gene products in Rous sarcoma virus-infected cells has recently been accomplished by McClain et al. (96). These workers demonstrated that cytoplasmic extracts obtained from cells infected with a temperature-sensitive *src* gene mutant of Rous sarcoma virus, when injected into mouse 3T3 cells, caused a temperature-sensitive disruption of cytoskeletal integrity.

While the transcription and translation of viral nuclei acids microinjected into eukaryotic cells can often be easily assessed by evaluating the ability of the virus in question to replicate or the viral products to induce gross morphological changes within the host cell (91-97), the use of this system to study translation of purified nonviral messages suffers from the difficulty of obtaining translation products sufficiently radiolabeled to permit standard biochemical analysis. In the past, experiments designed to determine the ability of differentiated eukaryotic cells to translate microinjected mRNAs have utilized the nonquantitative technique of fluorescent-antibody labeling of newly synthesized peptides (98,99). Recently, however, Huez et al. (100) have shown that translation of rabbit α and β globin mRNA microinjected into HeLa cells can be monitored by labeling the proteins produced with ^{35}S-methionine followed by gel analysis of the extracted proteins. As few as 150 cells were used in these experiments. This methodology permits accurate quantitation of the cells' transitional capacity, but it should be noted that to promote effective protein labeling, the addition of 3 mCi of isotope per milliliter of culture medium is required.

The microinjection technique has also been employed to convert TK^- cells to TK^+ cells efficiently with plasmids containing either the herpesvirus TK gene alone (101) or coupled to the gene for human β globin (102). In the latter instance the production of human β globin mRNA could be detected in the TK^+ clones.

Red Cell Ghost-Mediated Insertion of Macromolecules into Eukaryotic Cells

The desire to understand the interaction of polynucleotides and proteins with the constituents of a foreign cell led to the resurrection of the microinjection techniques originally developed in the early twentieth century (69,103) and briefly summarized above. However, as alluded to previously, the direct syringe-type microinjection, while providing a mechanism for answering several fundamental questions, is limited by the fact that large numbers of cells cannot be employed. This precluded a substantial amount of biochemical investigation. The need for a methodology that would allow for the mass insertion of macromolecules into cells led to an advance in the microinjection technology. In 1974, Furusawa et al. (104) demonstrated that a variety of macromolecules could be entrapped within human erythrocytes by the hypotonic swelling of the red cell in the presence of the molecules to be sequestered. These lysed red cells were termed ghosts. Once entrapped within the ghost the macromolecules were delivered into the cytoplasm of target cells by Sendai virus-mediated fusion of the ghost to the recipient cell.

The degree of sequestration of proteins, nucleic acids, and particulate matter by the red cell ghosts may depend on the size of the substance under investigation, although this is by no means certain due to contradictory data reported in the literature. For instance, Loyter et al. (105) reported that substances as large as latex spheres and T2 bacteriophage can be entrapped within red cell ghosts while Schlegel and Rechsteiner claimed that T4 bacteriophage is totally refractile to encapsulation (106). When proteins are used, their reported concentration within the ghost relative to their concentration in the hypotonic buffer is extremely variable, ranging from a low of 0.5% for *Escherichia coli* β-galactosidase (mol w 520,000) (107) to a high of 83% for albumin (mol w 62,000) (108). In the case of polynucleotides, tRNAs can be sequestered within the red cell ghost efficiently (109), whereas larger 18S and 28S rRNAs cannot (106).

Unlike the syringe microinjection techniques, the efficiency of delivery of red cell ghost-entrapped macromolecules into target cells depends on a large number of variables. Generally, data on the efficiency of cellular insertion are reported as the percentage of target cells containing the molecules originally sequestered within the ghost. These values range from 15 to 80% of the treated cells (110, 111). Although this efficiency is high enough to perform studies effectively using easily obtainable macromolecules, it presents difficulties when molecules available in only limited quantities are employed. This problem can best be illustrated by example. In work reported by Furusawa et al. (104), 2 ml of a 1.5% (v/v) suspension of human red blood cells were added to 1×10^5 human amniotic membrane cells. Since 1 ml of packed human red blood cells contains 6×10^9 cells, 2 ml of a 1.5% (v/v) suspension contains 1.8×10^8 red cells ghosts. Assuming that each target cell fuses with an average of one ghost (112), only 0.055% of the added entrapped material is utilized. In other words, 99.95% of this material is lost. Considering that in many cases as few as 15% of the target cells actually fuse with the ghosts, only 0.008% of the sequestered material is eventually utilized.

Another major problem is that the use of Sendai virus as a fusogen renders the ghost leaky to the sequestered material. Schlegel and Rechsteiner (112) have shown that as many as 70% of the added red cell ghosts lyse in the presence of the inactivated virus. Recently, however, an improvement in the technology of fusion of red cell ghosts to target cells has been reported by Kaltoft and Celis (113), who found that, by using polyethylene glycol (PEG) as a fusogen, they could deliver sequestered material into 80% of the target cells while drastically reducing the leakiness of the ghosts. Unfortunately, the concentration of PEG needed to obtain high fusion rates between the ghosts and target cells has been found to be extremely cytotoxic (114). This problem has been partially ameliorated by Mercer and Schlegel (114), who demonstrated that lower concentrations of PEG can be used when the target cells are treated with phytohemagglutinin prior to PEG addition.

Red cell ghost-mediated insertion of macromolecules has been used primarily to insert proteins and small nucleic acids into cells. Among the proteins delivered into cells by this technique are ferritin (105), thymidine kinase (112), immunoglobulins (110,115) β-galactosidase (108), nonhistone proteins (115), and SV40 T antigen (116). In view of the difficulty of sequestering high molecular weight RNAs (106), to date, most of the work done with nucleic acids has investigated the function of tRNA within a variety of cell types (117-120). By utilizing this methodology Kaltoft et al. (117) reported a preferential degradation of *E. coli* tRNA within murine C_3H cells compared to the degradation of similarly inserted yeast tRNA. If true, this observation would imply the existence of a mechanism within eukaryotic cells which could discriminate between very similar RNA molecules. Recently, these data have been challenged by Schlegel et al. (119), who, by using both mouse 3T3 and TC7 African green monkey kidney cells, found no difference between the intracellular processing of red cell ghost-delivered prokaryotic and eukaryotic tRNAs.

Other Methods of Cellular-Insertion of Macromolecules into Eukaryotic Cells

Although the techniques discussed previously have been most widely used by molecular and cellular biologists, other methods have also been reported.

Fusion of "Loaded" Chicken Erythrocytes with Mammalian Cells

One difficulty inherent in the use of human red cell ghosts as a vehicle for the insertion of molecules into cells is that the identification of the cells which have actually incorporated the desired macromolecule is dependent on cytotoxic fluorescent techniques, Jonak and Mora (121) have developed a procedure whereby nucleated chicken erythrocytes are induced to sequester compounds by a method similar to that used for human red blood cell ghosts. Polyethylene glycol-induced fusion of these erythrocytes with target cells results in the formation of

heterokaryons which can be microscopically identified. The use of this process
enables the investigator to distinguish cells that do not express the desired biological effect due to a lack of macromolecular insertion from those cells which
are unable to respond for biochemical reasons.

Human Red Cell "Loading" Induced by High-Voltage Dielectric Breakdown of Red Cell Membranes

An interesting variation on the red cell ghost technology was developed by Zimmermann and his co-workers (122,123). They found that the application of 12
kV/cm for 40 μsec to human red blood cells induces a dielectric breakdown (electrically induced permeabilization) of the red cell membrane. If protein is added
to the cell suspension during the electrical pulse, a rapid entry of the protein into
the ghost is obtained. Following a 10-min resealing process the "loaded" ghosts
can be used in standard fusion experiments.

Permeabilization of Target Cell Membranes with Sendai Virus

In view of the fact that Sendai virus-induced fusion of red blood cell ghosts to
target cells induces a substantial loss of cell membrane integrity (112), it was felt
that this same process could be employed to enhance the diffusion of macromolecules across nonpermeable membranes (124). It was found that addition of protein to cells cultured in the presence of Sendai virus resulted in the uptake of the
protein by the target cells. However, this uptake occurred for only 15 min and
yielded a very low intracellular accumulation of the protein. Furthermore, this
process has been shown to be applicable only to proteins.

Sendai Virus Encapsulation of Protein

Uchida et al. (125) have found that sonication of Sendai virus in the presence of
protein results in the sequestration of the protein within the virus. Fusion of
these altered virions with cells inserts the entrapped protein into the cytoplasmic
compartment. Like the previous methodology, this procedure cannot be used to
transfer polynucleotides.

DELIVERY OF MACROMOLECULES INTO EUKARYOTIC CELLS BY MEANS OF LIPOSOMES

We have previously discussed three major techniques for the delivery of macromolecules into cells: coculture of the molecule with the cell, direct syringe-type
microinjection of the molecule into the cell, and red cell ghost-mediated cellular
insertion facilitated by a fusogen such as Sendai virus or PEG. Problems exist
which limit the utility of all these techniques. (a) Direct addition of macromolecules to cells, both in vivo and in vitro, often leads to a rapid extracellular degradation of the molecule. Furthermore, when proteins are used in vivo (an ap-

plication that has been widely, if not successfully, tried), antibody production
against these proteins is observed. (b) Although direct syringe-type microinjection techniques ensure that the molecule being studied is delivered into the cell,
the physical manipulation required for this method severely limits the number of
cells that can be employed in each experiment. In addition, in vivo application
is precluded. (c) Methodologies that employ red cell ghosts as vehicles for delivery of macromolecules into cells have been shown to be unsuitable when large
molecules are employed (111). This failure has been attributed to the inability
of the large molecule to penetrate the swollen membrane of the ghost. The red
cell ghost procedure is also not applicable to in vivo experiments.

What follows is a review of the work done to date using liposomes as a means
of inserting macromolecules into eukaroytic cells. Utilization of liposomes, both
multilamellar vesicles (MLVs) and large unilamellar vesicles (LUVs) (see Chap. 1
for a detailed description of these procedures), allows one to sequester macromolecules and subcellular components as large as metaphase chromosomes (126)
and eukaryotic cell nuclei (127) and deliver them into large numbers of cultured
cells by endocytosis and/or fusion. The major advantage inherent in the use of
liposomes as a tool for microinjection of macromolecules stems from the fact
that the net charge, fluidity, permeability, and size of the vesicles can be controlled. Thus, innovative design of the liposome can, in many cases, enhance a
selected mode of cellular uptake in vitro (see Chap 3) or dictate a preferential
in vivo localization.

In this section we describe in detail the use of liposomes as a vehicle for introducing proteins, RNA, DNA, viruses, and some subcellular constituents into
eukaryotic cells. While most, if not all, of the work dealing with polynucleotides
and cellular fractions has been done in vitro, experiments using proteins have
been restricted primarily to the delivery of enzymes in vivo due to the appeal of
using liposomes to administer lysosomal enzymes to patients who are deficient
in that particular protein. Although this application of liposome technology has
clear and important clinical relevance, it falls beyond the scope of a chapter de-
voted to liposomes in molecular biology and will therefore not be reviewed here.
However, data relating to sequestration of proteins within liposomes as well as
the insertion of proteins into cells in vitro will be discussed.

Delivery of Proteins into Cells by Means of Liposomes

Sequestration

In 1968, Sessa and Weissmann (128), in a review of liposomes as models for
biological membranes, wrote: "In sum, lipid spherules constitute a valuable tool
with which to study many problems of membrane structure and function. Should
it prove possible to incorporate proteins, enzymes, or nucleic acids into these
models, this might constitute another step towards a true *in vitro* replica of mem-

branes of living systems," thus predicting the use of liposomes as carriers for
biologically significant macromolecules. In fact, two years later, the same authors
conclusively demonstrated, for the first time, the incorporation of a protein into
the aqueous interstices within multilamellar vesicles (129). In this work, egg
white lysozyme was inserted into liposomes composed of egg lecithin, cholesterol,
and either dicetylphosphate (DCP) for negatively charged liposomes or stearylamine (SA) for positively charged vesicles. To determine definitively that the enzyme was localized within the aqueous compartments of the liposomes the following criteria were established: (a) There must be a clear separation of free from
sequestered protein by gel filtration. With most proteins this separation can be
easily achieved on Sephadex G-200. (b) There must be no hydrophobic or charge-
charge interaction between the outermost vesicle bilayer and the protein since
this may result in a failure to achieve separation of the free protein from the liposomes by molecular sieving, thereby artificially increasing the apparent sequestration efficiency. To eliminate this possibility it must be shown that protein added
to a suspension of previously formed liposomes does not coelute with preformed
liposomes. (c) Disruption of gel-filtered liposomes by use of detergents or other
membrane perturbations must induce a shift in the gel filtration pattern of the
sequestered molecule from a position coincident with the liposome peak to one
that coelutes with free protein.

Fulfillment of these criteria permits one to conclude that liposome sequestration of the protein has occurred and that the protein is latent to the external
environment as represented graphically in Figure 1. In this experiment (130),
horseradish peroxidase and glucose were sequestered within MLVs composed of
lecithin, DCP, and cholesterol (molar ratio 7:2:1) and chromatographed on Sephadex G-200. When detergent was omitted, part of the enzyme, as well as the glucose (an aqueous compartment marker) coeluted with the liposomes in the void
volume. Addition of Triton X-100 to the purified liposome suspension resulted
in the release of the entrapped peroxidase and glucose, as evidenced by their separation from the void volume upon rechromatography.

Alternatively, latency can be evaluated by subjecting enzyme-containing
liposomes to proteolysis and assessing the residual liposome-associated enzyme activity. This technique is shown in Figure 2. Here, Fishman and Citri (131) incorporated L-asparaginase into MLVs composed of lecithin, cholesterol, and DCP
(molar ratio, 10:2:1). Treatment of the liposome suspension with trypsin did
not result in the loss of any enzyme activity as assessed by enzyme assay following
osmotic lysis of the liposomes. Conversely, when liposomes were osmotically
shocked prior to the addition of trypsin, 95% of the enzyme activity was lost.
This method can be used when catalytically active proteins are the sequestered
molecules. However, the latency of nonenzymatic proteins should be established
by the method of Sessa and Weissmann (129) outlined above. These general
criteria have been employed routinely by workers interested in using liposomes
as a vehicle for inserting proteins into cells.

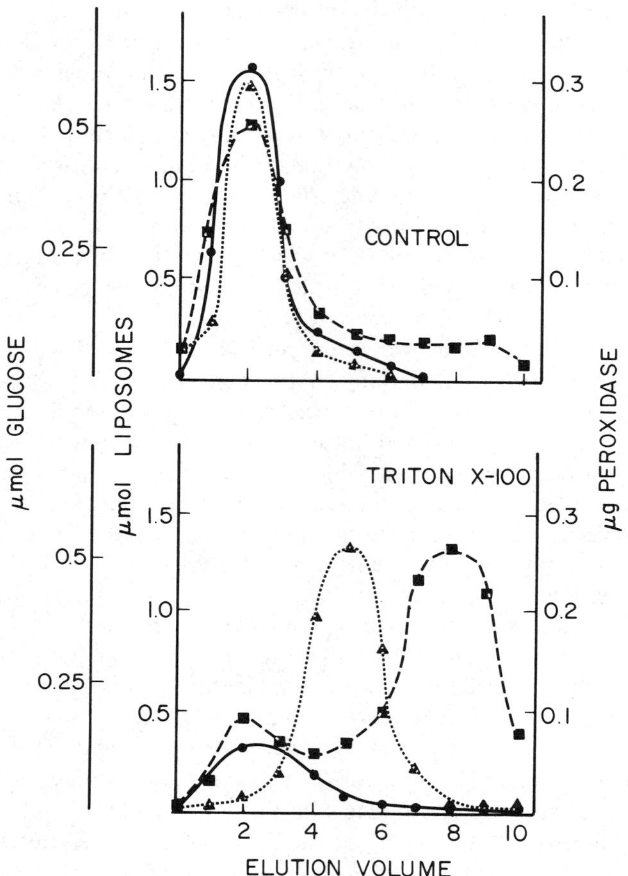

Figure 1 Sephadex G-200 chromatography of purified liposomes after incubation (37°C, 1 hr) with and without Triton X-100 (0.2% v/v) to demonstrate latency. Pooled liposome fractions (1.0 ml) eluted in 0.145 M NaCl-KCl, 0.05 M phosphate buffer (pH 7.4). Liposomes (●); HRP (▲); glucose (■). (From Ref. 130.)

In the past 10 years at least 30 different proteins have been sequestered within liposomes for use either in vitro, in vivo, or simply as models for enzyme latency. The efficiency of entrapment of these proteins is summarized in Table 1. It should be noted that the percent of the added protein sequestered within the liposomes varies dramatically. This fluctuation is due, at least in part, to the micromoles of lipid used and the volume of the reaction mixture. In an effort to normalize these values we have reported the data as percent entrapment per micro-

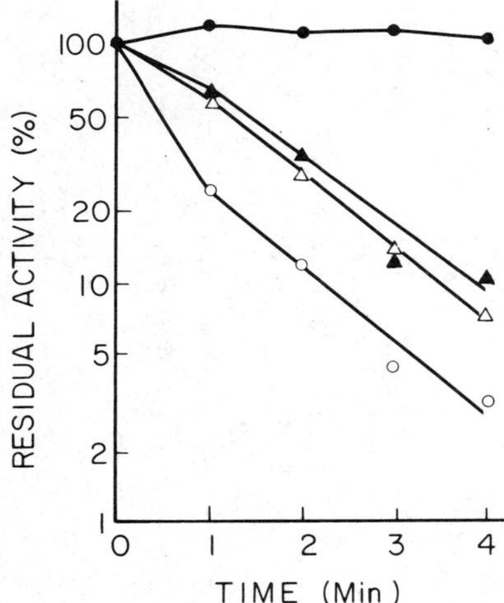

Figure 2 Resistance of entrapped L-asparaginase to proteolysis. The residual
activity of enzyme preparations incubated at 37°C with trypsin was assayed at
indicated time intervals (●) Intact asparaginase liposomes, (○) lysed asparaginase
liposomes, (△) asparaginase + empty liposomes, (▲) asparaginase alone. (From
Ref. 131.)

mole of lipid per milliliter. Although this formula does not completely eliminate
aberrant data, the range of values is considerably reduced. With some exceptions,
values range from 0.1 to 0.9. The exceedingly high values reported for β-gluco-
sidase (3.89, 2.98, 1.47) have been attributed to a hydrophobic interaction between the enzyme and the lipid vesicle (132), while most of the values below 0.10
represent sequestration of protein in small unilamellar vesicles (SUVs) which
have low trapping efficiency (133).

Other information can be obtained from this table. (a) There appears to be
no relationship between the efficiency of sequestration in MLVs and the size of
the protein used; for example, glucose oxidase, with a molecular weight of 150,000,
was entrapped in MLVs with an efficiency of approximately 0.35%/μmol lipid per
milliliter (134), while insulin, with a molecular weight of 12,000 was entrapped
with an efficiency of no greater than 0.33%/μmol lipid per milliliter (135). Trapping efficiency has also been shown to be unaffected by protein size when large
unilamellar liposomes are employed (133). However, as Adrian and Huang (133)
have demonstrated, when SUVs are employed, the percent entrapment decreases
as the size of the protein to be entrapped increases. (Lysozyme: mol wt 17,000,
percent entrapment 0.12; peroxidase: mol wt 40,000, percent entrapment 0.096;
conalbumin: mol wt 86,000, percent entrapment 0.069; α-amylase: mol wt

Table 1　Sequestration of Proteins Within Liposomes

Protein	Approximate mol wt.	Liposome type[a]	Liposome composition (molar ratio)	Percent sequestration[b]	Percent entrapment/ μmol lipid/ml[b,c]	Reference
Hexokinase	97,000	MLV	PC(2):DCP(0.2):C(1.5)	N.R.	N.R.	152
		MLV	SPH(2):DCP(0.2):C(1.5)	N.R.	N.R.	
G6PDH	205,000	MLV	PC(2):DCP(0.2):C(1.5)	8.0	N.R.	152
		MLV	SPH(2):DCP(0.2):C(1.5)	N.R.	N.R.	
β-Galactosidase	518,000	MLV	PC(2):DCP(0.2):C(1.5)	1.6	N.R.	152
		MLV	SPH(2):DCP(0.2):C(1.5)	N.R.	N.R.	
Soybean trypsin inhibitor	22,000	MLV	PC(7):DCP(2):C(1)	14	0.93	153
α_1-Antitrypsin	55,000	MLV	PC(7):DCP(2):C(1)	12	0.80	153
Amyloglucosidase	—	sMLV	PC(7):DCP(1):C(2)	6.5	0.45	154
		sMLV	PC(7):DCP(1):C(2)	6.5	0.45	167
		SUV	PC(7):PA(1):C(2)	N.R.	N.R.	161
Albumin	68,000	sMLV	PC(7):DCP(1):C(2)	10.6	0.74	154
		sMLV	PC(7):PA(1):C(2)	N.R.	N.R.	157
		sMLV	PC(7):PA(1):C(2)	N.R.	N.R.	163
		sMLV	PC(7):DCP(1):C(2)	10.6	0.74	167
β-Glucosidase	—	MLV	PC(3.5):C(1)	100	3.89	132
		MLV	PC(7):DCP(1):C(2)	25	0.88	
		MLV	PC(7):PA(1):C(2)	85	2.98	
		MLV	PC(7):SA(1):C(2)	42	1.47	

		MLV	PC(7):PA(1):C(2)	50	N.R.	176
		MLV	PC(3.5):C(1)	N.R.	N.R.	170
		MLV	PC(7):PA(1):C(2)	N.R.	N.R.	
Lysozyme	17,000	sMLV	PC(7):DCP(2):C(1)	18.6	1.12	141
		sMLV	PC(7):SA(2):C(1)	11.6	0.70	
		SUV	PC(13 μmoles)	0.32	0.12	133
		SUVet	PC(57 μmoles)	0.33	0.11	
		MLV	PC(13 μmoles)	0.32	0.12	
		LUV	PC(13 μmoles	0.33	0.25	
α-Amylase	97,000	SUV	PC (13 μmoles)	0.15	0.058	133
		SUVet	PC (57 μmoles)	0.16	0.053	
		MLV	PC (13 μmoles)	0.33	0.13	
		LUV	PC (13 μmoles)	0.33	0.25	
		sMLV	PC(N.R.)	26	N.R.	158
		sMLV	DPPC(N.R.)	70	N.R.	
Conalbumin	86,000	SUV	PC(13 μmoles)	0.18	0.069	133
		SUVet	PC(57 μmoles)	0.20	0.067	
		MLV	PC(13 μmoles)	0.33	0.13	
		LUV	PC(13 μmoles)	0.33	0.25	
α-Chymotrypsinogen	24,000	SUV	PC(13 μmoles)	0.32	0.12	133
		SUVet	PC(57 μmoles)	0.32	0.11	
		MLV	PC(13 μmoles)	0.33	0.13	
		LUV	PC(13 μmoles)	0.33	0.25	
Peroxidase	40,000	SUV	PC(13 μmoles)	0.25	0.096	133
		SUVet	PC(57 μmoles)	0.25	0.083	
		MLV	PC(13 μmoles)	0.33	0.13	
		LUV	PC(13 μmoles)	0.33	0.25	

Table 1 (continued)

Protein	Approximate mol wt.	Liposome type[a]	Liposome composition (molar ratio)	Percent seques- tration[b]	Percent entrapment/ μmol lipid/ml[b,c]	Reference
Peroxidase (contd)		MLV	PC(7):DCP(2):C(1)	5.8	0.39	130
		MLV	PC:(7):DCP(2):C(1)	17.5	1.16	156
		MLV	PC(4.5):DCP(2):C(3.5)	12.0	0.80	
		MLV	PC(6):DCP(2):C(1):gang(1)	14.0	0.93	
		MLV	DPPC(7):DCP(2):C(1)	4.6	0.30	
		MLV	SPH(7):DCP(2):C(1)	7.5	0.50	
		MLV	PC(7):C(1)	N.R.	N.R.	174
		MLV	PC(8):DCP(1):C(1)	N.R.	N.R.	
		MLV	PC(7):DCP(2):C(1)	N.R.	N.R.	
		sMLV	PC(7):DCP(2):C(1):Ly(0-0.5)	N.R.	N.R.	141
		sMLV	SPH(3):SA(1):C(1)	3.7	0.11	140
Human leukocyte interferon	—	MLV	PC(7):DCP(2):C(1)	11	0.73	155
L-Asparaginase	—	MLV	PC(10):DCP(1):C(2)	12	0.46	131
		MLV	PC(7):SA(1):C(2)	37	1.29	165
		sMLV	PC(7):PA(1):C(2)	37	1.29	
Macrophage activating factor	—	MLV	PC(4.95):PS(4.95):Ly(0.01)	4.8	0.24	146
		MLV	PC(4.95):PS(4.95):Ly(0.10)	N.R.	N.R.	175
Ricin	65,000	SUVet	PC(1):PA(0.07):C(1)	0.11	0.085	149
β-Galactosidase	520,000	MLV	PC(7):DCP(2):C(1)	12	N.R.	144

Glucose oxidase	150,000	sMLV	PC(7):DCP(2):C(1)	8	0.35	134
		sMLV	PI(23 μmoles)	52	N.R.	169
		sMLV	PC(7):PA(1):C(2)	48	N.R.	
Hexosaminidase A	—	MLV	PC(7):DCP(2):C(1)	6.9	0.46	136
Adenovirus type 12 T antigen	—	LUV	PS(3.2 μmoles)	6.0	0.23	151
Insulin	12,000	sMLV	PC(1):C(10)	N.R.	N.R.	171
		sMLV	(PC(7):C(2)	10	0.33	135
		sMLV	DOPC(7):C(2)	10	0.33	
		sMLV	DPPC(7):C(2)	10	0.33	
		sMLV	DSPC(7):C(2)	10	0.33	
		sMLV	DMPC(7):C(2)	1.6	0.053	
		sMLV	DLPC(7):C(2)	1.6	0.053	
		sMLV	PC(7):C(2)	10.9	0.15	173
		sMLV	DPPC(7):C(2)	10.9	0.15	
		sMLV	PC(7):C(2):gang(1)	10.9	N.R.	
Immunoglobulin	150,000	MLV	SPH(6):DCP(1):C(2)	13	0.68	177
		MLV	SPH(6):SA(1):C(2)	14	0.74	
Dextranase	—	MLV	PC(7):PA(1):C(2)	17.5	N.R.	159
		MLV	PC(7):PA(1):C(2)	22	N.R.	164
Carbonic anhydrase	29,000	sMLV	DPPC(8):PA(1):C(2)	44	0.63	160
Invertase	—	sMLV	PC(7):PA(1):C(2)	2.8	0.20	165
		sMLV	PC(7):PA(1):C(2)	N.R.	N.R.	163
		sMLV	PC(7):PA(1):C(2)	N.R.	N.R.	137
β-Glucuronidase	—	sMLV	DPPC(7):PA(1):C(2)	N.R.	N.R.	172
		sMLV	DPPC(7):SA(1):C(2)	5.0	0.27	162
		sMLV	DPPC(7):PA(1):C(2)	4.8	0.26	

Table 1 (continued)

Protein	Approximate mol wt.	Liposome type[a]	Liposome composition (molar ratio)	Percent sequestration[b]	Percent entrapment/ μmol lipid/ml[b,c]	Reference
Neuraminidase	—	sMLV	PC(7):PA(1):C(2)	5.7	0.20	166
α-Manosidase	—	sMLV	PC(7):PA(1):C(2)	N.R.	N.R.	168

[a]MLV, multilamellar vesicles; sMLV, sonicated multilamellar vesicles (these may in some cases be small unilamellar vesicles); SUV, small unilamellar vesicles; SUVet, small unilamellar vesicles produced by the ethanol injection procedure; LUV, large unilamellar vesicles; PC, phosphatidylcholine; DCP, dicetylphosphate; C, cholesterol; SPH, sphingomyelin; PA, phosphatidic acid; gang, ganglioside; SA, stearylamine; DPPC, dipalmitoylphosphatidylcholine; Ly, lysolecithin; PS, phosphatidylserine; PI, phosphatidylinositol; DOPC, dioleophosphatidylcholine; DSPC, distearoylphosphatidylcholine; DMPC, dimyristoylphosphatidylcholine; DLPC, dilauroylphosphatidylcholine.

[b]N.R., not reported.

[c]Percent entrapment/μmol lipid/ml represents the percent sequestration divided by the micromoles of lipid per milliliter of reaction volume.

97,000, percent entrapment 0.058.) (b) No consistent correlation can be made between the efficiency of entrapment and the lipid composition of the vesicle membrane.

Cellular Delivery of Proteins

As detailed in Chapter 3, liposome-sequestered molecules can be inserted into a cell by either endocytosis or fusion. In the first mode of liposome-cell interaction it has been shown that most of the endocytosed liposomes are merged with the lysosomes of the target cells (136). In these instances molecules within the aqueous spaces of the liposome are deposited within the lysosomal apparatus, a phenomenon that is deleterious to the cellular insertion of RNA or DNA but that may be exploited if one wishes to deliver naturally occurring lysosomal enzymes. The fusion mechanism of intracellular delivery of liposome-sequestered molecules involves the intercalation of the vesicle lipids into the cell membrane with the concomitant injection of the sequestered material into the cytosol. When MLVs are used, the contents of the outermost aqueous space are inserted into the cytoplasm, while the remaining concentric bilayers remain in the cytoplasm intact where they will slowly leak or be engulfed by the lysosome. If LUVs are used and fusion occurs, the entire content of the aqueous space is injected into the cytoplasmic compartment. Clearly, if one is studying the function of nonlysosomal enzymes, DNA, RNA, or subcellular constituents within the target cell, fusion between LUVs and the cell membrane would be advantageous. Maximization of each mode of liposome-cell interaction has been discussed previously (Chap. 3) and will therefore not be addressed here.

Delivery by Endocytosis   Perhaps the first evidence that liposome-sequestered protein could be inserted into cells in vitro was provided in 1973 by Gregoriadis and Buckland (137) who felt that addition of MLVs to phagocytic cells would result in their endocytosis followed by delivery of the liposomal contents into lysosomes. To test this hypothesis they used an in vitro model system composed of two invertase-negative cell types, mouse peritoneal macrophages and Chinese hamster fibroblasts. Culture of these cells in media containing a mixture of nonradioactive and radiolabeled sucrose led to the accumulation of sucrose within the lysosomes, causing them to appear as phase-lucent vacuoles in the perinuclear region of the cells. Addition of invertase (a lysosomal enzyme that hydrolyzes sucrose to glucose and fructose) to the sucrose-containing cells resulted in the hydrolysis of the stored sucrose and the elimination of the phase-lucent vacuoles (138). When invertase was sequestered in MLVs composed of lecithin, cholesterol, and phosphatidic acid (molar ratio, 7:2:1) and added to sucrose-filled macrophages or fibroblasts, a rapid elimination of the vacuoles could be observed. This phenomenon could not be observed when either liposome-encapsulated heat-inactivated invertase or culture medium alone was added to the cells. The intracellular activity of the liposome-delivered invertase was further supported by the demonstration that only those cells treated with active enzyme trapped within liposomes released significant quantities of radiolabeled glucose and fructose into the culture fluid. While these experiments clearly demonstrated the capacity of liposomes to deliver protein into cells, they failed to show that this mode of cellular insertion of enzyme provides any advantage over the direct addition of the protein to the cultured cells. However, as these authors pointed out, in an in vivo situation, liposome encapsulation of the protein would eliminate the problem of the production of antienzyme antibody.

Due to histochemical considerations (139), the most extensively studied enzyme with regard to its intracellular fate after in vitro delivery via liposomes has been horseradish peroxidase (130,140,141). In 1974, Magee et al. (140) incorporated peroxidase into MLVs composed of sphingomyelin, stearylamine, and cholesterol (molar ratio 3:1:1). These cationic liposomes appeared to sequester approximately 5% of the added enzyme. Addition of these liposomes to HeLa cells resulted in a considerable amount of interaction between the positive vesicles and negatively charged cells as measured by either the presence of ^{125}I-peroxidase or by histochemical analysis. Up to 25% of the radioactively labeled liposome-bound enzyme became cell-associated, but due to the electrostatic interactions, quantitation of the actual internalization of liposome-sequestered enzyme was not reported. However, electron microscopic examination of diaminobenzidine-stained HeLa cells which had been treated with peroxidase-containing liposomes revealed some peroxidase-positive material within the cytoplasm of the cells. These cytoplasmic inclusions appeared to be membrane bound, indicating that the liposomes were probably taken up by phagocytosis. In addition to the cytoplasmic inclusions, regions of the HeLa cell membrane appeared electron dense, which was interpreted by the authors to suggest that a fraction of the liposome-associated peroxidase may have become embedded within the vesicle bilayer and eventually transferred to the target cell membrane via a fusion mechanism. Addition of free enzyme to the HeLa cells did not result in significant incorporation of peroxidase into the cell membrane or any cytoplasmic constituent.

While endocytosis of liposomes by HeLa cells can be demonstrated, these cells do not normally engage in a high level of phagocytosis. Therefore, if the goal of one's research is to replace missing lysosomal enzymes, a more favorable model system is needed. Such a system was devised by Weissmann and his colleagues (130), who utilized phagocytes obtained from smooth dogfish (*Mustelus canis*) as target cells for liposome-entrapped peroxidase. These cells are ideal for liposome delivery studies since they are normally highly phagocytic, and most important, they are deficient in lysosomal myeloperoxidase. Furthermore, in addition to using cells that had a strong propensity for engulfing extracellular particulate matter, stimulation of phagocytosis was boosted by coating the liposomes with heat-aggregated isologous immunoglobulin. The rationale for immunoglobulin coating of liposomes was based on two facts. (a) It was shown that leukocytes endocytose immune precipitates formed in heat-inactivated serum due to the interaction of exposed Fc portions of the immunoblobulin with Fc receptors on the leukocytes (142). (b) Aggregated immunoglobulins were shown to become associated with liposomes by electrostatic and hydrophobic interactions (143). In the enzyme replacement experiments, horseradish peroxidase was incorporated into liposomes composed of lecithin, DCP, and cholesterol (molar ratio 7:2:1), followed by coating of the vesicle with heat-aggregated isologous immunoglobulin. Addition of the coated liposomes to phagocytes in vitro resulted in as high as 50% of the vesicle-associated enzyme being taken up by the cells as opposed to 1% when free enzyme was used. Liposomes not coated with aggregated immunoglobulin were far less effective in delivering enzyme into the target cells. Since negatively charged liposomes were used in these experiments, artificially elevated uptake values due to electrostatic interactions between the liposomes and the cells were minimal. This was surpported by the fact that inhibition of the phagocytic mechanism by addition of cytochalasin B at 4°C resulted in an almost complete lack of cell-associated peroxidase. If vesicles were simply bound to the cell surface, these conditions would have not effectively reduce liposome-cell association.

The same strategy used for the liposome-mediated insertion of peroxidase into dogfish phagocytes was used by Cohen et al. (136) to deliver human hexosaminidase A into polymorphonuclear leukocytes obtained from patients with Tay-Sachs disease. These cells have a total level of hexosaminidase activity in their lysosomes equivalent to that of normal leukocytes but lack the A isoenzyme. Addition of MLV liposomes coated with heat-aggregated human immunoglobulin and containing purified placental hexosaminidase A to leukocytes from Tay-Sachs patients resulted in the accumulation of hexosaminidase A activity in the lysosomes of those cells (Fig. 3). As in the previous experiment (130), the uptake of hexosaminidase A was eliminated by cytoskeletal disruption induced by cytochalasin B and required aggregation of the coating immunoglobulin to stimulate substantial phagocytosis. In addition, it was demonstrated that only a small fraction of the total enzyme uptake could be attributed to a mechanism by which the enzyme leaks out of the liposome and is carried into the cell when the liposome is phagocytized. When a mixture of free enzyme and buffer-loaded, immunoglobulin-coated liposomes was added to the leukocytes, the level of the A isoenzyme found within the cell was eightfold less than when the enzyme was added in its liposome-entrapped form. More precisely, when the enzyme was incorporated into the liposomes, 15.8% of the total hexosaminidase activity of the cells was due to the A isoenzyme, while only 2% was found when the mixture of free enzyme and buffer-loaded liposomes was used.

Attempts have been made to repair other lysosomal enzyme deficiencies. Reynolds et al. (144), using MLVs containing β-galactosidase, were able to restore the ability of feline skin G_{M1} gangliosidosis fibroblasts lacking this enzyme to

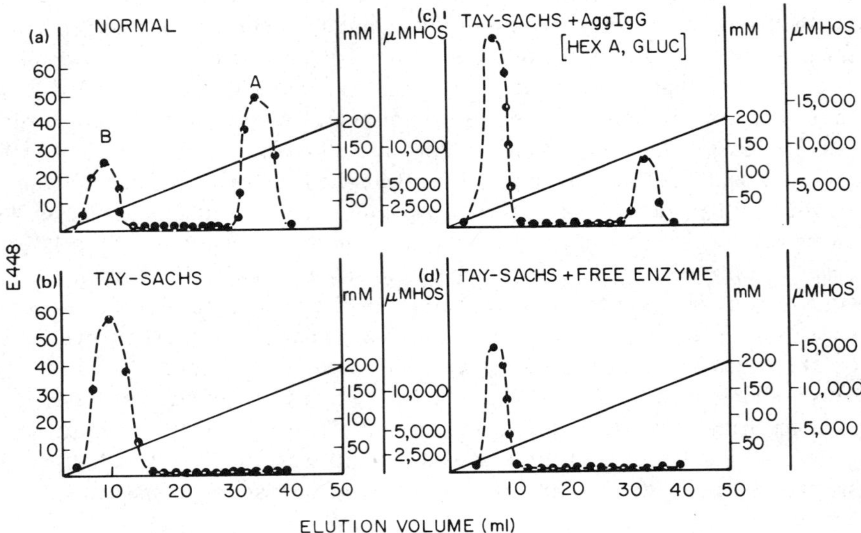

Figure 3 DEAE-cellulose column chromatography of leukocyte cell lysates. (a) Isozyme distribution in polymorphonuclear leukocytes (PMN) from a normal subject. (b) Isozyme distribution in PMN leukocytes of Tay-Sachs patients. (c) Isozyme distribution in PMN leukocytes of Tay-Sachs patients exposed to aggregated immunoglobulin-coated liposomes containing hexosaminidase A and glucose. (d) Isozyme distribution in Tay-Sachs leukocytes exposed to free hexosaminidase A. HOS, hexosaminidase concentration in micromoles; mM refers to the concentration of NaCl in elution gradient; E, fluorometric measurement of methylumbelliferone cleaved from specific substrate. (From Ref. 136. Reprinted with permission from Biochemistry 15:452. Copyright 1976 American Chemical Society.)

cleave [^{14}C] galactose from a variety of glucopeptides. This ability was maintained by the cells for up to 10 days.

Another in vitro model system for enzyme replacement therapy was developed by Ismail et al. (134). It had been previously demonstrated that active glucose oxidase within the lysosomes of leukocytes plays a role in the killing of catalase-positive microorganisms (e.g., *Staphylococcus aureus*) (145). The killing is accomplished, in part, by the production of hydrogen peroxide. Leukocytes obtained from patients with chronic granulomatous disease lack glucose oxidase and are therefore unable to kill *S. aureus* effectively. Addition of immunoglobulin-coated

MLV liposomes containing glucose oxidase to enzyme-deficient leukocytes resulted in (a) the generation of hydrogen peroxide due to the metabolism of added glucose, and (b) a modest stimulation in the bactericidal ability of the treated cells. Mixing of liposome-treated cells with *S. aureus* resulted in a twofold stimulation of the killing activity of the leukocytes compared to enzyme-deficient cells not treated with liposomes.

While the phagocytic mode of liposome uptake by cells has been most consistently exploited as a means of lysosomal enzyme replacement, other practical uses of this mechanism have been proposed. Sone et al. (146) have recently investigated the use of liposomes as a vehicle for the insertion of macrophage activating factor (MAF) into rat alveolar macrophages. MAF is a chemically poorly defined glycoprotein (lymphokine) which is released into the culture fluid by lymphocytes that have been stimulated with concanavalin A (147). Addition of this supernatant fluid to macrophages renders them cytotoxic to transformed cells of syngeneic, allogeneic, and xenogeneic origin but not to any nontumor cell line. Although free MAF is effective in the stimulation of the tumoricidal activity of macrophages, Sone et al. (146) have demonstrated that encapsulation of the MAF within MLVs allows one to achieve the same cytoxicity with lymphokine concentrations 1.6×10^4 times lower than when free MAF is employed. The interpretation of these data is difficult for two reasons. (a) It is believed that the normal fate of phagocytized liposomes is to eventually become localized in lysosomes. Since MAF is not a naturally occurring lysosomal constituent, the mechanism by which it is protected from proteolysis is not clear. (b) It is believed that the first step in the activation of macrophages by MAF involves the binding of the molecule to a membrane-bound receptor (148). By sequestering the MAF within liposomes this mechanism is bypassed; thus an alternative explanation for the macrophage activation may need to be invoked. Since it has not been demonstrated that the MAF is not exposed on the liposome surface, one possible explanation for the marked enhancement of MAF activity resulting from its liposome encapsulation could be that the MAF is associated with the outside of the liposome and, therefore, accessible to binding by receptors on the macrophage. Presentation of the lymphokine as part of a membrane structure may enhance its ability to bind and activate macrophages without the necessity for its internalization by the cell.

Delivery by Fusion Unlike endocytosis, which is a generally accepted and prevalent mode of liposome uptake by many cells, fusion between liposomes and cells represents a less understood mechanism. In fact, actual fusion between liposomes and cells is difficult to prove directly and unequivocally. However, data presented within the last three years indicate that fusion does occur, although at a level considerably below the uptake of liposomes by phagocytic cells. As alluded to earlier, fusion is the preferred mechanism of vesicle-cell interaction if one wishes to insert a functional macromolecule into the cytoplasm of a cell while avoiding lysosomal association. While the fusion mechanism has been exploited primarily to deliver RNA and DNA into cells, proteins have also been used. The first evidence that protein could be inserted into cells in vitro by a nonphagocytic mechanism was presented by Weissmann et al. (141) in 1977. In this work, MLVs composed of lecithin, DCP, and cholesterol (molar ratio 7:2:1) were prepared containing horseradish peroxidase and added to three nonphagocytic murine cell lines in the presence and absence of cytochalasin B. It was found that between 0.18 and 0.64% of the liposome-sequestered enzyme became cell associated when cytoskeletal function was inhibited. Addition of the fusogen lysolecithin to the liposome membrane increased, by as much as fourfold, the amount of enzyme taken up by the cells (Table 2). Since this level of uptake is quite low, the possibility that liposomes have simply adhered to the cells must be eliminated. Electron microscopic observation of the liposome-treated cells subsequent to washing revealed at least one peroxidase-reactive vesicle per cell. The majority of the liposomes appeared to be localized intracytoplasmically, although some were seen adhering to the plasma membrane. Under these circumstance, it is difficult to attribute all the cell-associated enzyme to vesicle-cell fusion. However, due to the low number of vesicles seen bound to the cell membrane, the level of enzyme associated with the cell cannot be totally accounted for by adherence. (It is important to note that since MLVs were used, the expected configuration of the liposome within the cytoplasm following a fusion event would be in the form of concentric bilayers with the outermost bilayer depleted. This would account for the observation of what appear to be complete liposomes within the cytoplasm in the presumed absence of endocytosis.)

Further evidence that proteins can be inserted into the cytoplasm of cells in vitro via a liposome-cell fusion mechanism was provided by Gardas and Macpherson (149). These authors entrapped the lectin ricin (a 65,000-dalton protein which inhibits protein synthesis by catalytic inactivation of the 60S ribosomal subunit) within small unilamellar vesicles prepared by the ethanol injection procedure (150) and added them to mutant baby hamster kidney (BHK) cells deficient in the receptor for ricin transport. After a 24-hr liposome-cell incubation, 99% of protein synthesis had been terminated. Addition of saline-containing liposomes or 50 μg/ml free ricin did not affect the cell's ability to produce proteins. Of particular interest is the fact that when ricin was entrapped in MLVs and added to cells, no inhibition of protein synthesis was observed. From these data it was concluded that the most probable mechanism of ricin injection was by direct fusion of liposomes to cell membranes. However, the reason MLVs were totally ineffective remains unclear, since one would expect that fusion of these vesicles with cells would result in the delivery of the contents of their outermost aqueous space into the cytoplasm and hence be available for ribosome binding.

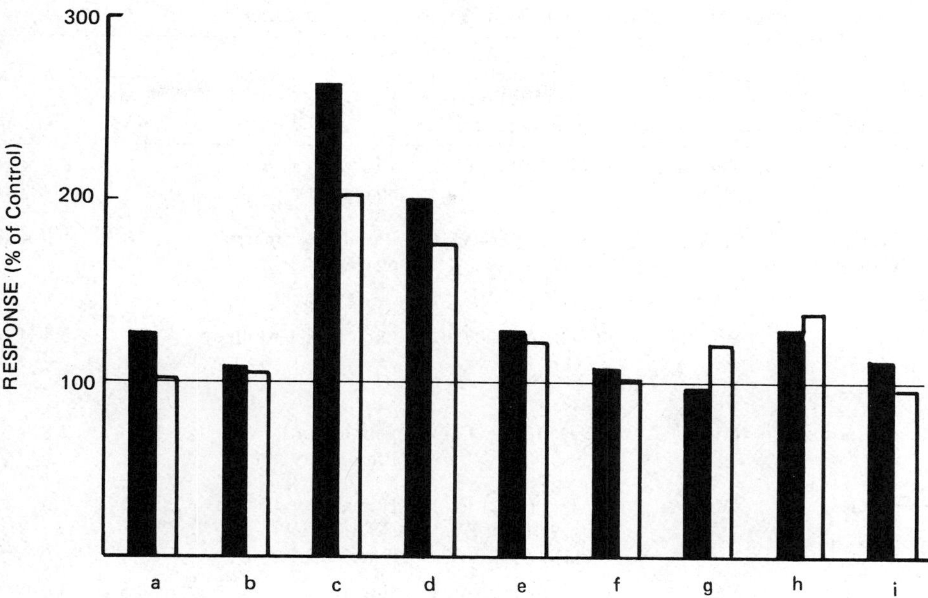

Figure 4   Effect of T637 and BHK21 cell extracts on induction of DNA synthesis in G_1-arrested BHK21 cell cultures. The filled columns represent the response
relative to the appropriate controls (autoradiography data). The open columns
show response relative to appropriate controls measured as total acid-insoluble
incorporation of radioactive thymidine. The encapsulated BHK21 cell extract
eluting from a double-stranded (ds)-DNA cellulose column at pH 8.0 is shown in
(a) and at pH 6.2 in (b). The encapsulated T637 cell extract eluting from ds-
DNA cellulose at pH 8.0 is shown in (c), from single-stranded (ss)-DNA cellulose
in the nonbinding fraction in (d), and eluting from ds-DNA cellulose at pH 6.2 in
(e). Nonencapsulated T637 cell ds-DNA cellulose eluate obtained at pH 8.0 (f)
or pH 6.2 (g) added together with preformed vesicles to test cultures and the elu-
ates obtained at pH 8.0 (h) or pH 6.2 (i) added alone to test cultures are indicated. (From Ref. 151.)

   Recently, Allebach et al. (151) have utilized LUVs to insert protein into cells.
Phosphatidylserine cochleate vesicles were produced containing adenovirus type
12 T antigen, which, when added to G_1-arrested BHK21 cells, caused a threefold
increase in the level of DNA synthesis (Fig. 4). This increase could not be induced
when free T antigen or buffer-loaded liposomes were employed. Since T antigen
functions by binding directly to cellular DNA, protein uptake by liposome fusion
probably accounts for most of the activity of cell-associated T antigen. However,

Table 2 Insertion of Proteins into Cells in Vitro with Liposomes[a]

Protein	Liposome type[b] (molar ratio)	Liposome composition (molar ratio)	Cell type	Time (hr)	μmol lipid
Peroxidase	aggIgM-MLV	PC(7):DCP(2):C(1)	*Mustelus canus* phagocytes	1	6.9×10^4
Macrophage activating factor	MLV	PC(4.95):PS(4.95): Ly(0.01	Rat alveolar macrophages	24	1.0×10^{-1}
Ricin	SUVet	PC(1):PA(0.007): C(1)	Baby hamster kidney	24	8.4×10^{-1}
β-Galactosidase	MLV	PC(7):DCP(2):C(1)	Feline skin Fibroblasts	6	2.8
Peroxidase	sMLV	PC(7):DCP(2):C(1) PC(6.9):DCP(2): C(1):Ly(0.1)	Human NB 45 lymphoid	0.5	3.0
	sMLV	PC(7):DCP(2):C(1) PC(6.9):DCP(2): C(1):Ly(0.1)	Human NB 20 lymphoid	0.5	3.0
	sMLV	PC(7):DCP(2):C(1) PC(6.9):DPC(2): C(1):Ly(0.1)	Human CRL-1121 fibroblast	0.5	3.0
Glucose oxidase	aggIgG-MLV	PC(7):DCP(2):C(1)	Human leukocytes	1	N.R.
Peroxidase	sMLV	SPH(3):SA(1):C(1)	HeLa	0.5	1.5
Hexosaminidase A	aggIgI-MLV	PC(7):DCP(2):C(1)	Human leukocytes	1	N.R.
Adenovirus type 12 T antigen	LUV	PS(3.2 μmoles)	Baby hamster kidney	3	N.R.
Invertase	sMLV	PC(7):PA(1):C(2)	Mouse peritoneal macrophages	3	1.0
			Human embryo lung fibroblasts	24	1.0

[a]N.R., not reported.

[b]aggIgM-MLV, multilamellar vesicles coated with heat-aggregated immunoglobulin; MLV, multilamellar vesicles; SUVet, small unilamellar vesicles produced by ethanol injection; LUV, large unilamellar vesicles.

[c]a, autologous serum; c, calf serum; f, fetal calf serum.

[d]PBS, phosphate buffered saline.

| Coculture conditions | | | | Probable intracellular localization | Percent cellular association | Reference |
Number of cells	Percent serum[c]	Temperature (°C)	Media[d]			
2.5×10^7	44a	30	Ringer's solution, pH 7.4	Lysosomes, phagosomes	50	130
1×10^5	5f	37	Eagle's minimal essential medium	N.R.	N.R.	146
5×10^5	0.2c	N.R.	Dulbecco's modified Eagle's medium	Cytoplasm	5	149
N.R.	20f	37	McCoy's 5a	Lysosomes	N.R.	144
5×10^6	0	37	PBS	Cytoplasm	0.18 0.82	141
5×10^6	0	37	PBS	Cytoplasm	0.64 0.60	
5×10^6	0	37	PBS	Cytoplasm	0.59 0.95	
5×10^6	17a	37	PBS	Lysosomes	N.R.	134
2.5×10^6	0	37	Hank's balanced salt solution	Ubiquitous	17	140
4×10^6	9a	37	PBS	Lysosomes	N.R.	136
N.R.	0	37	Ca^{2+}-Mg^{2+}-free basal minimal Eagle's	Cytoplasm and nucleus	N.R.	137
2×10^6	20f	37	Medium 199	Lysosomes	N.R.	137
N.R.	10c	37	Eagle's basal medium	Lysosomes	N.R.	

the precise mechanism involved with the penetration of the protein into the nucleus is not known.

Delivery of RNA into Cells by Means of Liposomes

The incorporation of intact RNA molecules into eukaryotic cells by direct cell-RNA incubation (see preceding sections) presents several difficulties, due primarily to the inevitable presence of nucleases in the incubation buffer. One obvious way to protect RNA from enzymatic degradation during its interaction with cells and possibly also to increase its level of incorporation would be to sequester the RNA molecules within liposomes to take advantage of the putative ability of liposomes to fuse with cell membranes and discharge their content into the cytosol of the target cell. Thus, the biological activity and the fate of RNA molecules in a predetermined, unaltered cellular environment might be studied. However, the development of a successful liposome technology to achieve an efficient cellular insertion of intact RNA molecules requires the solution of several technical problems. In fact, the use of this technology can be envisioned only if evidence is provided that (a) RNA is efficiently sequestered within liposomes, (b) RNA sequestered in liposomes is intact and maintains its biological activity, and (c) liposome-sequestered RNA is delivered efficiently to cells.

Sequestration

When liposomes are formed in a buffer containing RNA, some of the RNA molecules are sequestered into the aqueous space within the liposomes. Proof that RNA can indeed be sequestered by liposomes requires experimental evidence very similar to the criteria established for the entrapment of proteins (see the preceding section). By fulfilling these criteria, Ostro et al. (178) and Dimitriadis (179,180) have shown that radioactive RNA molecules can be sequestered by LUV liposomes formed by the ether infusion technique (178) and by the phosphatidylserine cochleate method (179,180). Both laboratories used negatively charged lipids to avoid nonspecific, electrostatic association between the liposomes and the RNA molecules. The suspensions obtained after liposome production containing both free and liposome-entrapped radioactively labeled RNA was subjected to RNAse treatment followed by gel filtration. Two peaks of radioactivity (RNA) were obtained; one was associated with the liposomes in the column void volume (liposome-sequestered RNA), while the other eluted much later, (Fig. 5b). As a control, radioactive RNA was added to preformed, buffer-loaded liposomes and the mixture was then subjected to RNAse treatment and sieve chromatography. In this instance the radioactivity eluted as a single peak corresponding to degraded RNA (Fig. 5a). This experiment showed that liposome-associated RNA was internalized and not bound to the surface of the vesicle. When RNA was extracted from liposomes recovered from the column, it was found to be undegraded. Figure 6a shows the sedimentation profile

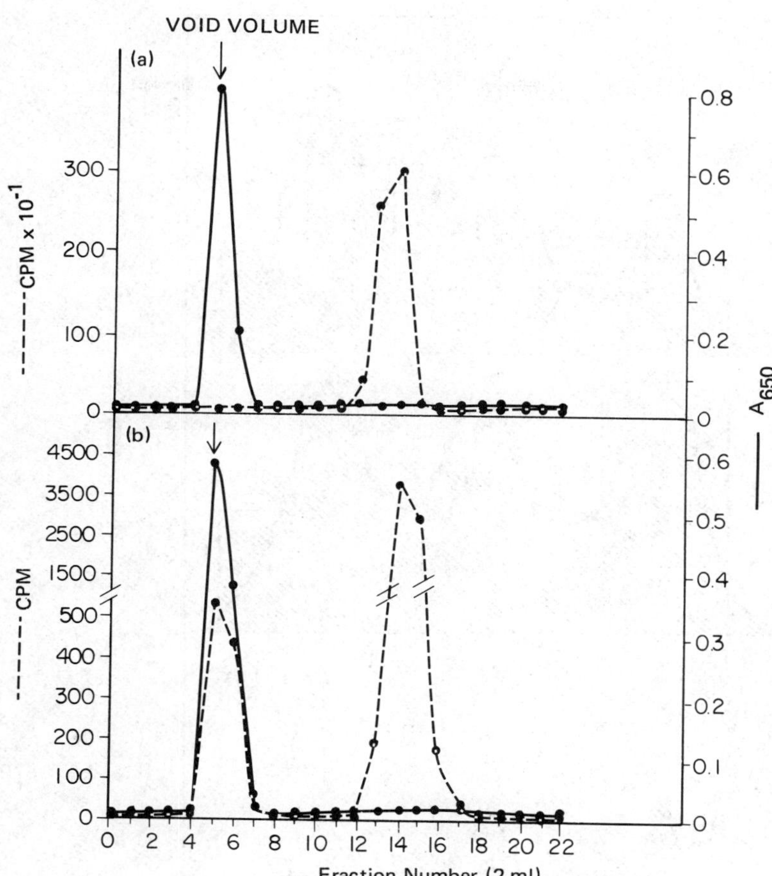

Figure 5 Elution profile of liposomes and RNA from a Sepharose 4B column. (a) Ribonuclease-treated suspension of empty liposomes and exogenously added [^{3}H] RNA. (From Ref. 178.) (b) Ribonuclease-treated suspension of liposomes formed in the presence of [^{3}H] RNA.

of intact unlabeled RNA on a sucrose gradient. Alternatively, the latency of poly-nucleotides entrapped within liposomes can be demonstrated by showing that RNAse treatment of gel-filtered liposomes does not reduce the amount of acid precipitable RNA (182).

More rigorous proof that RNA has not been damaged during its sequestration into liposomes can be provided if one shows that the sequestered RNA maintains its full biological potential upon reextraction. In our laboratory we have

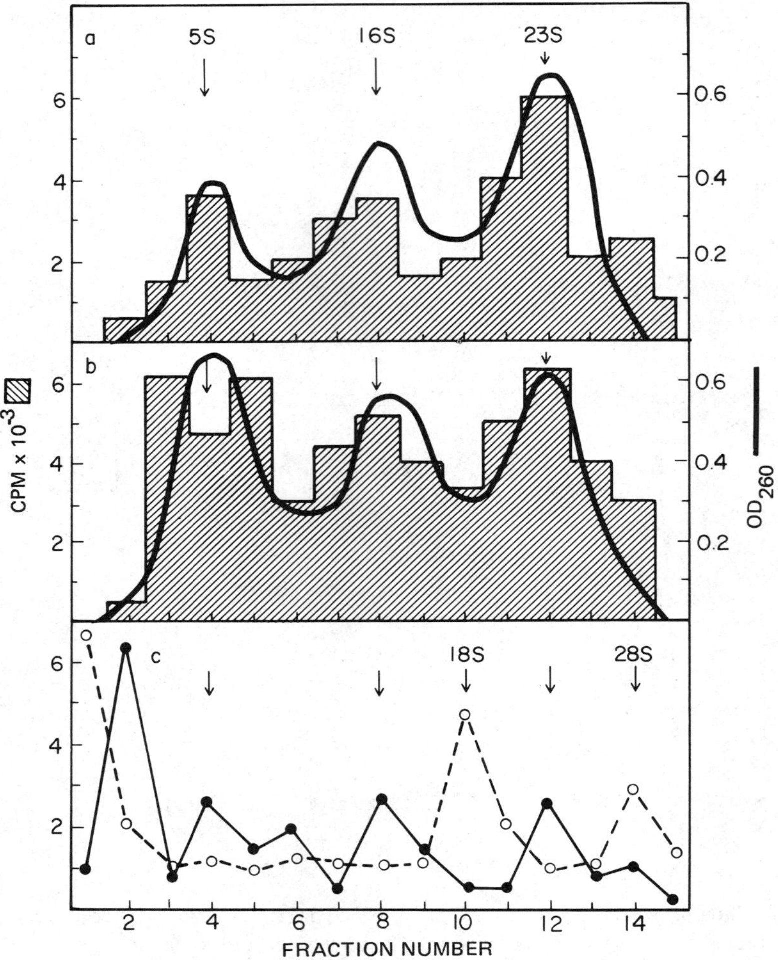

Figure 6 Sucrose density gradient analysis of *E. coli* [³H] RNA. RNA was extracted from (a) RNase- and pronase-treated, gel-filtered liposomes; (b) the supernatant obtained following harvesting of HEp-2 cells which had been treated with liposome-sequestered [³H] RNA; (c) washed pellet of HEp-2 cells treated with liposome-sequestered [³H] RNA (●) or an equal amount of "naked" RNA (○). The solid lines in (a) and (b) represent the A₂₆₀ profile of nonradioactive *E. coli* RNA added during the extraction process. (From Ref. 181.)

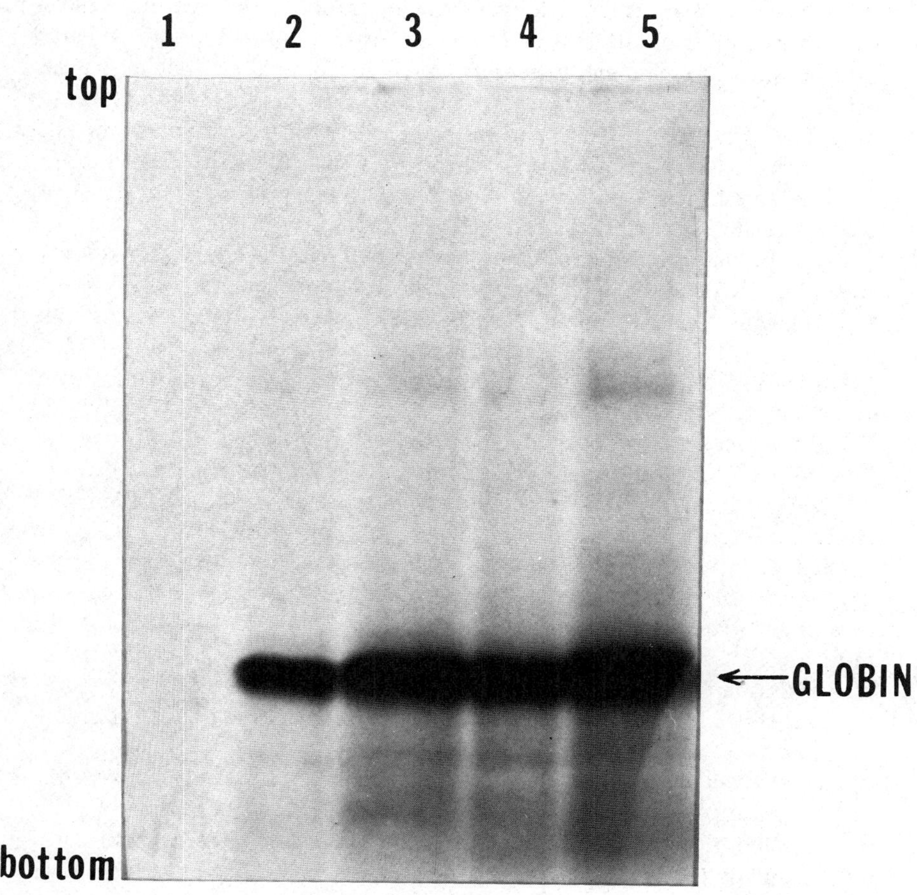

Figure 7 SDS gel electrophoresis of the translation products of rabbit globin mRNA generated in a wheat germ cell-free system. 1, No message added. 2, 0.5 μg of globin mRNA. 3 and 4, globin mRNA extracted after sequestration in liposomes. 5, ^{125}I-labeled globin standard.

sequestered rabbit globin mRNA in ether infusion liposomes composed of lecithin and DCP (molar ratio 8:2). Following gel filtration, the RNA was extracted from the liposomes and added to a wheat germ cell-free protein synthesizing system to assess its ability to direct the synthesis of globin. Figure 7 shows that the protein synthesized is not distinguishable in its electrophoretic properties from purified rabbit globin.

Table 3 summarizes the efficiency of sequestration of different RNA molecules by liposomes of different type and composition. As in Table 1, the data are expressed as percent entrapment per micromole of lipid per milliliter of suspension. Variations in the efficiency of sequestration are apparent but not remarkable. Efficiencies of entrapment between 0.25 and 1 were found when negatively charged liposomes were used, values very similar to those reported for proteins in Table 1. Positively charged liposomes containing stearylamine (SA) became associated with 10 times more RNA than was associated with negatively charged liposomes, a phenomenon probably due to electrostatic interactions between the RNA and the lipids comprising the liposome membrane (180).

Some investigators have reported that the efficiency of RNA sequestration is a function of the size of the molecule. Dimitriadis (180) observed that 5% of tRNA but only 1.5% of rRNA was liposomally sequestered. Also, Wreshner et al. (183) found that rRNA released from the liposomes by treatment with Triton X-100 contained proportionally more 5S and 18S than 28S rRNA. However, these investigators used either hybrid liposomes (183) or those produced from cochleate cylinders (180), two methods of preparation that involve fusion of either small liposomes or cochleate cylinders into larger vesicles. It is therefore possible that geometric factors might reduce the sequestration efficiency of the large RNA molecules. When RNA was sequestered by liposomes prepared by the ether infusion technique of Deamer and Bangham (184), all populations of bacterial RNA were sequestered with equal efficiency (Fig. 6a). The concentration of the RNA in the liposome formation buffer did not affect the efficiency of RNA sequestration when either phosphatidylserine cochleate or ether infusion liposomes were used (181,185).

To ascertain definitively whether liposomes prepared in different ways differ in their ability to sequester RNA, we have compared the sequestration of *E. coli* RNA by liposomes prepared by the ether infusion, reverse-phase evaporation (REV), and by the phosphatidylserine cochleate techniques. The results shown in Table 4 demonstrate that the REV technique allows one to sequester the greatest amount of RNA, followed in efficiency by the ether infusion and cochleate methods, respectively.

Liposome membranes appear to be impermeable to sequestered high molecular weight RNA. In our laboratory (181) *E. coli* [^{3}H] RNA was sequestered within liposomes of different composition prepared by the ether infusion technique. The radioactivity retained in the liposomes after 1 and 24 hr at 24°C was assessed by repeating the liposome purification procedure. Regardless of the lipid composition, no more than 10% of the radioactivity had leaked out of the liposomes during the 24-hr period, a value certainly within the limits of the experimental error of the procedure. Conversely, when [^{3}H] uridine was the sequestered molecule, 50-90% of the radioactivity leaked out of the liposomes in the first hour.

Table 3 Sequestration of RNA into Liposomes[a]

Type of liposome	Liposome composition (molar ratio)	Type of RNA	Sequestration (%)	% Entrapment (μmol lipids/ml)	Notes	Reference
LUV (cochleate)	PS(7):PC(2):C(1) and PS(7):PC(2):SA(1)	Rat rRna and globin mRNA	7–11	N.R.	Low molecular weight RNA is sequestered more efficiently	183
LUV (ether infusion)	PS, PC, C, DCP, Ly[b]	*E. coli*	6–8	0.35–0.55	Sequestration independent of RNA concentration and size	181
LUV (ether infusion)	PC(8):DCP(2)	*E. coli*	10–15	0.50–0.75	—	186
LUV (ether infusion)	PC(7):DCP(2):C(1)	*E. coli*	5	0.25	Independent of the size of RNA	178
sMLV	PC(4):C(1.5):SA(0.25)	Poly U	14–26	2.5–5.0	Sequestration lower when RNA concentration is increased	182
LUV (cochleate)	PS	Poliovirus	5–10	0.5–1.0	Independent of RNA concentration	185
LUV (cochleate)	PS	Rabbit globin mRNA	3	0.33	—	179
LUV (cochleate)	PS	5S, 18S, 28S, and globin mRNA	1.5–5	N.R.	Low molecular weight RNA is sequestered more efficiently	180
sMLV	PC(6):C(2):DCP(1)	Poly(I)$\cdot$poly(C)	3	0.17	—	190
sMLV	PC(6):C(2):SA(2)	Poly(I)$\cdot$poly(C)	37	2.0	RNA is associated with liposomes by electrostatic interaction	

[a]Abbreviations as in Table 1.
[b]Different lipid compositions were used.

Table 4 Comparative Efficiency of RNA Sequestration and Delivery into Cells by Three Types of LUVs

Type of liposome[a]	Volume of aqueous phase (ml)	Percent RNA sequestered (range)	Percent Entrapment (μmoles lipid/ml)	Delivery into cells[b] (% of entrapped RNA)	
				HEp-2	P3
LUV (cochleate)	1.0	2 (0.5-3.0)	0.15	6.9 ± 1.50[c]	4.92 ± 2.53[c]
LUV (ether infusion	2.0	5 (2-8)	0.80	6.24 ± 0.61	4.35 ± 2.40
LUV (reverse evaporation	0.3	21 (15-40)	0.50	5.39 ± 1.49	3.00 ± 1.29

[a] All liposomes were made with 13 μmol of phosphatidylserine.

[b] 20×10^6 cells were incubated with liposomes (2 μmol) for 60 min at 37°C in 2 ml of RPMI 1640 medium.

[c] Mean ± SD.

Cellular Delivery of RNA

Cocultivation of liposomes containing radioactive RNA with cells results in part of the radioactivity becoming cell associated. Evidence that this radioactivity is the result of a direct liposome-cell interaction and that radioactive RNA is internalized by the cells can be obtained by demonstrating the following: (a) radiolabeled RNA extracted from RNAse-treated, purified liposomes and from the supernatant left after the harvesting of liposome-treated cells is undegraded, and (b) [³H] RNA extracted from the washed cell pellet shows signs of degradation. If the [³H] RNA remains intact during sequestration within liposomes and during their subsequent incubation with cells and appears degraded only when extracted from the cell pellet, one can conclude that the degradation occurred intracellularly.

Figure 6 represents the sedimentation pattern in a sucrose density gradient of *E. coli* [³H] RNA extracted from (a) ribonuclease-treated, gel-filtered liposomes; (b) the supernatant following the harvest of human epithelial carcinoma (HEp-2) cells that have been incubated with liposomes for 60 min at 37°C; and (c) washed HEp-2 cells after a 60-min treatment with liposome-encapsulated [³H] RNA or with equivalent amounts of "naked" [³H] RNA (181). RNA extracted from purified liposomes (Fig. 6a) and from the supernatant after cell harvest (Fig. 6b) appears undegraded. This, coupled with the fact that unprotected RNA added directly to cells in culture results in its rapid breakdown, indicates that RNA does not leak out of liposomes during liposome-cell cocultivation. However, as shown in Figure 6c, the *E. coli* [³H] RNA extracted from the HEp-2 cell pellet following liposome-mediated cellular insertion appears to be partially degraded. Since RNA sequestered within liposomes is protected from nuclease attack during the sequestration process as well as during the liposome-cell cocultivation, it can be reasonably concluded that the observed RNA degradation has occurred intracellularly. Cells incubated with "naked" *E. coli* [³H] RNA contained eukaryotic (18S and 28S) [³H] RNA as well as large amounts of degraded RNA at the top of the gradient, indicating a severe breakdown of the RNA during incubation with HEp-2 cells followed by a reutilization of the degradation products (181).

Further evidence that radioactive RNA associated with target cells is internalized and not due to liposomes adhering to the membrane was provided by a scanning electron microscopy of murine myeloma cells following their incubation with liposomes as well as after repeated washings. Following liposome-cell incubation many liposomes appear to adhere to the cells. However, after four washings, the number of bound liposomes is reduced to a level insufficient to explain the amount of cell-associated RNA routinely achieved in the laboratory (D. Lavelle, M. J. Ostro, and D. Giacomoni, unpublished observations.)

Table 5 summarizes the efficiency of delivery of liposome-sequestered RNA

Table 5 Delivery of Liposome-Sequestered RNA into Cells

Type of liposome	Composition of liposomes (molar ratio)	RNA used	Cell type	Incubation conditions (in 60 min)		RNA delivered (%)	Reference
				μmol lipids	Number of cells		
LUV (cochleate)	PS	Reticulocyte	Mouse L929	0.1	1×10^6	2.5-3.2	180
LUV (ether infusion)	PC(8):DCP(2):Ly(0.4)	*E. coli*	HEp-2	1.2	2×10^6	4.5	181
LUV (ether infusion)	PC(8):DCP(2):Ly(0.4)	*E. coli*	Rabbit spleen	1.2	2×10^6	1.8	181
LUV (ether infusion)	PC(8):DCP(2):Ly(0.4)	*E. coli*	Carrot protoplasts	1.2	2×10^6	17.0	186
LUV (ether infusion)	PC(8):DCP(2)	*E. coli*	HEp-2	1.2	2×10^6	3.8	181
LUV (ether infusion)	PC(8):DCP(2)	*E. coli*	Rabbit spleen	1.2	2×10^6	1.2	181
LUV (ether infusion)	PC(8):DCP(2)	*E. coli*	Carrot protoplasts	1.2	2×10^6	11.7	186
LUV (ether infusion)	PC(7):DCP(2):C(1)	*E. coli*	HEp-2	1.2	2×10^6	3.1	181
LUV (ether infusion)	PC(7):DCP(2):C(1)	*E. coli*	Rabbit spleen	1.2	2×10^6	1.4	181
LUV (ether infusion)	PC(7):DCP(2):C(1)	*E. coli*	Carrot protoplasts	1.2	2×10^6	6.0	186
LUV (ether infusion)	PC(4):C(4):DCP(2)	*E. coli*	HEp-2	1.2	2×10^6	5.0	181
LUV (ether infusion)	PC(4):C(4):DCP(2)	*E. coli*	Rabbit spleen	1.2	2×10^6	2.1	181
LUV (cochleate)	PS	Poliovirus	HeLa, CHO, mouse L	10^{-3} to 10^{1}	1×10^6	N.R.	185

into cells. In most of these studies large unilamellar lipid vesicles obtained by the ether infusion method (184) were used. It appears that, regardless of the composition of the liposomes, carrot protoplasts can incorporate more RNA (6-17%) than ether HEp-2 cells (4-5%) or rabbit spleen cells (1-2%), suggesting that efficiency of liposome-cell interaction depends largely on the nature of the target cell (181,186).

Evidence that RNA delivered into target cells via liposomes is able to express its biological function was first provided by Ostro et al. (187). These authors sequestered rabbit globin mRNA into liposomes composed of lecithin and DCP (molar ratio 8:2) and incubated them with HEp-2 cells. The treated cells were cultivated overnight in medium containing radioactive amino acids. Following the culture period, protein extracts obtained from the cells were enriched in putative globin by immunoabsorbent chromatography and then subjected to electrophoresis on 10% polyacrylamide gels in the presence of a rabbit globin standard. The gel profile of radioactively labeled protein revealed a protein species that comigrated with the globin standard but that was absent from the protein extract isolated from cells treated with buffer-loaded liposomes and exogenously added globin mRNA (Fig. 8a). When the experiment was carried out with cells which had been inhibited in their ability to synthesize RNA by pretreatment with actinomycin D, the peak of radioactivity that comigrated with the globin standard was enhanced (Fig. 8b). Once again, addition of buffer-loaded liposomes and free globin mRNA to the cells did not result in the production of any globin-like protein. These data, as well as the ability of part of the newly synthesized proteins to bind selectively to purified antiglobin antiserum, provided evidence that synthesis of a foreign protein had occurred in HEp-2 cells following liposome-mediated mRNA delivery (187). Independently, Dimitriadis demonstrated the translation of rabbit globin mRNA in mouse lymphocytes after liposome-mediated cellular insertion (188).

More recently, Wilson et al. (185) have shown that LUVs can sequester poliovirus RNA and deliver it into mammalian cells in an infectious form. The infectious titer of RNA in the vesicles was roughly 1 in 10^4 genome equivalents, a value lower than that of an intact virus preparation (1 in 10^2 genome equivalents) but about 100-fold higher than that of "naked" poliovirus RNA inserted into cells by transfection in the presence of DEAE-dextran and DMSO (185). These authors attributed the relatively low infectivity (relative to intact virus) of liposome-sequestered RNA to inefficient delivery of the molecules rather than to the fact that only very few molecules retained their infectivity. Liposomes have also been used to deliver RNA into lymphocytes in order to transform them into cells with an immune competence specified by the RNA. Magee et al. (189) inserted RNA-rich splenic extracts isolated from guinea pigs immunized against a hepatocarcinoma cell line into normal lymphocytes via liposomes. Antilymphocyte immunoglobulin was also present during liposome formation. After incubation

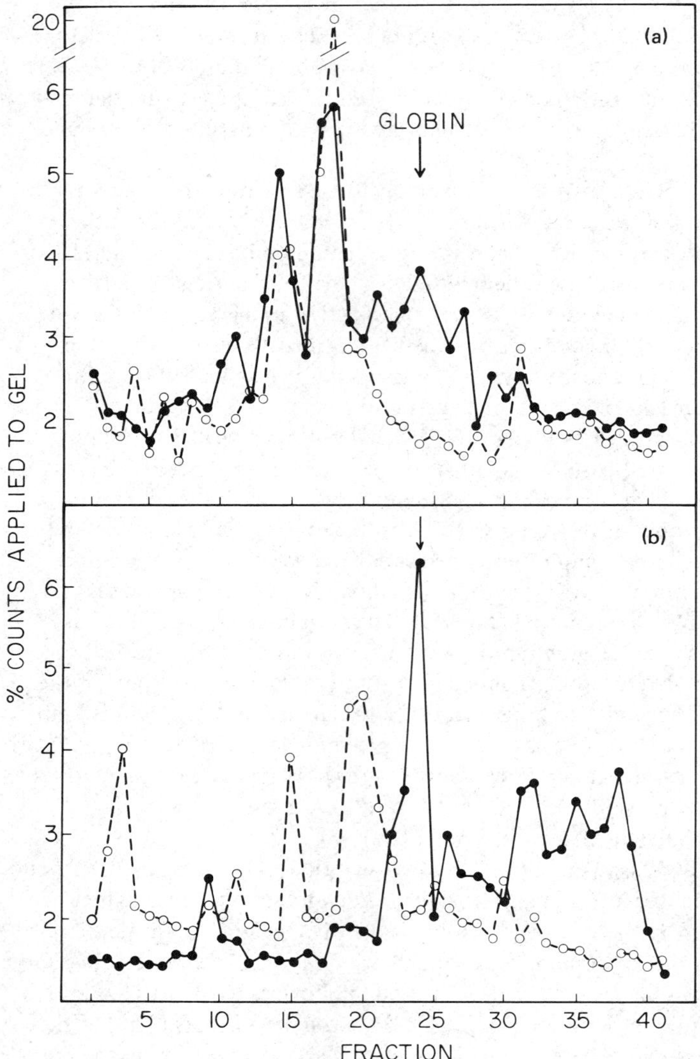

Figure 8   SDS polyacrylamide gel electrophoresis of a mixture of radioactive HEp-2 cell protein isolates purified by immunoadsorbent chromatography and rabbit globin standard. (a) (●) Radioactivity profile of proteins from HEp-2 cells treated with liposome-sequestered globin mRNA: (○) radioactivity profile of proteins from untreated HEp-2 cells. (b) (●) Radioactivity profile of proteins from HEp-2 cells grown in the presence of 5 μg/ml actinomycin D and treated with liposome-sequestered globin mRNA; (○), radioactivity profile of HEp-2 cells treated with empty liposomes and 100 μg of exogenously added globin mRNA. (From Ref. 187.)

with these liposomes, the normal lymphocytes became cytotoxic to the carcinoma cells, presumably because of phenotypic changes induced by the "immune RNA" carried by the liposome. When liposomes were prepared in the absence of immunoglobulins the lymphocytes did not develop cytotoxicity, indicating that the specific immunoglobulins may play a role in "homing" of the liposomes to the target cells. However, since cationic liposomes were used in these experiments, the possibility exists that the role of the immunoglobulin was to prevent salt bridge formation between the negatively charged RNA and the positively charged liposome membrane, thereby permitting the efficient release of the RNA from the liposome during the vesicle-cell interaction.

In the only reported in vivo experiment involving liposome-entrapped RNA, Magee et al. (190), in an attempt to induce an interferon response, injected liposomes containing poly(I)·poly(C) into mice. When multilamellar liposomes containing the polynucleotides were injected intravenously, a level of interferon production was achieved that was 20 times higher than when free poly(I)·poly(C) was administered.

Delivery of DNA into Cells by Means of Liposomes

The problems associated with the sequestration of DNA molecules into liposomes and their delivery into cells are not very different from the problems one encounters when working with RNA or proteins. However, DNA must be delivered into the nucleus to express its genetic potential. Excellent evidence has been provided in recent years that liposomes can sequester DNA without damaging it physicochemical or biological properties. Fraley et al. (191) sequestered DNA isolated from the plasmid pBR322 (a plasmid that carries genetic determinants for resistance to tetracycline and ampicillin) into large unilamellar liposomes. When the DNA-liposome suspension was chromatographed on a Sepharose column, part of the DNA comigrated with the liposomes. Liposome-associated DNA, after treatment of the liposomes with DNAse, was reextracted and electrophoresed on an agarose gel (Fig. 9). The migration pattern obtained showed that the DNA molecules had suffered very slight physical alterations since they appeared to have shifted to the open circular form. These experiments strongly suggest that liposome-associated DNA was internalized by the liposomes and had become latent to the action of hydrolytic enzymes. More important, the reextracted DNA has maintained most of its biological activity, as demonstrated by its ability to transfer tetracycline resistance to *E. coli* cells, although with only one-tenth the efficiency of a standard plasmid preparation.

Table 6 summarizes the data on the efficiency of sequestration of DNA by liposomes reported by different laboratories. In most cases 4-10% of the DNA used was sequestered by the liposomes, a value similar to that reported for liposomal sequestration of proteins and RNA (Tables 1 and 3). When REV liposomes were used, the sequestration was enhanced five-fold (192). Hoffman et al. (193)

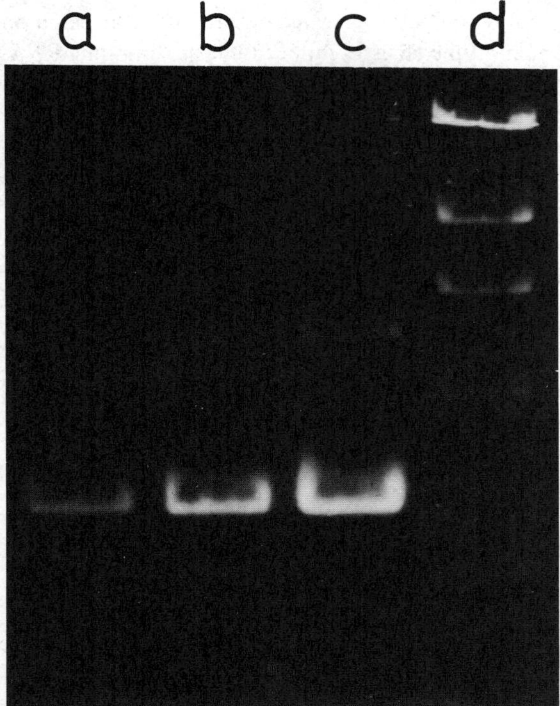

Figure 9 Agarose gel profiles of control and liposome-extracted SV40 DNA.
Lane a, SV40 DNA extracted from liposomes (0.3 μg); lane b, extracted control
SV40 DNA (0.5 μg); lane c, control SV40 DNA (0.5 μg); lane d, λDNA (Hind
III digest). (From Ref. 191.)

observed that when liposomes were formed in a low-ionic-strength buffer (0.02
M Tris, 1 mM MgCl$_2$ pH 7.5) 95% of the DNA became associated with the lecithin bilayer. When the buffer contained 0.14 M NaCl, only 8% of the DNA became liposome-associated. However, at low ionic strength the DNA was not
latent to nuclease degradation, whereas when the ionic strength of the buffer
was raised, little if any liposome-associated DNA could be digested by DNAse.
They concluded that at low ionic strength, electrostatic interactions occur that
cause association of DNA with the liposome membrane but not its sequestration.
Mannino et al. (194) observed that the efficiency of sequestration of DNA into
LUVs prepared by the phosphatidylserine cochleate method depended on the
size of the DNA. When the molecular weight was lower than 10^6, 8-10% of the
DNA was sequestered, wheras only 1% of the DNA with a molecular weight of

Table 6 Sequestration of DNA into Liposomes

Type of liposome	Liposome composition (molar ratio)	Type of DNA	Sequestration (%)	Notes	Reference
LUV (cochleate)	PS	T7 bacteriophage (restriction fragments)	10	Low molecular weight DNA sequestered more efficiently	194
LUV (cochleate)	PS	Plasmid pMB9 (mol wt 4.7×10^6) and *E. coli* DNA	6-8	Independent of size of DNA	199
LUV (ether infusion)	PC(10):PG(1)	Plasmid pBR322 (mol wt 2.6×10^6)	2-6	—	191
MLV	PG	SV40 λ bacteriophage Ehrlich ascites cells	4-8	At low ionic strength DNA also associates with PG by electrostatic forces	193
MLV	PC	Plasmids pBR322 (mol wt 2.6×10^6) and pCR1 (mol wt 8.6×10^6)	4-8	—	196
MLV	PC(7):C(2)	Plasmids pBR322 and pCR1	4-8	—	196
LUV (REV)	PS	SV40	30-40	—	192

4-6 $\times$ 10^6 was entrapped. As mentioned earlier, this dependence on size was also described for RNA sequestration by the same type of liposomes (Table 3) and indicates a possible limitation on the use of liposomes obtained by this method.

Liposomes have been used to deliver DNA molecules into mammalian and plant cells. Following an observation by Cassels (195) that lipid vesicles could fuse with tomato protoplasts and deliver their contents into the cells, Lurquin (196) demonstrated that plasmid DNA could be inserted into protoplasts by this method. When multilamellar liposomes containing plasmid DNA were incubated with cowpea protoplasts for 45 min, 5% of the sequestered DNA became associated with the protoplasts. When the incubation was performed in the presence of PEG, the value was raised to 25%. Relatively intact plasmid DNA could be reextracted from the nuclei of these cells, whereas the DNA associated with the cytoplasm appeared to be degraded. Successful insertion of biologically active DNA into mammalian cells has been reported by Fraley et al. (192). These authors sequestered SV40 DNA into REV liposomes and incubated them with African green monkey kidney cells. Functional insertion of DNA was assessed by the plaque assay used for viral infections. In a first series of experiments it was found that liposome-sequestered SV40 DNA had a frequency of infection of 1.5 $\times$ 10^3 PFU/ng DNA, while control DNA transfected by the DEAE-dextran method had an infectivity of 4-5 $\times$ 10^6 PFU/ng DNA. However, the relative infectivity of liposome-sequestered DNA is probably higher than these numbers suggest. In fact, while 0.01 ng of DNA was used for the DEAE-dextran transfection, a much larger amount of liposome-sequestered DNA was employed (10 ng). It is therefore possible that multiple strands of DNA inserted into the cell by each liposome could have created a situation whereby excess DNA within the cell was wasted, thus artificially reducing the PFU/ng values obtained. Of obvious interest is the fact that liposome-mediated infection was not affected by the presence of DNAse. Higher infectivity of liposome-delivered SV40 DNA was achieved when (a) phosphatidylserine:cholesterol liposomes were used, and (b) cells were treated with 40% glycerol following liposome-cell interaction. Under these conditions 2-3 $\times$ 10^5 PFU/ng DNA was obtained, whereas incubation of cells with naked DNA without DEAE-dextran showed very low infectivity.

Recent experiments by Wong et al. (197) suggest the possibility of transferring functional prokaryotic genes into animal cells via liposomes. These authors isolated a restriction fragment encoding β-lactamase from the pBR322 plasmid and sequestered it within sonicated multilamellar liposomes composed of lecithin and phosphatidylserine (molar ratio 9:1). Upon incubation of these liposomes with HeLa cells (4 μmol of lipids per 10^7 cells) for 2 hr at 37°C, 2.5% of the DNA was internalized by the cells. The cells so treated produced measurable amounts of the bacterial β-lactamase. The data reported do not allow one to determine whether a permanent change has occurred at the genetic level. Also, the

intracellular location at which the transcription of DNA took place was not investigated. If confirmed, this type of experiment could be of great importance
in the understanding of prokaryotic and eukaryotic gene regulation. Liposomes
have also been used to deliver pBR322 DNA into a strain of *E. coli* (SF8) susceptible to tetracycline (191), thus resulting in the appearance of clones of *E.
coli* resistant to that antibiotic. The transformation efficiency was low (about
1% of that achieved when free plasmid DNA was used) but was insensitive to
DNAse added to the incubation mixture.

Delivery of Particulate Biological Materials into Cells
by Means of Liposomes

The versatility of liposomes as microsyringes has prompted a few scientists to use
them to inject particulate biological materials into cells. Mukherjee et al. (126)
have shown that chromosomes associated with liposomes can be transferred into
foreign cells, where they can induce changes in the cells' genetic makeup more
efficiently than can isolated chromosomes alone. Metaphase chromosomes from
a murine x human somatic cell hybrid expressing hypoxanthine guanine phos-
phorybosil transferase (HPRT) of the human type were suspended in an ether-
chloroform mixture, added to a chloroform solution containing lecithin and
cholesterol (molar ratio, 8:2), and followed by evaporation of the organic solvents. The lipid-chromosome mixture was suspended in buffer, leading to the
formation of "lipochromosomes." The lipochromosomes were incubated with
the murine cell line A9 lacking HPRT. These cells cannot grow in HAT medium
unless they acquire the ability to produce this enzyme. In three different experiments these authors observed that incubation of A9 (HPRT$^-$) cells with HPRT$^+$
lipochromosomes yielded clones of cells that could grow in HAT medium. An
average of 50 clones per 5×10^6 cells developed. When metaphase chromosomes
alone were used, only 2-5 clones of HPRT$^+$ cells were observed. To prove that
the HPRT produced by these clones was of the donor type (human), thereby
eliminating the possibility that the clones obtained were genetic revertants, the
HPRT produced by the clones was analyzed by an electrophoretic system able
to distinguish human HPRT from the slower migrating mouse form. Figure 10
shows that the murine A9 clones produced human HPRT, proving that a gene
transfer had indeed occurred.

More recently, Kondorosi and Duda (127) sequestered nuclei from chicken
red cells into phosphatidylserine cochleate vesicles and incubated them with
Chinese hamster ovary cells that could not grow in media lacking proline (CHO/
pro$^-$). Cells so treated were cultivated in medium lacking proline so that only
those cells that had acquired the ability to make proline endogenously could sur-
vive. Proline-positive clones grew with a frequency of $4\text{-}6 \times 10^{-5}$. When nuclei
alone were used the transformation frequency was not higher than the spontane-

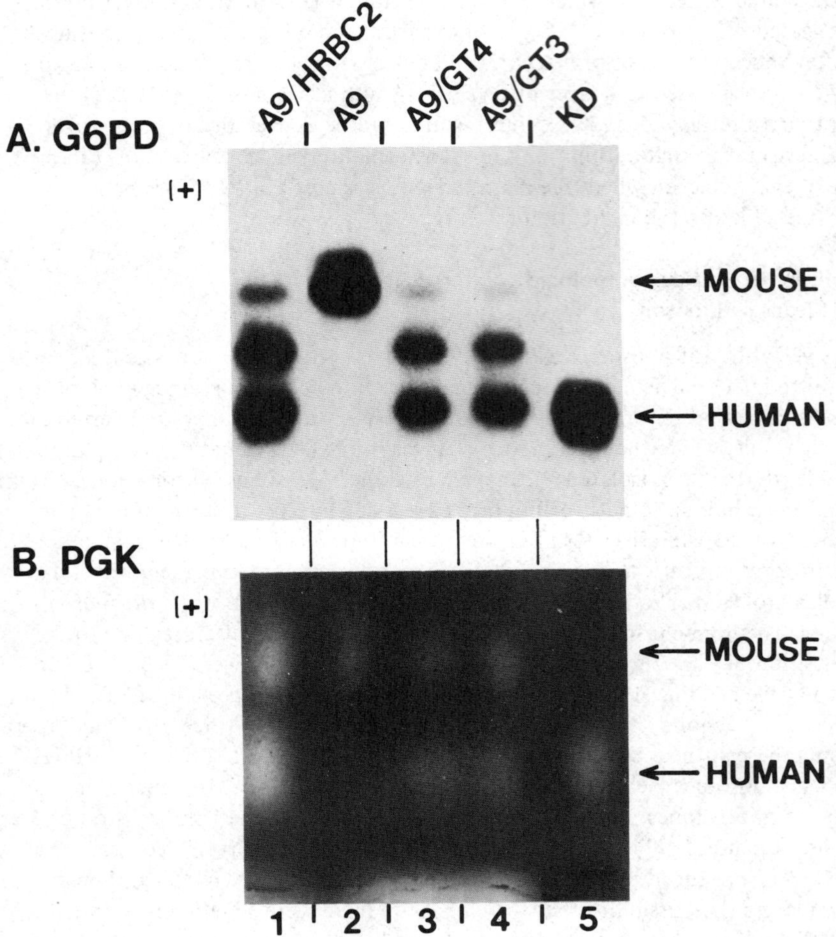
A9/HRBC2
A9
A9/GT4
A9/GT3
KD
A. G6PD
(+)
MOUSE
HUMAN
B. PGK
(+)
MOUSE
HUMAN
1   2   3   4   5

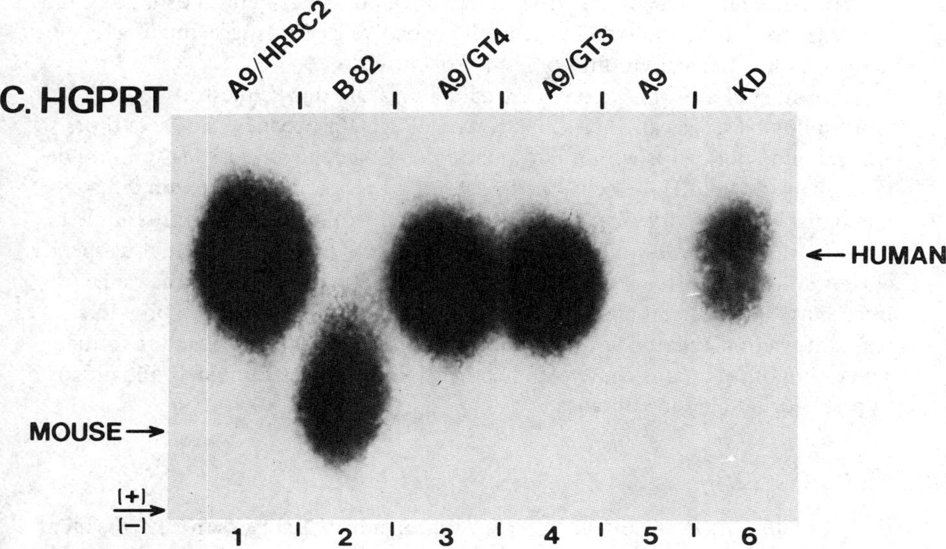

Figure 10 (A) Electrophoresis of glucose-6-phosphate dehydrogenase (G6PD). Note the three bands of G6PD in the HAT-resistant clones (channels 3 and 4). The heteropolymer is seen between the mouse and the human forms of the enzyme. The recipient A9 cells have only the mouse G6PD activity. (B) Electrophoresis of PGK. Human PGK migrates slower than the mouse enzyme. A9 has only the mouse form, whereas the HAT-resistant clones (channels 3 and 4) have both mouse and human forms of the enzyme. (C) Electrophoresis of HGPRT. Note the human form of the enzyme in the HAT-resistant clones (channels 3 and 4). The recipient cell line A9 does not have any HGPRT activity. B82 is a thymidine kinase-deficient mouse line derived from L cells. This cell line is positive for HGPRT, G6PD, and PGK enzymes. (From Ref. 126.)

ous reversion rate $(6\text{-}10 \times 10^{-7})$. Interestingly, no chicken chromosomes or fragments thereof were observed in the proline-positive clones, suggesting that a gene insertion into the host chromosomes had occurred.

Wilson et al. (198) have reported data indicating that sequestration of poliovirus in lipid vesicles can alter some of its biological properties. These authors encapsulated poliovirus within phosphatidylserine cochleate vesicles and found that when added to HeLa cells, the liposome-sequestered viruses exhibited a much higher infectivity than that which resulted from addition of virus alone. Also, only 90% of this infectivity was lost following incubation with virus-specific antiserum, whereas free poliovirus was totally inactivated. To test the range of infectivity of the encapsulated virus, Chinese ovary cells, normally poliovirus resistant, were incubated with the "lipovirus." Such treatment resulted in the generation of infectious centers, indicating that liposomes can help poliovirus bypass "forbidden" membranes.

REFERENCES

1. Purchio, A. F., E. Erickson, J. S. Brugge, and R. L. Erickson. 1978. Identification of a polypeptide encoded by the avian sarcoma virus *src* gene. Proc. Natl. Acad. Sci. USA 75:1507.
2. Stacey, D. W. 1979. Messenger activity of virion RNA for avian leukosis viral envelope glycoprotein. J. Virol. 29:949.
3. Stacey, D. W. and H. Hanafusa. 1978. Nuclear conversion of micro-injected avian leukosis virion RNA into an envelope-glycoprotein messenger. Nature 273:779.
4. Gierer, A. and G. Schramm. 1956. Infectivity of RNA from tobacco mosaic virus. Nature 177:702.
5. Westphal, O., O. Luderitz, and F. Bister. 1955. Über die Extraktion von Bacterien. Dtsch. Z. Verdauungskr. 15:170.
6. Fraenkel-Conrat, H., B. Singer, and R. C. Williams. 1957. Infectivity of viral nucleic acid. Biochim. Biophys. Acta 25:87.
7. Colter, J. S., H. H. Bird, A. W. Myer, and R. A. Brown. 1957. Infectivity of RNA isolated from virus infected tissues. Virology 4:522.
8. Alexander, H. E., G. Koch, I. M. Mountain, K. Sprunt, and O. Van Damme. 1958. Infectivity of RNA of poliovirus on HeLa cell monolayers. Virology 5:172.
9. Colter, J. S. and K. A. O. Ellem. 1961. Structure of viruses. Annu. Rev. Microbiol. 15:219.
10. Alexander, H. E., G. Koch, I. M. Mountain, and O. Van Damme. 1958. Infectivity of RNA from polio virus in human cell monolayers. J. Exp. Med. 108:493.
11. Holland, J. J., L. C. McLaren, and J. T. Sylverton. 1959. The mammalian cell virus relationship. IV. Injection of naturally insusceptible cells with enterovirus RNA. J. Exp. Med. 110:65.

12.   Holland, J. J., L. C. McLaren, B. H. Hoyer, and J. T. Sylverton. 1960.
      Enteroviral ribonucleic acid. II. Biological, physical and chemical studies.
      J. Exp. Med. 812:841.
13.   Pagano, J. S. and A. Vaheri. 1965. Enhancement of infectivity of polio-
      virus RNA with DEAE-dextran. Arch. Gesamte Virusforsch. 17:456.
14.   Tatemoto, K. K. and P. Fabish. 1963. Influence of acid polysaccharides
      on plaque formation by influenza A2 and B virus. Proc. Soc. Exp. Biol.
      Med. 114:811.
15.   Bachrach, H. L. 1966. RNA of foot and mouth disease virus, an ultrasensi-
      tive plaque assay. Proc. Soc. Exp. Biol. Med. 123:939.
16.   Towell, D. R. and J. S. Colter. 1967. Observations on the assay of infec-
      tious viral RNA: effect of DMSO and DEAE-dextran. Virology 32:84.
17.   Amstey, M. S. and P. D. Parkman. 1966. Enhancement of polio RNA in-
      fectivity by DMSO. Proc. Soc. Exp. Biol. Med. 123:438.
18.   Bhargava, P. M. and G. Shanmugam. 1971. Uptake of nonviral nucleic
      acids by mammalian cells. In *Progress in Nucleic Acid Research and Mole-
      cular Biology,* Vol. 11, J. N. Davidson and W. E. Cohn (Eds.). Academic
      Press, New York, p. 103.
19.   Chin, P. H. and M. S. Silverman. 1967. Studies on the transfer of myeloma
      protein synthesis with RNA isolated from the C3H plasma cell tumor. I.
      Uptake of isotopically labelled tumor RNA by lymphoid cells. J. Immunol.
      99:476.
20.   Wang, S. R., D. Giacomoni, and S. Dray. 1973. Physical and chemical
      characterization of RNA incorporated by rabbit spleen cells. Exp. Cell.
      Res. 78:15.
21.   Shanmugan, G. and P. M. Bhargava. 1969. Uptake of *E. coli* RNA by rat
      liver cells in suspension. Indian J. Biochem. 6:64.
22.   Bishop, D. C. and P. Abramoff. 1969. In vitro uptake of isotopically
      labelled RNA by mammalian spleen cells. Proc. Soc. Exp. Biol. Med. 120:
      378.
23.   Ellem, K. A. O. and J. S. Colter. 1961. The interaction of infectious RNA
      with mammalian cells. III. Comparison of infection and RNA uptake in
      the HeLa cell-polio RNA and L cell-Mengo RNA systems. Virology 15:113.
24.   Herrera, F., R. H. Adamson, and R. C. Gallo. 1970. Uptake of transfer
      RNA by normal and leukemic cells. Proc. Natl. Acad. Sci. USA 67:1943.
25.   Gallagher, R. E., C. A. Walter, and R. C. Gallo. 1972. Uptake and amino-
      acylation of exogenous tRNA by mouse leukemia cells. Biochem. Biophys.
      Res. Commun. 49:782.
26.   Niu, M. C., C. C. Cordova, and L. C. Niu. 1961. RNA-induced changes in
      mammalian cells. Proc. Natl. Acad. Sci. USA 42:1689.
27.   Niu, M. C. and C. C. Cordova. 1964. [14]C as marker for the activity of a
      RNA-induced enzyme. Fed. Proc. 23:317.
28.   Zimmermann, E., M. Zoller, and F. Turba. 1963. Induction of synthesis
      of serum albumin in mouse ascite cells by means of nucleolar RNA from
      liver. Biochem. Z. 339:53.

29. Tuohimaa, P., S. J. Segal, and S. S. Koide. 1972. Induction of avidin synthesis by RNA obtained from chick oviduct. Proc. Natl. Acad. Sci. USA 69: 2814.

30. Villee, C. A. 1973. Effects of exogenous polynucleotides on uterine enzymes. In *The Role of RNA in Reproduction and Development*, M. C. Niu and S. J. Segal (Eds.). North-Holland, Amsterdam.

31. Mroczkowski, B., H. P. Dym, E. J. Siegel, and S. M. Heywood. 1980. Uptake and utilization of RNA by myogenic cells in culture. J. Cell Biol. 87: 65.

32. Sterzl, J. and M. Hrubesova. 1956. The transfer of antibody formation by means of nucleoprotein fractions to nonimmunized recipients. Folia Biol. 2:21.

33. Fishman, H. 1961. Antibody formation in vitro. J. Exp. Med. 114:837.

34. Friedman, H. 1964. Antibody plaque formation by normal mouse spleen cell cultures exposed to RNA from immune mice. Science 146:934.

35. Cohen, E. P. and J. J. Parks. 1964. Antibody production by nonimmune spleen cells incubated with RNA from immunized mice. Science 144:1012.

36. Cohen, E. P., R. W. Newcomb, and I. K. Crosby. 1965. Conversion of nonimmune spleen cells to antibody forming cells by RNA: strain specificity of the response. J. Immunol. 95:583.

37. Campbell, D. H. and J. S. Garvey. 1957. The retention of ^{35}S labeled bovine serum albumin in normal and immunized rabbit liver tissue. J. Exp. Med. 105:361.

38. Campbell, D. H. and J. S. Garvey. 1963. Nature of retained antigen and its role in immune mechanism. Adv. Immunol. 3:261.

39. Chin, P. H. and M. S. Silverman. 1967. Studies on the transfer of myeloma protein synthesis with RNA isolated from the C_3H plasma cell tumor (X5563). II. In vivo and in vitro culture of RNA-treated cells. J. Immunol. 99:489.

40. Adler, F. L., M. Fishman, and S. Dray. 1966. Antibody formation initiated in vitro. III. Antibody formation and allotypic specificity directed by RNA from peritoneal exudate cells. J. Immunol. 97:554.

41. Bell, C. and S. Dray. 1969. Conversion of nonimmune spleen cells by RNA of lymphoid cells from an immunized rabbit to produce antibody of foreign light chain allotype. J. Immunol. 103:1190.

42. Schafer, A. E., M. Fishman, and F. L. Adler. 1974. Studies on antibody induction in vitro. III. Cellular requirements for the induction of antibody synthesis by solubilized T_2 phage and immunogenic RNA. J. Immunol. 112:1981.

43. DiMayorca, G. A., B. E. Eddy, S. E. Stewart, W. S. Hunter, C. Friend, and A. Bendich. 1959. Isolation of infectious DNA from polyoma infected tissue cultures. Proc. Natl. Acad. Sci. USA 45:1805.

44. Weil, R. 1961. A quantitative assay for an infective agent related to polyoma viruses. Virology 14:46.

45. Gerber, P. 1962. An infectious DNA derived from vacuolating virus (SV40). Virology 16:96.

46. Black, P. H. and W. P. Rowe. 1965. Induction of SV40 T antigen with
     SV40 DNA. Virology 27:436.
47. McCutchan, J. H. and J. S. Pagano. 1968. Enhancement of the infectivity
     of simian virus 40 DNA with DEAE-dextran. J. Natl. Cancer Inst. 41:351.
48. Warden, D. and H. V. Thorne. 1969. The infectivity of polyoma virus
     DNA from mouse embryo cells in the presence of DEAE-dextran. J. Gen.
     Virol. 3:371.
49. Nicholson, M. O. and R. M. McAllister. 1968. Infectivity of human adeno-
     virus-1 DNA. Virology 48:14.
50. Graham, F. L. and A. J. Van der Eb. 1973. A new technique for the assay
     of infectivity of human adenovirus 5 DNA. Virology 52:456.
51. Laudo, D. and M. L. Ryhimer. 1973. Pouvoir infectieux du DNA d'herpes-
     virus en culture cellulaires. C. R. Acad. Sci. Paris Ser. D, 269:527.
52. Graham, F. L., G. Veldhuisen, and N. M. Wilkie. 1973. Infectious herpes
     virus DNA. Nature New Biol. 245:265.
53. Szybalska, E. H. and W. Szybalski. 1962. Genetics of human cell lines.
     IV. DNA mediated heritable transformation of a biochemical trait. Proc.
     Natl. Acad. Sci. USA 48:2026.
54. Graham, F. L., P. J. Abrahams, C. Mulder, H. L. Heijmeker, S. O. Warmaar,
     F. A. J. de Vries, W. Fiers, and A. J. Van der Eb. 1974. Studies on in vitro
     transformation by DNA and DNA fragments of human adenovirus or SV40
     DNA. Cold Spring Harbor Symp. Quant. Biol. 39:637.
55. Maitland, N. J. and J. K. McDougall. 1977. Biochemical transformation of
     mouse cells by fragments of HSV-DNA. Cell 11:231.
56. Wigler, M., S. Silverstein, L. S. Lee, A. Pellicer, Y. C. Cheng, and R. Axel.
     1977. Transfer of purified herpes virus thymidine kinase gene to cultured
     mouse cells. Cell 11:223.
57. Wigler, M., R. Sweet, G. K. Siru, B. Wold, A. Pellicer, A. Lacy, T. Maniatis,
     S. Silverstein, and R. Axel. 1979. Transformation of mammalian cells
     with genes from prokaryotes and eukaryotes. Cell 16:777.
58. Buetti, E. and H. Diggelmann. 1981. Cloned MMTV is biologically active
     in transfected mouse cells and its expression is stimulated by glucocorticoid
     hormones. Cell 23:335.
59. Wigler, M., A. Pellicer, S. Silverstein, and R. Axel. 1978. Biochemical
     transfer of single copy eukaryotic genes using total cellular DNA as donor.
     Cell 14:725.
60. Porter, M. T., A. L. Fluharty, and H. Kihama. 1971. Correction of abnor-
     mal cerebroside sulfate metabolism in cultured metachromatic leucodys-
     tryphy fibroblasts. Science 172:1263.
61. Wiesman, U. N., E. E. Rossi, and N. N. Herschkowitz. 1972. Treatment
     of metachromatic leukodystrophy in fibroblasts by enzyme replacement.
     N. Engl. J. Med. 284:672.
62. O'Brien, J. S., A. L. Miller, A. W. Loverde, and M. L. Veath. 1973. San-
     filippo disease type B: Enzyme replacement and metabolic correction in
     cultured fibroblasts. Science 181:753.

63. Lagunoff, D., D. M. Nicol, and P. Pritzl. 1973. Uptake of β glucuronidase by deficient human fibroblasts. Lab Invest. 29:449.

64. Dawson, G., R. Matalon, and Y. T. Li. 1973. Correction of enzymatic defect in cultured fibroblasts from patients with Fabry's disease: treatment with purified α galactosidase from ficin. Pediatr. Res. 7:684.

65. Bach, G., R. Friedman, B. Weissmann, and E. F. Neufeld. 1972. The defect in the Hurler and Scheie syndromes: deficiency of α-L-iduronidase. Proc. Natl. Acad. Sci. USA 69:2048.

66. Baudhuin, P., H. G. Hers, and H. Loeb. 1964. An E. M. and biochemical study of type II glycogenesis. Lab. Invest. 13:1139.

67. Lauer, R., T. Mascanines, A. S. Racela, and A. M. Diehl. 1968. Administration of a mixture of fungal gluconidases to a patient with type II glycogenesis (Pompe's disease). Pediatrics 42:672.

68. Desnick, R. J., S. R. Thorpe, and M. B. Fiddler. 1976. Toward enzyme therapy for lysozomal storage diseases. Physiol. Rev. 56:57.

69. Barber, M. A. 1911. A technique for the inoculation of bacteria and other substances and of microorganisms into the cavity of the living cell. J. Infect. Dis. 8:348.

70. Briggs, R. and T. J. King. 1952. Transplantation of living nuclei from blastula cells into enucleated frogs' eggs. Proc. Natl. Acad. Sci. USA 38: 455.

71. Elsdale, T. R., J. B. Gurdon, and M. Fischberg. 1960. A description of the technique for nuclear transplantation in *Xenopus laevis*. J. Embryol. Exp. Morphol. 8:437.

72. Lane, C. D., G. Marbaix, and J. B. Gurdon. 1971. Rabbit hemoglobin synthesis in frog cells: the translation of reticulocyte 9S RNA in frog oocytes. J. Mol. Biol. 61:73.

73. Lane, C. D. and C. M. Gregory. 1973. Duck haemoglobin synthesis in frog cells. The translation and assay of reticulocyte 9S RNA in oocytes of *Xenopus laevis*. Eur. J. Biochem. 34:219.

74. Berns, A. J. M., M. Kraaikamp, H. Bloemendal, and C. D. Lane. 1972. Calf crystallin synthesis in frog cells: the translation of lens-cell 14S RNA in oocytes. Proc. Natl. Acad. Sci. USA 69:1606.

75. Jilka, R. L., R. L. Cavalieri, L. Yaffe, and S. Pestka. 1977. Synthesis and glycosylation of the MOPC-46B immunoglobulin kappa chain in *Xenopus laevis* oocytes. Biochem. Biophys. Res. Commun. 79:625.

76. Kindas-Mugge, I., G. Kreil, and C. D. Lane. 1973. Insect protein synthesis in frog cells: the translation of honey bee promellitin messenger RNA in *Xenopus* oocytes. J. Mol. Biol. 87:451.

77. Stevens, R. and A. Williamson. 1972. Specific IgG mRNA molecules from myeloma cells in heterogeneous nuclear and cytoplasmic RNA containing poly A. Nature 239:143.

78. Labarca, C. and K. Paigen. 1977. mRNA-directed synthesis of catalytically active mouse β-glucuronidase in *Xenopus* oocytes. Proc. Natl. Acad. Sci. USA 74:4462.

79. Lebleu, B., E. Hubert, J. Content, L. DeWit, I. A. Braude, and E. DeClercq. 1978. Translation of mouse interferon mRNA in *Xenopus laevis* oocytes and in rabbit reticulocyte lysates. Biochem. Biophys. Res. Commun. 82: 665.

80. Reynolds, F. H., Jr., E. Prekumar, and M. P. Pitha. 1975. Interferon activity produced by translation of human interferon messenger RNA in cell-free ribosomal systems and in *Xenopus* oocytes. Proc. Natl. Acad. Sci. USA 72:4881.

81. Huez, G., G. Marbaix, D. Gallwitz, E. Weinberg, R. Devos, E. Hubert, and Y. Cleuter. 1978. Functional stabilization of HeLa cell histone messenger RNA's injected into *Xenopus* oocytes by 3'-OH polyadenylation. Nature 271:572.

82. Ghysbael, J., E. Hubert, M. Travnicek, D. P. Bolognesi, A. Burny, Y. Cleuter, G. Huez, R. Kettmann, G. Marbaix, D. Portetelle, and H. Chantrenne. 1977. Frog oocytes synthesize and completely process the precursor polypeptide to virion structural proteins after microinjection of avian myeloblastosis virus RNA. Proc. Natl. Acad. Sci. USA 74:3230.

83. Bloemendal, H., and J. B. Gurdon. 1972. Unpublished observation.

84. Wu, T., S. Gilmour, G. Dixon, and J. B. Gurdon. 1972. Unpublished observation.

85. Brown, D. D. and J. B. Gurdon. 1977. High fidelity transcription of 5S DNA injected into *Xenopus* oocytes. Proc. Natl. Acad. Sci. USA 74:2064.

86. Brown, D. D. and J. B. Gurdon. 1978. Cloned single repeating units of 5S DNA direct accurate transcription of 5S RNA when injected into *Xenopus* oocytes. Proc. Natl. Acad. Sci. USA 75:2849.

87. Kressmann, A., S. G. Clarkson, V. Pirrotta, and M. L. Birnstiel. 1978. Transcription of cloned tRNA gene fragments and subfragments injected into the oocyte nucleus of *Xenopus laevis*. Proc. Natl. Acad. Sci. USA 75: 1176.

88. DeRobertis, E. M. and M. V. Olson. 1979. Transcription and processing of cloned yeast tyrosine tRNA genes injected into frog oocytes. Nature 278: 137.

89. Mertz, J. E. and J. B. Gurdon. 1977. Purified DNA's are transcribed after microinjection into *Xenopus* oocytes. Proc. Natl. Acad. Sci. USA 74:1502.

90. Diacumakos, E. G., S. Holland, and P. Pecora. 1970. A microsurgical methodology for human cells in vitro: evolution and applications. Proc. Natl. Acad. Sci. USA 65:911.

91. Graessmann, A., M. Graessmann, E. Hoffmann, J. Niebel, G. Brandner, and N. Mueller. 1974. Inhibition by interferon of SV40 tumor antigen formation in cells injected with SV40 cRNA transcribed in vitro. FEBS Lett. 39: 249.

92. Graessmann, M. and A. Graessmann. 1976. "Early" simian-virus-40 specific RNA contains information for tumor antigen formation and chromatin replication. Proc. Natl. Acad. Sci. USA 73:366.

93. Tijan. R., G. Fey, and A. Graessmann. 1978. Biological activity of puri-

fied simian virus 40 T-antigen proteins. Proc. Natl. Acad. Sci. USA 75: 1279.

94. Mueller, C., A. Graessmann, and M. Graessmann. 1978. Mapping of early SV40-specific functions by microinjection of different early viral DNA fragments. Cell 15:1579.

95. Graessmann, A., M. Graessmann, and C. Mueller. 1977. Regulatory function of simian virus 40 DNA replication for late viral gene expression. Proc. Natl. Acad. Sci. USA 74:4831.

96. McClain, D. A., P. F. Maness, and G. M. Edelman. 1978. Assay for early cytoplasmic effect of the *src* gene product of Rous sarcoma virus. Proc. Natl. Acad. Sci. USA 75:2750.

97. Gershey, E. L. and E. G. Diacumakos. 1978. Simian virus 40 production after viral uncoating in the CV-1 cell nucleus. J. Virol. 28:415.

98. Anderson, W. F., M. C. Willing, D. E. Axelrod, T. V. Gopalakrishnan, and E. G. Diacumakos. 1978. A new approach in the search for globin gene regulatory factors. In *Cellular and Molecular Regulation of Hemoglobin Switching,* G. Stamatoyannapoulos and A. W. Nienhuis (Eds.). Grune & Stratton, New York, p. 779.

99. Stacey, D. W. and V. G. Allfrey. 1976. Microinjection studies of duck globin messenger RNA translation in human and avian cells. Cell 9:725.

100. Huez, G., C. Bruck, and Y. Cleuter. 1981. Translational stability of native and deadenylated rabbit globin mRNA injected into HeLa cells. Proc. Natl. Acad. Sci. USA 78:908.

101. Capecchi, M. R. 1980. High efficiency transformation by direct microinjection of DNA into cultured mammalian cells. Cell 22:479.

102. Anderson, W. F., L. Killos, L. Sanders-Haigh, P. J. Kretshmen, and E. G. Diacumakos. 1980. Replication and expression of thymidine kinase and human globin genes microinjected into mouse fibroblasts. Proc. Natl. Acad. Sci. USA 77:5399.

103. Chambers, R. 1922. New apparatus and methods for the dissection and injection of living cells. Anat. Rec. 24:1.

104. Furusawa, M., T. Nishimura, M. Yamaizumi, and Y. Okada. 1974. Injection of foreign substances into single cells by cell fusion. Nature 249:449.

105. Loyter, A., N. Zakai, and R. G. Kulka. 1975. Ultramicroinjection of macromolecules or small particles into animal cells. J. Cell Biol. 66:292.

106. Schlegel, R. A. and M. C. Rechsteiner. 1978. Red cell-mediated microinjection of macromolecules into mammalian cells. Methods Cell Biol. 20:341.

107. Wille, W. and K. Willecke. 1976. Retention of purified proteins in resealed human erythrocyte ghosts and transfer by fusion into cultured murine recipient cells. FEBS Lett. 65:59.

108. Schlegel, R. A. and M. C. Rechsteiner. 1975. Microinjection of thymidine kinase and bovine serum albumin into mammalian cells by fusion with red blood cells. Cell 5:371.

109. Capecchi, M. R., R. A. Von der Haar, N. E. Capecchi, and M. M. Sveda.

1977. The isolation of a suppressible nonsense mutant in mammalian cells. Cell 12:371.

110. Yamaizumi, M., T. Uchida, Y. Okada, and M. Furuzawa. 1978. Neutralization of diphtheria toxin in living cells by microinjection of antifragment A contained within resealed erythrocyte ghosts. Cell 12:227.

111. Wille, W. and K. Willecke. 1976. Retention of purified proteins in resealed human erythrocyte ghosts and transfer by fusion into cultured murine recipient cells. FEBS Lett. 65:59.

112. Schlegel, R. A. and M. C. Rechsteiner. 1975. Microinjection of thymidine kinase and bovine serum albumin into mammalian cells by fusion with red blood cells. Cell 5:371.

113. Kaltoft, K. and J. E. Celis. 1978. Ghost mediated transfer of human hypoxanthine-guanine phosphoribosyltransferase into deficient Chinese hamster ovary cells by means of polyethylene glycol-induced fusion. Exp. Cell Res. 115:423.

114. Mercer, W. E. and R. A. Schlegel. 1979. Phytohemaglutinin enhancement of cell fusion reduces polyethylene glycol cytotoxicity. Exp. Cell Res. 120:417.

115. Furusawa, M., M. Yamaizumi, T. Nishimura, T. Uchida, and Y. Okada. 1976. Use of erythrocyte ghosts for injection of substances into animal cells by cell fusion. In *Methods in Cell Biology*, Vol. 14, D. M. Prescott (Ed.). Academic Press, New York, p. 73.

116. Kriegler, M. P., J. D. Griffin, and D. M. Livingston. 1978. Phenotypic complementation of the SV40 mutant defect in viral DNA synthesis following microinjection of SV40 antigen. Cell 14:983.

117. Kaltoft, K., J. Zeuthen, F. Engback, P. W. Piper, and J. E. Celis. 1976. Transfer of tRNA's to somatic cells mediated by Sendai virus-induced fusion. Proc. Natl. Acad. Sci. USA 73:2793.

118. Celis, J. E. 1977. Injection of tRNA's into somatic cells. Search for in vivo systems to assay potential nonsense mutations in somatic cells. Brookhaven Symp. Quant. Biol. 29:178.

119. Schlegel, R. A., P. Iversen, and M. C. Rechsteiner. 1978. The turnover of tRNA's microinjected into animal cells. Nucl. Acid Res. 5:3751.

120. Celis, J. E., K. Kaltoft, A. Celis, R. Fenwick, and C. T. Caskey. 1979. Microinjection of tRNA's into somatic cells. In *Nonsense Mutations and tRNA Suppressions*, J. E. Celis and J. D. Smith (Eds.). Academic Press, London, p. 254.

121. Jonak, G. J. and M. Mora. 1980. Rapid identification of microinjected mammalian cells by fusion with loaded chicken erythrocytes. In *Introduction of Macromolecules into Viable Mammalian Cells*, R. Baserga, C. Croce, and G. Rovera (Eds.). Alan R. Liss, New York, p. 157.

122. Zimmermann, V., G. Pilwat, C. Holzapfel, and K. Rosenheck. 1976. Electrical hemolysis of human and bovine red blood cells. J. Membr. Biol. 30:135.

123. Zimmermann, V., F. Riemann, and G. Pilwat. 1976. Enzyme loading of

electrically homogenous human red blood cells ghosts prepared by dielectric breakdown. Biochim. Biophys. Acta 436:460.

124.   Yamaizumi, M., T. Uchida, and Y. Okada. 1979. Macromolecules can penetrate the host cell membrane during the early period of incubation with HVJ (Sendai virus). Virology 95:218.

125.   Uchida, T., M. Yamaizumi, and Y. Okada. 1977. Reassembled HVJ (Sendai virus) envelopes containing nontoxic mutant proteins of diphtheria toxin show toxicity to mouse L cells. Nature 266:839.

126.   Mukherjee, A. B., S. Orloff, J. D. Butler, T. Triche, P. Lally, and J. Schulman. 1978. Entrapment of metaphase chromosomes into phospholipid vesicles (lipochromosomes): carrier potential in gene transfer. Proc. Natl. Acad. Sci. USA 75:163.

127.   Kondorosi, E. and E. Duda. 1980. Introduction of foreign genetic maternal into cultured mammalian cells by liposomes loaded with isolated nuclei. FEBS Lett. 120:37.

128.   Sessa, G. and G. Weissmann. 1968. Phospholipid spherules (liposomes) as a model for biological membranes. J. Lipid Res. 9:310.

129.   Sessa, G. and G. Weissmann. 1970. Incorporation of lysozyme into liposomes. A model for structure-linked latency. J. Biol. Chem. 245:3295.

130.   Weissmann, G., D. Bloomgarden, R. Kaplan, C. Cohen, S. Hoffstein, T. Collins, A. Gotlieb, and D. Nagle. 1975. A general method for the introduction of enzymes, by means of immunoglobulin-coated liposomes, into lysosomes of deficient cells. Proc. Natl. Acad. Sci. USA 72:88.

131.   Fishman, Y. and N. Citri. 1975. L-Asparaginase entrapped in liposomes: preparation and properties. FEBS Lett. 60:17.

132.   Braidman, I. P. and G. Gregoriadis. 1977. Rapid partial purification of placental glucoceribroside β-glucosidase and its entrapment in liposomes. Biochem. J. 164:439.

133.   Adrian, G. and L. Huang. 1979. Entrapment of proteins in phosphatidylcholine vesicles. Biochemistry 18:5610.

134.   Ismail, G., L. A. Boxer, and R. L. Baehner. 1979. Utilization of liposomes for correction of the metabolic and bactericidal deficiencies in chronic granulomatous disease. Pediatr. Res. 13:769.

135.   Dapergolas, G. and G. Gregoriadis. 1977. The effect of liposomal lipid composition on the fate and effect of liposome-entrapped insulin and tubocurarine. Biochem. Soc. Trans. 5:1383.

136.   Cohen, C. M., G. Weissmann, S. Hoffstein, Y. C. Awasthi, and S. K. Srivastava. 1976. Introduction of purified hexosaminidase A into Tay-Sachs leukocytes by means of immunoglobulin-coated liposomes. Biochemistry 15:452.

137.   Gregoriadis, G. and R. A. Buckland. 1973. Enzyme-containing liposomes alleviate a model for storage disease. Nature 244:170.

138.   Cohn, Z. A. and B. A. Ehrenreich. 1969. Uptake, storage, and intracellular hydrolysis of carbohydrates by macrophages. J. Exp. Med. 129:201.

139.   Graham, Jr., R. C., and M. J. Karnovsky. 1966. The early stages of ab-

sorption of injected horseradish peroxidase in the proximal tubules of mouse kidney. Ultrastructural cytochemistry by a new technique. J. Histochem. Cytochem. 14:291.

140. Magee, W. E., C. W. Goff, J. Schoknecht, M. D. Smith, and K. Cherian. 1974. The interaction of cationic liposomes containing entrapped horseradish peroxidase with cells in culture. J. Cell Biol. 63:492.

141. Weissmann, G., C. Cohen, and S. Hoffstein. 1977. Introduction of enzymes by means of liposomes, into non-phagocytic human cells in vitro. Biochim. Biophys. Acta 498:375.

142. Weissmann, G., R. B. Zurier, P. J. Spieler, and I. M. Goldstein. 1971. Mechanisms of lysosomal enzyme release from leukocytes exposed to immune complexes and other particles. J. Exp. Med. 134:149s.

143. Weissmann, G., A. Brand, and E. C. Franklin. 1974. Interaction of immunoglobulin with liposomes. J. Clin. Invest. 53:536.

144. Reynolds, G. D., H. J. Baker, and R. H. Reynolds. 1978. Enzyme replacement using liposome carriers in feline G_{M1} gangliosidosis fibroblasts. Nature 275:754.

145. Klebanoff, S. J. 1975. Antimicrobial mechanisms in neutrophilic polymorphonuclear leukocytes. Semin. Hematol. 12:117.

146. Sone, S., G. Poste, and I. J. Fidler. 1980. Rat alveolar macrophages are susceptible to activation by free and liposome-encapsulated lymphokines. J. Immunol. 124:2197.

147. North, R. J. 1978. The concept of the activated macrophage. J. Immunol. 121:806.

148. Poste, G., R. Kirsh, W. Fogler, and I. J. Fidler. 1979. Activation of tumoricidal properties in mouse macrophages by lymphokines encapsulated in liposomes. Cancer Res. 39:881.

149. Gardas, A. and I. Macpherson. 1979. Microinjection of ricin entrapped in unilamellar liposomes into a ricin-resistant mutant of baby hamster kidney cells. Biochim. Biophys. Acta 584:538.

150. Bartzri, S. and E. D. Korn. 1973. Single bilayer liposomes prepared without sonication. Biochim. Biophys. Acta 298:1015.

151. Allebach, E. S., R. J. Mannino, W. A. Strohl, and K. Raska, Jr. 1980. Stimulation of DNA synthesis by an adenovirus type 12 T antigen fraction containing protein kinase activity and encapsulated in liposomes. Virology 107:240.

152. Kataoka, T., J. R. Williamson, and S. C. Kinsky. 1973. Release of macromolecular markers (enzymes) from liposomes treated with antibody and complement. An attempt at correlation with electron microscopic observations. Biochim. Biophys. Acta 298:158.

153. Finkelstein, M. C., J. Maniscalco, and G. Weissmann. 1978. Entrapment of soy bean trypsin inhibitor and α_1-antitrypsin by multilamellar liposomes. Anal. Biochem. 89:400.

154. Gregoriadis, G., P. D. Leathwood, and B. E. Ryman. 1971. Enzyme entrapment in liposomes. FEBS Lett. 14:95.

155. Anderson, P., J. Vilcek, and G. Weissmann. 1981. Entrapment of human leukocyte interferon in the aqueous interstices of liposomes. Infect. Immun. 31:1099.

156. Finkelstein, M. C. and G. Weissmann. 1979. Enzyme replacement via liposomes. Variations in lipid composition determine liposomal integrity in biological fluids. Biochim. Biophys. Acta 587:202.

157. Van Rosijen, N. and R. Van Nieuwmegen. 1979. Attempts to study the localization of liposomes and liposome-entrapped antigen in the spleen. Acta Histochem. 65:41.

158. Galzigna, L., L. Garbin, and A. Burlina. 1979. Liposome-incorporated enzymes: studies on amylase. Clin. Biochem. 12:267.

159. Colley, C. M. and B. E. Ryman. 1976. The use of liposomally entrapped enzyme in the treatment of an artificial storage condition. Biochim. Biophys. Acta 451:417.

160. Sengupta, S. and S. Rous. 1978. Inhibition of the hyperglycemic effect of cAMP by intravenous injection to mice of carbonic anhydrase entrapped in liposomes. Biochem. Biophys. Res. Commun. 8:795.

161. Tyrrell, D. A., B. E. Ryman, B. R. Keeton, and V. Dubowitz. 1979. Use of liposomes in treating type II glycogenosis. Br. Med. J. 2:88.

162. Steger, L. D. and R. J. Desnick. 1977. Enzyme therapy. VI. Comparative in vivo fates and effects on lysosomal integrity of enzyme entrapped in negatively and positively charged liposomes. Biochim. Biophys. Acta 464:530.

163. Gregoriadis, G. and E. D. Neerunjun. 1974. Control of the rate of hepatic uptake and catabolism of liposome-entrapped proteins injected into rats. Possible therapeutic applications. Eur. J. Biochem. 47:179.

164. Colley, C. M. and B. E. Ryman. 1974. A model for a lysosomal storage disease and a possible method of therapy. Biochem. Soc. Trans. 2:871.

165. Neerunjun, E. D. and G. Gregoriadis. 1976. Tumour regression with liposome-entrapped asparaginase: some immunological advantages. Biochem. Soc. Trans. 4:133.

166. Gregoriadis, G., D. Putman, L. Louis, and D. Neerunjun. 1974. Comparative effect and fate of non-entrapped and lysosome-entrapped neuraminidase injected into rats. Biochem. J. 140:323.

167. Gregoriadis, G. and B. E. Ryman. 1972. Fate of protein-containing liposomes injected into rats. An approach to the treatment of storage diseases. Eur. J. Biochem. 24:485.

168. Patel, H. M. and B. E. Ryman. 1974. α-Mannosidase in zinc-deficient rats: possibility of liposomal therapy in mannosidosis. Biochem. Soc. Trans. 2:1014.

169. Dapergolas, G., E. D. Neennjun, and G. Gregoriadis. 1976. Penetration of target areas in the rat by liposome-associated bleomycin, glucose oxidase and insulin. FEBS Lett. 63:235.

170. Belchetz, P. E., J. C. W. Crawley, I. P. Braidman, and G. Gregoriadis. 1977. Treatment of Gaucher's disease with liposome-entrapped glucocerebroside:β-glucosidase. Lancet 2:116.

171. Patel, H. M., N. G. Harding, F. Logue, C. Kesson, A. C. MacCuish, J. C. MacKenzie, B. E. Ryman, and I. Scobie. 1978. Intrajejunal absorption of liposomally entrapped insulin in normal man. Proc. Biochem. Soc. Trans. 6:784.

172. Hudson, L. D. S., M. B. Fiddler, and R. J. Desnick. 1979. Enzyme therapy. X. Immune response induced by enzyme and buffer-loaded liposomes in C3H/HeJ mice. J. Pharmacol. Exp. Ther. 208:507.

173. Dapergolas, G. and G. Gregoriadis. 1976. Hypoglycaemic effect of liposome-entrapped insulin administered intragastrically into rats. Lancet 2: 824.

174. Weissmann, G., C. Cohen, and S. Hoffstein. 1976. Introduction of missing enzymes into the cytoplasm of cultured mammalian cells by means of fusion prone liposomes. Trans. Acad. Am. Physicians 89:171.

175. Fogler, W. E., A. Raz, and I. J. Fidler. 1980. In situ activation of murine macrophages by lymphokine containing liposomes. Cell. Immunol. 53: 214.

176. Braidman, I. and G. Gregoriadis. 1976. Preparation of glucocerebroside β-glucosidase for entrapment in liposomes and treatment of patients with adult Gaucher's disease. Biochem. Soc. Trans. 4:259.

177. Magee, W. E. and O. V. Miller. 1972. Liposomes containing antiviral antibody can protect cells from virus infection. Nature 235:339.

178. Ostro, M. J., D. Giacomoni, and S. Dray. 1977. Incorporation of high M.W. RNA into large artificial lipid vesicles. Biochem. Biophys. Res. Commun. 76:836.

179. Dimitriadis, G. J. 1978. Entrapment of RNA in liposomes. FEBS Lett. 86:289.

180. Dimitriadis, G. J. 1978. Introduction of RNA into cells by means of liposomes. Nucl. Acid Res. 5:1381.

181. Ostro, J. J., D. Lavelle, W. Paxton, B. Matthews, and D. Giacomoni. 1980. Parameters affecting the liposome mediated insertion of into eukaryotic cells in vitro. Arch. Biochem. Biophys. 201:392.

182. Kulpa, C. F. and T. J. Tinghitella, 1976. Encapsulation of poly U in phospholipid vesicles. Life Sci. 19:1879.

183. Wreshner, D. H., G. Gregoriadis, D. B. Gunner, R. R. Dourmashkin. 1978. Entrapment of mRNA and rRNA in large monolamellar liposomes derived from hybrid small monolamellar liposomes. Biochem. Soc. Trans. 6:930.

184. Deamer, D. W. and A. D. Bangham. 1976. Large volume liposomes by an ether evaporation method. Biochim. Biophys. Acta 443:629.

185. Wilson, T., D. Papahadjopoulos, and R. Taber. 1979. The introduction of poliovirus RNA into cells via lipid vesicles. Cell 17:77.

186. Matthews, B., S. Dray, J. Widholm, and M. Ostro. 1979. Liposome mediated transfer of bacterial RNA into carrot protoplasts. Planta 145:37.

187. Ostro, M. J., D. Giacomoni, D. Lavelle, W. Paxton, and S. Dray. 1978. Evidence for translation of rabbit globin mRNA after liposome mediated insertion into a human cell line. Nature 274:921.

188. Dimitriadis, G. J. 1978. Translation of rabbit globin mRNA introduced
      by liposomes into mouse lymphocytes. Nature 274:923.
189. Magee, W. E., J. H. Cromenberger, and D. E. Thor. 1978. Marked stimu-
      lation of lymphocyte mediated attack on tumor cells by target-directed
      liposomes containing immune RNA. Cancer Res. 38:1173.
190. Magee, W. E., M. L. Talcott, S. X. Staub, and C. Y. Vriend. 1976. A com-
      parison of negatively and positively charged liposomes containing entrapped
      poly(I)· poly(C) for interferon production in mice. Biochem. Biophys.
      Acta 451:610.
191. Fraley, R. T., C. S. Fornary, and S. Kaplan. 1979. Entrapment of a bac-
      terial plasmid in phospholipid vesicles: potential for gene transfer. Proc.
      Natl. Acad. Sci. USA 76:3348.
192. Fraley, R. T., S. Subramani, P. Berg, and D. Papahadjopoulos. 1980.
      Introduction of liposome-encapsulated SV40 DNA into cells. J. Biol.
      Chem. 255:10431.
193. Hoffman, R. M., L. M. Margolis, and L. D. Bergelson. 1978. Binding and
      entrapment of high M.W. DNA by lecithin liposomes. FEBS Lett. 93:365.
194. Mannino, R. J., E. I. Allerbach, and W. A. Strohl. 1979. Encapsulation
      of high M.W. DNA in large unilamellar phospholipid vesicles. FEBS Lett.
      101:229.
195. Cassels, A. C. 1978. Uptake of charged lipid vesicles by isolated tomato
      protoplasts. Nature 275:760.
196. Lurquin, P. F. 1979. Entrapment of plasmid DNA by liposomes and their
      interaction with plant protoplasts. Nucl. Acid Res. 6:3773.
197. Wong, T. K., C. Nicolau, and P. H. Hofschneider. 1980. Appearance of
      β lactamase activity in animal cells upon liposome mediated-gene transfer.
      Gene 10:87.
198. Wilson, T., D. Papahadjopoulos, and R. Taber. 1977. Biological proper-
      ties of poliovirus encapsulated in lipid vesicles: antibody resistance and in-
      fectivity in virus resistant cells. Proc. Natl. Acad. Sci. USA 74:3471.
199. Dimitriadis, G. J. 1979. Entrapment of plasmid DNA in liposomes. Nucl.
      Acid Res. 6:2697.

6

Immunologic Aspects of Liposomes

Carl R. Alving and Roberta L. Richards / Walter Reed Army Institute of
Research, Washington, District of Columbia

INTRODUCTION

The field of immunology contains immensely complicated systems of proteins
and protein-cell interactions. The important roles of lipids in immunological
phenomena, which were largely ignored in the past, have recently become topics
of major interest. Part of this awakened interest about lipids has occurred because of the growing realization, among both immunologists and membrane
chemists, of the immense utility of *liposomes* as a model and as a practical tool
for studying numerous immunological topics related to membranes.

As shown by the scope of this book, liposome theory and technology can be
utilized in virtually any biological discipline, but in the early years of liposome
development, as judged by the number of publications, immunological questions
were not a major focus. In surveying the literature for this review, we believe
that we were able to find more than 90% of all articles, even peripheral articles,
published in the field of immunologic aspects of liposomes. The total number

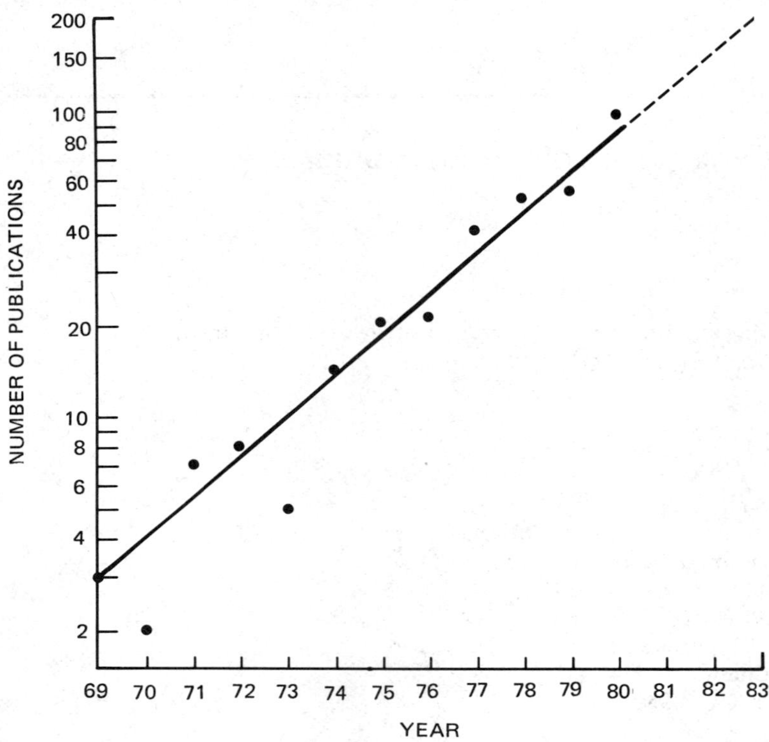

Figure 1 Publications per year in the field of immunologic aspects of liposomes. Only full papers and reviews are included.

initially collected amounted to more than 350 articles from more than 100 laboratories. As shown in Figure 1, the field is growing logarithmically. According to our survey, 8 articles were published in 1972, 95 were published in 1980, 150 are predicted for 1982, and 200 for 1983. More than 75% of the literature has appeared in the past four years. This review is based on the literature that we discovered from 1968 up to approximately mid-April 1981.

Numerous other reviews, some having particular frames of interest, have been published in this field in the past four years (1-12). The reader is referred to a previous comprehensive review by one of us in 1977 (13). Papers from a recent symposium on liposomes and immunobiology have been published (14).

Definition of Liposomes

Prior to the use of liposomes, immunological studies on lipid haptens commonly employed lipid emulsions consisting of empirically derived combinations of "auxiliary lipids," usually lecithin and cholesterol, as carriers (reviewed in Refs. 13 and 15).

The methods employing auxiliary lipids as supporting frameworks for immunological reactions were useful, but they suffered from the disadvantage that lipid emulsions having different molar ratios of lecithin and cholesterol and glycolipid have markedly different physical and geometric properties. The structure of auxiliary lipids containing high concentrations of cholesterol, as visualized microscopically, is irregular and does not lend itself easily to geometric analysis (16). Differences of surface area obviously could influence the number, and possibly the density, of epitopes for antibody binding; and relatively subtle changes of lipid composition and fluidity can have profound effects on complement fixation (17-22).

The effects of auxiliary lipids on immune reactions were unpredictable. The combination of lipids that worked with antibodies against one lipid hapten might not work with another, and the optimum lipid formulation had to be worked out in advance (13,15). Although it could not have been recognized in the past, certain formulations of auxiliary lipids (e.g., lecithin:cholesterol in equimolar concentrations) now might be referred to legitimately as liposomes. Other auxiliary lipid formulations, such as those that lack phospholipids entirely, would not qualify, at least not for the purposes of this review, as conventional liposomes. Other examples of *nonliposomal* auxiliary lipid combinations are lecithin:cholesterol in ratios ranging from 1:3 (w/w) to 1:9 (w/w) (or approximately as high as 1:18 by molar ratio). Auxiliary lipid emulsions still are widely employed for helping to characterize antiglycolipid antibodies (e.g., Refs. 23-28) and antiphospholipid antibodies (e.g., Refs. 29 and 30).

In order to minimize confusion with work on auxiliary lipids, in this review we have decided arbitrarily that the authors should *recognize* first, that their work fits into the domain of liposome technology; second, that the lipid mixture must contain a phospholipid, such as lecithin or sphingomyelin; third, that the molar ratio of phospholipid to cholesterol of the liposomes generally should not be less than 1:2; fourth, that the lipid suspension should contain at least 50% water, by weight, compared to the phospholipid; and fifth, that there should be essentially no organic solvent in the final preparation. In rare instances, we included discussions of preparations that had a small amount of organic solvent.

Fortunately, for reviewers who cannot always evaluate lipid dispersions at a distance, if an author does not recognize when (or if) he is using liposomes, his work usually generates much less interest from a liposome standpoint. In most cases, the concept of liposomes is almost as important as the fact of liposomes. Unfortunately, because of the long history of auxiliary lipids in immunology and the popular emergence of liposomes serving similar functions, some authors now refer to virtually any lipid emulsion as liposomes, even if the emulsion lacks phospholipids. In our opinion, it is useful to restrict the term "liposomes" to lipid dispersions that meet the criteria listed above.

ANTIBODIES AGAINST LIPOSOMAL ANTIGENS

Numerous substances have been employed as antigens in liposomes. Table 1 lists nearly every example and nearly every reference that we could find in which each type of liposome-associated substance was used in any capacity as an antigen.

Glycolipids

Glycolipids have been recognized as important cellular constituents since the discovery of galactocerebroside (galactosyl ceramide) and sulfatide (sulfogalactosyl ceramide), by Thudichum in the nineteenth century (31). Forssman antigen (pentahexosyl ceramide) was discovered as the major antigen of sheep erythrocytes in 1911 (32). In view of this long history, it seems satisfying and appropriate that glycosphingolipids, particularly Forssman antigen and galactocerebroside, were the first antigens employed in studies on immunological properties of liposomes (17,33-37).

The use of glycolipids as liposomal antigens was fortuitous. These molecules are widely distributed in nature and a sizable literature exists on their chemistry and biology. Many glycolipids are easily extracted and purified (38,39). Several brain glycolipids, including galactocerebroside, mono- and digalactosyl diglycerides, sulfatide, gangliosides, and related compounds and even synthetic glycolipids, can be purchased. Antisera against glycolipids are readily obtained (40), and anti-Forssman (anti-sheep erythrocyte serum, also known as hemolysin) is widely available commercially for complement fixation assays. Glycolipids may be important for intercellular recognition (41,42) and they may serve as cellular receptors for numerous proteins (43) and as blood group antigens (44-46). Antibodies against certain glycolipids (particularly gangliosides) can have effects on brain function, development, and behavior (47-50). Finally, and perhaps most important, a large literature exists on the immunology of glycolipids (reviews in Refs. 13, 15, 51, and 52).

Numerous techniques have been developed in the past 70 years for measuring antibody activity against glycolipids (see the reviews cited above). A substantial literature has now also developed on the use of liposomes for routine assay of antiglycolipid antibodies (see the references in Table 1). Many laboratories, ours

Table 1 Substances Utilized as Antigens in Liposomes

Substance	References
Glycolipids	
Galactosyl ceramide (galacto-derebroside)	17, 18, 36, 37, 65, 66, 72, 75, 77, 79, 180, 213a, 236, 440
Lactosyl ceramide [ceramide dihexoside (CDH)]	3, 27, 37, 65, 72, 75, 180
Ceramide trihexoside (CTH)	65, 67, 68, 72
Globoside I	3, 36, 37, 55, 56, 67, 72, 73, 162, 167, 168, 253, 441
Paragloboside	442
Cytolipin R	180
Forssman	3, 18, 33–37, 55, 56, 59, 72, 73, 155, 157–159, 161–163, 172, 179, 184, 205, 208, 216, 253, 254, 257, 284, 285, 441, 443–445
Galactosyl diglycerides	13, 17, 71, 72
Sulfoglycolipids	78, 446
Blood type A glycolipid	56
Human white matter lipids	71
Virus glycolipid	435
Schistosomal glycolipids	447
Gangliosides and asialogangliosides	
Brain	17, 79, 180, 192, 445, 448
Erythrocyte	442
Sea urchin spermatozoa	445, 448
G_{M1}	70, 72, 85, 380
G_{D1a}	70
G_{D1b}	72, 380
G_{M2}	72, 173, 445
Asialo G_{M2}	3, 55, 58, 72, 73
G_{M3}	3, 445
Phospholipids and phospholipid derivatives	
Cardiolipin	19, 165, 181, 185, 188, 311
Phosphatidylcholine or sphingomyelin	57, 74, 81–85

Table 1 (continued)

Substance	References
Phospholipids and phospholipid derivatives (continued)	
DNP-PE derivatives	93-98, 101, 148, 154, 156, 169, 175, 177, 193, 200, 315, 316, 356-358, 363-365, 398, 399, 412, 416, 421-423, 426-428, 430, 431, 449
Hydroxyiodonitrophenyl PE derivative	450
Spin-label-PE derivatives	175, 315, 316, 364
Fluorescein-PE derivatives	95, 145, 422, 428, 430
ABA-Tyr-PE derivatives	425, 426, 428, 429, 449
ABS-Tyr-PE derivatives	450
Phosphocholine-PE derivatives	98, 399
Theophylline-PE derivative	191
Lipopolysaccharides and lipid A	
LPS	99, 100, 162, 183, 404
Lipid A	81, 82, 84, 85, 100, 102
Miscellaneous lipids	
Cholesterol and cholesterol derivatives	84, 87-89
Dicetylphosphate	84
MDP-fatty acid derivatives	85
Proteins	
Albumin	11, 335, 366, 402, 405, 406, 408, 414, 415, 450-454
Gamma globulins, immunoglobulins, or antibodies	11, 94, 116, 149, 153, 414, 415
Dansylated proteins	145, 149, 154
Enzymes	105, 109, 112-114, 366
Glycophorin	117
Erythrocyte glucose transporter	118
Thyroglobulin	160
Myelin proteins	118, 455
Diphtheria toxoid	104, 106, 401

Table 1 (continued)

Substance	References
Proteins (continued)	
Cholera toxin	70, 84, 85, 152
Malaria antigen	394, 407, 409
Virus antigens	107, 108, 120–127, 132, 290, 291, 293, 295, 297–300, 305, 306, 456–462
Histocompatibility antigens	125, 128–135, 137, 138, 150, 290–294, 296–307, 461–463

included, rarely study purified glycolipids as antigens outside of liposomes. The major rationale for this has been that most simple glycolipids rarely, if ever, occur naturally outside of a lipid bilayer membrane.

Beginning with studies on purified Forssman glycolipid (35), the importance of the oligosaccharide moiety of the liposomal antigen on the specificity of antibody binding was established (17,35–37). Forssman (ceramide pentahexoside) and globoside I (ceramide tetrahexoside) are nearly identical glycolipids, differing only in the presence of a fifth sugar, α-N-acetylgalactosamine (GalNAc), at the terminal end of Forssman oligosaccharide, while the terminal sugar of globoside is β-GalNAc (53) (Fig. 2). Despite the similarity of these glycolipids, in using a double diffusion technique, cross-reactivity of anti-Forssman or antigloboside serum was not detected, respectively, with purified globoside or Forssman glycolipid (53). Double diffusion techniques using lipids can be unreliable due to the insoluble or micellar nature of the antigens (54). However, when liposomes containing purified glycolipids were used, absence of immune cross-reactivity between Forssman and globoside was confirmed (36,37,55,56). Purified antibodies against asialo G_{M2} (terminal β-GalNAc, Fig. 2) did not cross-react with either liposomal Forssman or globoside (3). This indicates that the terminal sugar is not necessarily the sole determinant of antiglycolipid specificity. The role of the terminal sugar of liposomal glycolipid was carefully studied by Young et al. using monoclonal IgM and IgG3 antibodies against asialo G_{M2} (58).

Liposomes also were used by Alving et al. to characterize the specificity of a monoclonal IgM Waldenström macroglobulin (McG) having anti-Forssman specificity (37,59). The McG paraprotein did not cross-react with globoside in liposomes (37). In contrast, when tested at $4°C$ with ficin-treated human erythrocytes, Naiki and Marcus found that McG agglutinated the erythrocytes, and agglutination was inhibited equally by Forssman or globoside suspended in sodium taurocholate (60). The latter authors correctly raised a note of caution in inter-

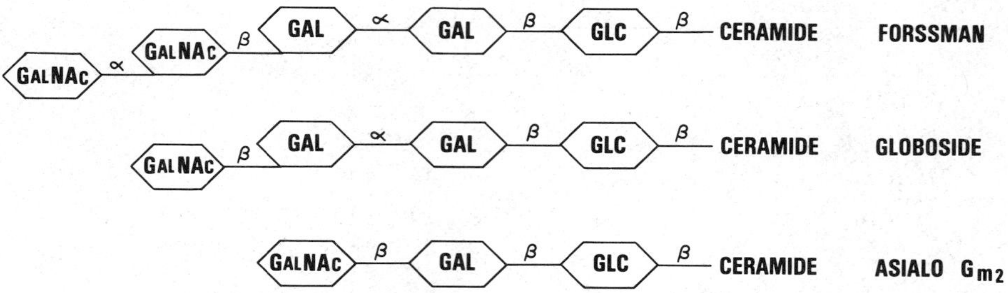

Figure 2 Structures of three glycolipids having terminal N-acetylgalactosamines. GalNAc, N-acetylgalactosamine; GAL, galactose; GLC, glucose.

preting the specificity of the McG antibody. It seems that the GalNAc of globoside might be presented differently when globoside is in liposomes at room temperature (where it *does not* bind McG, Ref. 37) than when the globoside is in ficin-treated erythrocytes at 4°C or in sodium taurocholate (where it *does* bind McG, Ref. 60). As a matter of interest, a human monoclonal anti-Forssman hybridoma antibody, produced after fusion of human lymphocytes with mouse myeloma cells, did not cross-react with globoside in a solid-phase radioimmunoassay (61). Rabbit anti-Forssman antibodies do not ordinarily react with globoside in trypsinized human erythrocytes (3,55).

From the disparate reports on anti-Forssman and antigloboside antibodies cited above it might appear that studies of antibody specificity performed with cells give fundamentally different results compared to those obtained with liposomes or with other noncellular techniques. This conclusion is not correct. Numerous factors, including temperature, epitope density, and steric hindrance by neighboring lipids or proteins, can influence antibody binding to glycolipids, and this can apply equally to cells or liposomes. To illustrate this point, it may be instructive to examine the immunological system in which ficin-treated erythrocytes are employed at 4°C for detecting anti-P blood group antibodies, as described by Marcus and colleagues. Anti-P antibodies are human cold agglutinins that react only at low temperature with human cells containing globoside (46,62). By this criterion, one type of antibody, Donath-Landsteiner antibodies, could be called "anti-P." The Donath-Landsteiner antibodies are cold agglutinins that are responsible for paroxysmal cold hemoglobinuria. They apparently have specificity for binding to either Forssman or globoside, or both (63). Donath-Landsteiner antibodies are assayed with normal (i.e., not enzyme-treated) erythrocytes at 4°C, and, depending on the antibody, binding can be inhibited best by either Forssman or globoside (63). Rabbit antibodies that are raised by active immunization against Forssman or globoside *do not* give positive results in the Donath-

Landsteiner test (63). In fact, rabbit antigloboside antibodies ordinarily react only with human erythrocytes that have been treated with a proteolytic enzyme (54). It is apparent that membrane-associated glycolipids can be expressed differently as antigens depending on the antibody and on the assay conditions employed, and perhaps on the physical state (i.e., fluidity or manner of orientation) of the glycolipid in the bilayer. It seems that despite the common ability to bind to globoside, there may be fundamental differences between conventional anti-P, Donath-Landsteiner, and rabbit antigloboside antibodies.

It has even been reported that the epitope density of a ganglioside (hematoside, or G_{M3}) that had been adsorbed to erythrocytes can determine whether the cells are agglutinated by a monoclonal anti-G_{M3} cold agglutinin at $0°C$ only, or at both 0 and $37°C$ (64). It is evident that regardless of whether cells or liposomes are being employed for immunological assay of glycolipids, the results must be interpreted in the light of current knowledge and theories relating to membrane immunology.

Several specificity studies with liposomes have also employed glycolipid antigens having a terminal galactose (Gal). These glycolipids include neutral glycolipids, such as galactosyl ceramide, lactosyl ceramide, ceramide trihexoside, and mono- and digalactosyl diglycerides; and gangliosides, such as mixed gangliosides, G_{M1}, and G_{D1b} (Fig. 3). Rabbit antisera, and even purified antibodies, against galactosyl ceramide [ceramide monohexoside (CMH)] cross-reacted readily with liposomal lactosyl ceramide [ceramide dihexoside (CDH)], mono- and digalactosyl diglycerides, and mixed gangliosides (17,65,66).

Purified rabbit anti-CMH or anti-CDH antibodies did not cross-react with liposomal ceramide trihexoside (CTH), and anti-CTH antibodies did not cross-react with CMH or CDH (65,67,68). The basis of this specificity, analogous to the differential specificity of Forssman and globoside, may have been anomeric because the terminal sugar of CMH and CDH is β-Gal and that of CTH is α-Gal. Anomericity is not the sole determinant of antigalactose specificity of antibodies against liposomal glycolipids. Antidigalactosyl diglyceride antibodies (terminal α-Gal) cross-reacted strongly with monogalactosyl diglyceride (terminal β-Gal), and as mentioned above, purified anti-CMH antibodies (terminal β-Gal) cross-reacted with digalactosyl diglyceride (17,65). However, antidigalactosyl diglyceride antiserum, as with anti-CTH, failed to react with liposomal CMH (17), and the absence of cross-reactivity in these cases might have been due to anomeric differences. Antibody specificities similar to some of those described above with liposomal glycolipids have also been reported with glycolipids in natural membranes (69).

When using liposomes, or even other techniques, for assaying specificities of antiglycolipid antisera, it is extremely easy to be confused or fooled by the presence of naturally-occurring antiglycolipid antibodies. Natural antibodies in rabbit or human sera have been described that reacted with liposomal CMH, CDH, CTH,

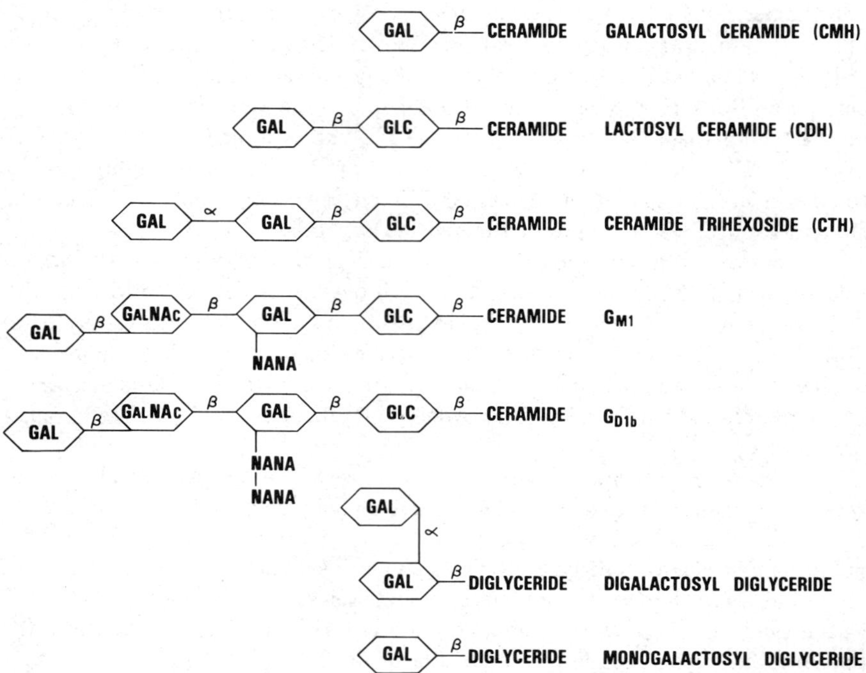

Figure 3   Structures of seven glycolipids having terminal galactoses. GAL, galactose; GalNAc, N-acetylgalactosamine; GLC, glucose; NANA, N-acetylneuraminic
acid (sialic acid).

globoside, Forssman, digalactosyl diglyceride, and gangliosides G_{M1}, G_{M2}, G_{M3},
G_{D1a}, and asialo G_{M2} (3,65,70-73). Adding to potential confusion, natural antiphospholipid ("antiliposome") antibodies may react with liposomes lacking glycolipid (72,74) (see "Phospholipids," p. 219).
     The frequency and titer of natural antiglycolipid antibodies differed between
species and individuals and depended on the glycolipid being tested. Natural
anti-CTH antibodies were not observed in rabbits by Hamers et al. (67). In contrast, in our experiments the most dramatic observations relating to frequency
and titer were that virtually all normal rabbit sera had high titer antibodies against
CDH and CTH but not against CMH (65). All normal human sera tested had very
high antibody activities against asialo G_{M2}, but not against G_{M2} (72). There was
no correlation between relative titers of anti-Forssman, antigloboside, anti-G_{M2},
and antiasialo G_{M2} in individual human sera, thus suggesting the absence of significant cross-reactivity between the natural antibodies (72). Purified natural anti-

Forssman antibodies from human sera did not react with liposomes containing globoside (56). All normal humans had detectable, but generally low level activity against globoside (72). The rarity of such activity against globoside (i.e., anti-P activity) in human erythrocytes at 4°C (46,62) illustrates again that assay of antibody activity against liposomal glycolipids at room temperature (or at higher temperatures) may yield results that are quite different from those obtained at low temperature or by different techniques. It should be noted that the antibody titer of rabbit antiserum against CMH was directly related to temperature when tested with liposomes (75).

Since all of the "naturally-occurring" antibody activities described above represent "autoantibodies," the significance of these discoveries may have broad implications in the field of immunology. Regardless of the reasons for, or biological roles, if any, for such antibodies, the natural antibodies play havoc with serological studies on glycolipids. Affinity chromatography methods for purifying antiglycolipid antibodies have been described (76) but the procedures for coupling the glycolipids may be inconvenient or time consuming. Because liposomal glycolipids are easily "coupled" to a particle (i.e., the liposome itself) we devised a method employing "affinity binding" of antibodies to liposomes and subsequent elution to purify antiglycolipid antibodies (77). The entire procedure, starting from preparation of liposomes and resulting in purified antibodies, requires only a few hours, and as much as a 3800-fold purification was obtained from whole antiserum (77). Antibodies obtained through affinity binding to liposomes have been useful and convenient for studying specificities of numerous antiglycolipid antibodies (3,65, 73,78). The purified antibodies have also been used for fluorescent localization of glycolipids on lymphocytes (79).

Phospholipids

All, or nearly all, phospholipids are antigenic, and this area has been the subject of numerous investigations (reviews in Refs. 13, 15, and 52). Because the membrane fluidity of the bulk liposomal phospholipids (such as lecithin), as well as the type, density, fluidity, and so on, of antigenic phospholipids (such as cardiolipin or synthetic phospholipid-hapten conjugates) can be manipulated readily by various means, phospholipids have been useful as antigens in liposome studies on complement mechanisms (discussed further in "Antibody-Dependent Classical Pathway," p. 228). The degree of antigenicity of individual phospholipids varies greatly. Cardiolipin was the first lipid antigen discovered (80) and it has been used widely as the antigen in routine screening tests for syphilis. Presumably because it is highly antigenic and has a large supporting literature, cardiolipin has been employed frequently as a liposomal antigen (see Table 1).

Other than cardiolipin, the only phospholipids from natural sources that have been widely used as antigens in liposomes are the choline phosphatides, lecithin and sphingomyelin (Table 1). The antibodies against these phospholipids have been referred to as "antiliposome" antibodies because they are directed against the bulk lipids of the liposomes (74,81-83); however, antilecithin antibodies have also been studied by nonliposomal methods (29). Antiliposome antibodies are readily produced in rabbits or mice by injecting liposomes containing lipid A derived from endotoxin (81,82,84,85). Antiliposome antibodies occur naturally in rabbit sera (72,74) and in certain human sera (57). The titers of these antibodies in rabbits can be raised nonspecifically by immunization with other antigens that do not contain choline phosphatides (74,82,83). The antibodies recognize phosphocholine, at least partially, and they can bind to liposomes containing only phosphatidylcholine but not to liposomes lacking a choline phosphatide (82). The antibodies are inhibited by high concentrations of soluble phosphocholine (81,86). A very low level of antibody activity against cholesterol also was detected by solid-phase radioimmunoassay (84). It has been claimed that anticholesterol antibodies are present in rabbit antisera either against human serum lipoproteins (87,88), or against cholesterol, or a cholesterol-protein conjugate (89).

Monoclonal antiliposome hybridoma antibodies have been obtained after immunizing mice with liposomes containing lipid A. The phospholipid antigen consisted either of dipalmitoylphosphatidylcholine (DPPC) or phosphatidylinositol monophosphate (86,90). The monoclonal antibodies had specificities that were similar to the original mouse antisera, or to rabbit antisera, and in each case they were inhibited by phosphocholine, but not by choline. An intriguing feature of the monoclonal antibody specificities is that the antibodies were also inhibited by several other phosphorylated compounds, for example adenosine monophosphate (AMP), and to a greater degree, adenosine triphosphate (ATP) and inositol hexaphosphate (but not inositol), and even to a very slight extent by sodium phosphate (86,90). The production of antibodies having a partial phosphocholine specificity induced by injection of liposomes containing lipid A apparently represents a unique phenomenon since the antibodies lacked both the TEPC-15 and V_HPC idiotypes found on virtually all other mouse antiphosphocholine antibodies (86).

At least one antiliposome antibody, antiphosphatidylinositol phosphate (PIP), can bind to cells. A monoclonal antibody against liposomes containing PIP did not react well with liposomes lacking an inositol phosphatide (91). The antibody also bound to Chinese hamster ovary cells and blocked the cytotoxicity of diphtheria toxin (91). Previous experience had shown that diphtheria toxin was a phosphate-binding protein that could bind to PIP in liposomes, and PIP was proposed as a cellular receptor for the toxin (92). The anti-PIP antibody had a specificity that was reminiscent of the binding specificity of diphtheria toxin in that it was blocked by AMP and, to a greater extent, by ATP (91).

Although antiliposome antibodies were produced by injecting liposomes containing lipid A, they were not induced by liposomes containing another adjuvant, muramyl dipeptide (MDP), conjugated to a fatty acid (85). The explanation for this might be that lipid A was a much stronger adjuvant than the MDP conjugate, and it is possible that more prolonged immunization with the MDP conjugate could have induced antiliposome antibodies. However, it must be emphasized that lipid A is a classic B-cell mitogen [it is the active moiety of lipopolysaccharide (LPS)], and lipid A would be expected to induce a polyclonal response. Bacterial cells treated with acid to remove the polysaccharide from LPS and expose the lipid A also induced antiliposome antibodies, even though such bacterial cells probably lacked any conjugated phosphocholine, including choline phosphatides (82). This observation, taken together with the occurrence of "natural" antiliposome antibodies (see above), leaves open the possibility that antiliposome antibodies result from a nonspecific proliferation of clones secreting antiphospholipid antibodies.

Regardless of the mechanism of induction of antiphospholipid or antiliposome antibodies, we have never seen any obvious signs of ill-effects among individual rabbits or humans having such antibodies—although the possibility of subtle effects, such as on aging or arteriosclerosis, has not been explored (57). It is quite an open (and interesting) question whether such antibodies are detrimental, beneficial, or of no consequence to the individual. Even if the antibodies do not cause tissue injury, it is clear that they could, at least potentially, disrupt liposomes injected for therapeutic purposes. The natural occurrence of antiliposome antibodies that we have observed in humans and the potential that might exist for producing antiliposome antibodies therefore pose theoretical problems that could affect the utility of liposomes as drug carriers or as vehicles for vaccines.

A fascinating series of synthetic antigens having haptenic groups covalently linked to phosphatidylethanolamine has been created, initially by Kinsky and coworkers, for various applications in liposomes (Table 1) (reviewed in Refs. 1, 5, and 13). The most popular of the conjugated phospholipid antigens are in the series having dinitrophenyl (DNP) as a hapten, particularly DNP-aminocaproyl-phosphatidylethanolamine (DNP-Cap-PE) (93-96). One reason for the widespread use of the DNP-lipid conjugates is the ready availability of anti-DNP antibodies produced by injection of DNP-protein conjugates. Early studies by Kinsky utilized these antigens in comparison with previous experiments employing antibodies against glycolipids (93,94), but the majority of later contributions from his laboratory utilized conjugated phospholipids as models to examine the role of liposomes in enhancing the antigenicity of the lipid antigens (see "Immunization with Liposomal Lipids," p. 251; also reviewed in Refs. 1, 6, 13, and 97).

Other derivatives, employing either azobenzenearsonic acid-tyrosine (ABA-Tyr), azobenzenesulfonic acid-tyrosine (ABS-Tyr), fluorescein, theophylline, nitroxide radical, or phosphocholine, have been synthesized for various purposes (see Table 1). With respect to phosphocholine, this is the terminal moiety of a

complicated hapten, 3-[3-(4-azophenylphosphocholine)-4-hydroxyphenyl]-N-
propionyl (APPC-PPr), attached to PE (APPC-PPr-PE) (98). Antibodies against
APPC-PPr-PE apparently had the typical TEPC-15 antiphosphocholine idiotype,
and were inhibited by phosphocholine, and even by choline. Anti-APPC-PPr-Pe
antibodies did *not* react with liposomes lacking APPC-PPr-PE (98). Since liposomes lacking APPC-PPr-PE still contained phosphocholine moieties on the phosphatides, it is evident that the ability to bind phosphocholine may be only one
factor in determining whether an antibody can bind to phosphatidylcholine or
sphingomyelin in liposomes.

Lipopolysaccharides and Lipid A

Smooth and rough lipopolysaccharides (LPS) from different strains of *Salmonella*
were incorporated as antigens in liposomes (99,100). Under the conditions used,
LPS that had been alkali-treated to remove esterified fatty acids was more effective as an antigen than was native LPS (99). Alkali-treated LPS could bind nonspecifically to erythrocytes and caused "sensitization" to immune hemolysis,
and it could similarly bind to, and sensitize, preformed liposomes that initially
lacked LPS, to immune damage by specific anti-LPS antibodies and complement
(99). Lipid A, the lipid moiety of LPS, also served as an antigen in liposomes (81,
100), and as with alkali-treated LPS, alkali-treated lipid A passively sensitized
liposomes to immune damage (100). When improved methods for incorporating
lipid A were used during liposome formation, untreated lipid A had activity in
liposomes that was approximately equal to alkali-treated lipid A (81). Conversion of lipid A to a chloroform-soluble form was the most efficient way to ensure
uniform mixing of lipid A with the bulk lipids during liposome formation (101,
102).

Antisera against LPS from various rough mutants were tested for specificity
with liposome-associated LPS (99). Some cross-reactivity was observed against
structurally similar LPS molecules from the mutants but universal cross-reactivity
was not observed (99). In contrast, antilipid A antiserum cross-reacted with every
LPS tested, including smooth (wild type) LPS. Presumably, the broad cross-
reactivities of antilipid A antibodies were caused by structural similarities, or
identities, of lipid A in the various LPS molecules from which lipid A was obtained.
Extensive cross-reactivity of antilipid A serum with smooth LPS was confirmed by
a passive hemolysis assay (103).

Lipid A (and therefore presumably LPS) consists of a heterogeneous group of
molecules (102) and extensive studies have been performed to test the abilities of
antilipid A sera to react with individual purified lipid A fractions, both in liposomes (102) and in a solid-phase radioimmunoassay (84).

Numerous studies have employed lipid A as an adjuvant to enhance the antigenicity of liposomal antigens, and this topic is discussed in the sections "Phospho-

lipids," p. 219, "Alteration of Lymphocyte Lipids and Effects on Function," p. 245, and "Liposomes as Adjuvants and Antigen Carriers," p. 249.

Proteins

One of the most commonly asked questions, in our experience, is whether proteins can be utilized as antigens in liposomes for studying antibody binding and complement fixation. The answer, of course, is yes, but considerable complexity or difficulty might be encountered depending on the individual protein. Obviously, if the protein is entirely encapsulated in the aqueous spaces inside the liposomes, it may be resistant to immunological attack by antibodies. This has been observed by several laboratories (104-108). Because of the potential protection that might be afforded to proteins by liposomes, liposome-encapsulation has been proposed as a method for avoiding allergic reactions due to antibody binding (serum sickness or Arthus reaction) (104-106,109). In contrast, if a protein has hydrophobic regions—even ones that are not readily detected under ordinary circumstances—then the protein, or at least a fraction of the protein molecules, may assume a transmembrane configuration during liposome formation. Cholera toxin did not bind at all to preformed liposomes lacking its receptor, ganglioside G_{M1} (110), but when attempts were made to "encapsulate" cholera toxin in liposomes lacking G_{M1}, approximately one-third of the toxin was detectable on the liposome surface by antitoxin antibodies and approximately one-third of the activity of the toxin itself was still detectable (111). Similar observations have been made in the course of attempts to encapsulate antibodies in liposomes (see "Targeting of Liposomes by Antibodies," p. 246). The inability to encapsulate certain proteins fully could have importance in determining whether liposomes might be suitable as carriers for certain enzymes for therapeutic use in humans. Furthermore, even if the fully encapsulated protein does not bind antibodies in vitro, it still might bind antibodies in vivo (112-114).

Numerous membrane proteins have large hydrophobic regions for which the preferred environment in liposomes, as in cells, may be the hydrophobic region of the phospholipid bilayer. Many such proteins have been "reconstituted" by incorporation into liposomes and some have been used as liposomal antigens that are partially exposed on the liposome surface and bind antibodies. Examples of hydrophobic liposomal protein antigens that bind antibodies are aggregated IgG antibodies (115,116), glycophorin (117), erythrocyte glucose transporter (118), myelin basic proteins, and sciatic nerve myelin P2 protein (119), various viral antigens (120-127), and many preparations of histocompatibility antigens (123, 125, 128-138). The ability of antibodies to react with hydrophobic proteins can be influenced by the liposomal lipid composition (119). Incorporation of histocompatibility antigens into liposomes, either with or without viral antigens, has been a popular and productive approach, and this topic is discussed again in the section "Immunity Associated with Histocompatibility Antigens," p. 237.

Numerous soluble proteins, plasma proteins, and others can bind nonspecifically to liposomes and potentially could serve as antigens for antibody binding studies (139; see also Chap. 5). Although proteins such as albumin (140) or heat-aggregated immunoglobulins (141) can bind nonspecifically to liposomes, this phenomenon has not been employed as a mechanism for attaching these protein antigens to the surface of liposomes to study specific antibody and complement interactions. Drawbacks to this approach include difficulties in regulating the amount and orientation of attached proteins and the fact that nonspecific interactions of proteins with liposomes frequently cause permeability changes in the liposomes (140,141; reviewed in Ref. 142).

Methods have been developed recently for attaching potential protein antigens, such as enzymes or immunoglobulins, to preformed liposomes while the liposomes remain intact (143-148) (see "Targeting of Liposomes by Antibodies," p. 246). An alternative approach to direct attachment of proteins to preformed liposomes is to attach a protein antigen to a lipid molecule, such as a fatty acid, and insert the lipid-protein conjugate into liposomes (149-151). The latter process can result in liposomes having attached antigen that can react with specific antibodies against the antigen (149,150). Finally, although it may be obvious, any protein that attaches to a liposomal constituent, regardless of the mechanism or specificity of attachment, could serve as a potential liposomal antigen. As an example, cholera toxin attached to its receptor, ganglioside G_{M1}, served as an antigen on the surface of liposomes. Anticholera toxin antibodies induced complement-dependent immune damage and glucose release from the liposomes (70,152). As another example, regardless of whether immunoglobulins were partially encapsulated within liposomes (153), inserted into liposomes as a fatty acid-immunoglobulin complex (149), or were specifically attached to DNP-Cap-PE antigen in liposomes (94), in each case a liposomal antigen was produced that reacted with appropriate specific anti-immunoglobulin antibodies.

Fluorescent liposomes having dansylated protein, either albumin or immunoglobulin, attached to the outer surface can react with appropriate antiprotein and antidansyl antibodies (145,149,154).

Liposomal Immunoassays for Antibodies

Many methods have been developed for detecting and assaying immune reactions of antibodies with liposomes. Several of these assays are based on measuring the release, due to immune damage, of a marker substance trapped inside the liposomes. Such an assay involves preparation of liposomes by swelling the dried lipids in a marker solution; removing the untrapped marker solution from the liposomes by dialysis, centrifugation, or column chromatography; reacting the liposomes with antibodies and complement (or, in antibody-dependent cell-mediated cytotoxicity, with antibodies and cytotoxic cells); and measuring the fraction of total trapped marker that is released.

Isotope Analysis

Radioactive substances, such as [^{3}H] glucose (155), ^{86}Rb (66,156-159), ^{51}Cr (138,158,159), or ^{99m}TcO$_4$ (160), have been used as trapped markers, but they can have several disadvantages. It is difficult and inconvenient to reduce the level of untrapped marker sufficiently to allow measurement of low levels of marker release caused by immune damage. In addition, once the immune reaction is complete, the liposomes must be removed from the incubation medium since the radioactivity both inside and outside the liposomes will be detected. Separation of released radioactive marker from the liposomes still containing unreleased marker is frequently inconvenient and usually results in a substantial loss of sensitivity of the assay. Furthermore, it does not allow continuous monitoring of results.

Spectrophotometry

Release of trapped marker also can be measured spectrophotometrically. These assays involve a chemical reaction of the trapped marker after its release from the liposomes, thus allowing measurement only of marker that is outside the liposomes. The most widely used marker for spectrophotometric assays is glucose, and the glucose is usually measured via a coupled enzyme system that causes increased A$_{340}$ due to reduction of NADP (33,161). Other substances that have been entrapped as internal liposomal markers include the enzymes of the glucose assay system itself (hexokinase and glucose-6-phosphate dehydrogenase) (162), galactose (163,164); CrO$_4^{-2}$ (141,165) horseradish peroxidase (131), and arsenazo III, a calcium-sensitive dye (166).

Agglutination of liposomes by antibodies has also been determined spectrophotometrically. The increased light scattering due to antibody binding has been measured at 340 nm (167) and 600 nm (168).

Fluorometry

A variety of methods using fluorescence have been developed for assaying antibody interactions with liposomes. Fluorescent markers that have been used in measuring marker release due to immune damage include umbelliferone phosphate, which is converted to umbelliferone after release from the liposomes (169,170); fluorescein (171); a 1:1 complex of 1-aminonaphthalene-3,6,8-trisulfonic acid with p-xylene-α-α'-bispyridinium bromide (172-174), which is quenched from fluorescing until it is released from the liposomes; and FITC-urea (87). Binding of fluorescent antibody to liposomes has been determined fluorometrically (175) or by fluorescence microscopy (176). Binding of liposomes to cells via antibodies (153,177) has also been determined by fluorescence microscopy. The amount of trapped liposomal fluorescent dye present in liposomes bound to cells can be determined by flow microfluorometry (177). Diffusion of antigen molecules in the liposomal membrane has been observed by following redistribution of fluorescence due to bound antibody after photobleaching (176,178).

Immunoprecipitation

Antibody-dependent agglutination of liposomes has been used as the basis for
assays of antibody binding (179). In addition to the spectrophotometric methods
mentioned above (167,168), assays have been described for determining agglutination visually (153,179-181). Agglutination can be measured as radioactivity
bound to glass fiber filters after interaction of radiolabeled liposomes with antibody (154). Binding of liposomes by antibody also has been used as the initial
step in a method for the purification of antilipid antibodies from serum (77).
This method involved interaction of antigen-containing liposomes with serum,
separating the antibody-coated liposomes by centrifugation and then eluting the
antibodies off the liposomes.

Radioimmunoassay

Radioimmunoassay techniques have been used in determining binding of antibodies to liposomes. One example of this type of assay involved reacting anti-
galactosyl ceramide serum with radiolabeled liposomes containing galactosyl
ceramide, adding anti-immunoglobulin antibody, centrifuging the liposomes, and
determining the fraction of the radioactivity remaining in the supernatant (180).
A similar assay has been described for determining antimyosin antibody covalently
attached to liposomes as the fraction of ^{125}I-myosin precipitated (144). Antibody binding also can be determined using a solid-phase radioimmunoassay where
antigen-containing liposomes are bound to microtiter plates and then reacted with
the specific antibody followed by ^{125}I-labeled anti-immunoglobulin. The label
bound to the wells is then counted (84,85).

Electron Spin Resonance Immunoassay

Both antibodies and lipid antigens have been determined by measuring release of
trapped spin-label marker (183-186; reviewed in Ref. 187). Binding of antibody
to liposomes has also been measured by observing the effects of unbound antinitroxide antibody on the electron spin resonance (ESR) spectrum of a soluble
nitroxide spin label (20,21). Other assays have been based on observation of perturbation by antibody of spin label incorporated in the liposomal lipid bilayer
(21,188,189).

Electron Microscopy

Freeze-etch electron microscopy has been used to visualize individual IgG and
Clq molecules bound to liposomes (190).

Electrode Assays

Specific electrodes have been used to measure release of a trapped marker after
immune damage to liposomes. In an assay for theophylline, release of trapped

enzymes has been measured as the rate of oxygen depletion in the presence of substrate using a Clark oxygen electrode (191). Complement and antibody levels in microliter amounts of serum have been determined using tetrapentylammonium ion-loaded liposomes and a tetrapentylammonium ion-selective electrode (192, 193).

Effects of Membrane Composition on Antibody Binding

The binding characteristics of antibodies against membrane-associated lipid antigens can often be strongly influenced by other lipids, or proteins, in the membrane. A classical example is the probable steric interference by membrane proteins that blocks the binding of antibodies against glycolipids or phospholipids on certain cell membranes (54; reviewed in Ref. 13).

Experiments with liposomal glycolipid antigens (galactosyl ceramide and Forssman) suggested that the bulk liposomal phospholipids could influence antibody binding to the haptenic polar region of a glycolipid in the bilayer (17,37). Adsorption studies showed that liposomes having longer hydrophobic regions [e.g., distearoylphosphatidycholine (DSPC) or sphingomyelin] bound fewer antiglycolipid antibodies than did those having shorter hydrophobic regions [e.g., dimyristoylphosphatidylcholine (DMPC)] (37,77). Glycolipids that had longer oligosaccharide groups bound antibodies more efficiently than did glycolipids having shorter oligosaccharides (65,75). The foregoing results were interpreted as indicating that the bulk phospholipids in liposomes could interfere with antibody binding to glycolipids in liposomes by steric hindrance (reviewed in Ref. 13). Indeed, space-filling models clearly showed that steric hindrance to antibody binding might be reasonably expected, and even predicted, with certain combinations of glycolipids and phospholipids (75,194). Steric hindrance to binding of antiglycolipid antibodies was partially overcome as the experimental temperature was raised from room temperature (as in most of the studies noted above) to 35°C (75). However, steric hindrance was also reported with other lipid combinations even at 32°C (20,21).

The dynamics of antibody interactions with phospholipid antigens at the surface of a liposomal membrane are complex. The numerous variables include size and type of hydrophilic and hydrophobic regions of the antigen, molecular motion both of antigen and of surrounding bulk lipids, relative kinetics of antibody binding compared to lateral diffusion of antigen, effects of antibody binding on lateral diffusion kinetics of the antigen, antibody type and affinity, distance between adjacent antigen molecules, and effects of cholesterol. All of these variables have been examined in a series of papers by McConnell and colleagues using a natural phospholipid (cardiolipin) or synthetic phospholipid-hapten conjugates (19-22,178,190,195-197; reviewed in Refs. 2 and 5).

COMPLEMENT AND LIPOSOMES

The first experiments utilizing liposomes containing a glycolipid antigen were
devised in order to have a convenient model for studying complement activation
and membrane damage (reviews in Refs. 1, 13, 198-200). It is now apparent that
liposomes can be useful as a model in nearly any aspect of complement research,
both immune and nonimmune.

Antibody-Dependent Classical Pathway

The classical pathway of immune complement activation (reviewed in Ref. 201)
generally begins with antibody binding to antigen. Through means that still have
not been completely elucidated, the first component of complement, C1 (consisting of subunits C1q, C1r, and C1s), is activated via the Fc portion of the antibody. Activation or binding of complement components then occurs sequentially
in the series C1, C4, C2, C3, C5, C6, C7, C8, C9. If complete complement activation occurs at the surface of a membrane, membrane damage and permeability
changes occur. All of the foregoing information was known in 1968 when liposomes were first employed in the complement field. At that time, and even now,
studies on the mechanism of complement-mediated cytolysis commonly utilized
a model hemolytic system originally discovered in the late nineteenth century
(see Refs. 202 and 203 for historical reviews). This remarkably useful and versatile but complex model system, jocularly dubbed the "immunological zoo" (204),
consists of sheep erythrocytes, rabbit (or other species) anti-sheep erythrocyte
serum (also called hemolysin), and unheated serum from guinea pig (or fresh
serum from nearly any species) as a complement source. The major antigen in
the sheep erythrocyte is a glycolipid, Forssman antigen (Fig. 2, p. 216).

Virtually all of the essential features of the classical complement-dependent
immune hemolytic system were systematically duplicated with liposomes containing purified Forssman antigen. The liposomes initially were prepared entirely from a protein-free sheep erythrocyte extract ("sheep fraction IIa") containing phospholipids, cholesterol, and Forssman (33-35). The basic elements of
the system were defined by showing that the liposomes could bind (adsorb) anti-
sheep erythrocyte antibodies and fix (inactivate) hemolytic guinea pig complement, and liposomes lacking antigen could be sensitized to antibody binding by
inclusion of a semipurified Forssman glycolipid fraction (34). With purified complement components it was shown that antibody-induced complement-dependent
membrane damage to liposomes was absolutely dependent on C2 and C8 and, as
in the erythrocyte model, complement damage was further stimulated by C9
(205).

A striking feature of classical complement-induced hemolysis is the appearance of characteristic small circular lesions observed by electron microscopy of
negatively stained erythrocyte membranes (206). Similar or identical lesions,

indicative of membrane damage induced by terminal complement components, were observed on liposomes (207-209). These fascinating lesions, which have generated considerable controversy, and considerable research, have been purified from erythrocytes and incorporated into liposomes (210,211). The lesions are discussed further in "Mechanism of Complement Damage," p. 233.

An interesting line of approach, initiated by McConnell and his colleagues, has demonstrated that membrane lipid composition and lateral mobility of lipid antigens can influence complement fixation mediated by antibodies. With DMPC liposomes, cardiolipin antigen, and human antibodies from syphilitic sera, antibody binding occurred, but complement fixation did not occur when the liposomes had less than 40 mol% cholesterol. Fixation was enhanced with increased cholesterol (19). Absence of complement fixation at less than 40 mol% cholesterol was attributed to lateral phase separation of the cardiolipin antigen in a highly fluid state separated from the more rigid state of the bulk phospholipid (DMPC). The lateral phase distribution of the antigen restricted its molecular motion in the bulk phospholipid. Increasing the cholesterol concentration to 40 mol% resulted in a homogeneous distribution of cardiolipin and also resulted in complement activation (19).

In further studies, a novel series of antigens, consisting of dipalmitoyl phosphatidylethanolamine conjugated to nitroxide spin label haptens, allowed direct measurement of antibody (antinitroxide antibody) binding, leading to changes of paramagnetic resonance spectra detected on an ESR spectrometer (20,21,212). In this system, initiation and extent of complement activation was influenced by phospholipid fatty acyl chain length, relative fluidity of the bulk and haptenic phospholipids, hapten concentration, antibody concentration, antibody class, liposome concentration, degree of liposome agglutination, and liposomal cholesterol concentration (20-22,212,213a). The detailed relationships described between these factors are complex (reviewed in Ref. 2), and they have also been studied on a smaller scale elsewhere (17,18,82,157,165,181,213b-215). It is evident that lateral hapten mobility, and fluidity of the bulk phospholipids, both of which can be influenced by numerous intrinsic or extrinsic variables, may have profound effects on classical pathway complement activation mediated by antibodies on membrane surfaces.

One extrinsic variable, temperature, has had surprisingly little emphasis. For the record it should be pointed out that the apparent titer of complement, as well as the apparent antibody titer, is strongly affected by temperature when tested with liposomes (75). The relative effects of temperature on apparent antibody and complement titers are also influenced by liposomal lipid composition (75,216).

In general, it may be stated that "fluid" liposomes (e.g., DMPC at 32°C) fix (i.e., bind or inactivate) complement more readily than do "solid" liposomes (e.g., DPPC at 32°C). This generalization does not hold true under all conditions; in any given system certain variables must be controlled, such as keeping the antibody concentration at a low level (20), in order to observe correlations between relative degree of complement fixation and membrane composition. Because the complement cascade is initiated by binding of Clq to antibody, the influence of membrane composition, and hapten (and hapten-antibody) mobility, on Clq binding was studied (196). Binding of Clq was not influenced by membrane fluidity or by hapten or hapten-antibody mobility (196), but the *rate of activation* of Cl to Cl̄ was greater with fluid than with solid liposomes (217,218).

Antibody-Independent Classical Pathway

Activation by C-Reactive Protein

C-reactive protein (CRP) is an acute-phase reactant of human serum that is able to activate complement by the classical pathway (219-223). CRP was first recognized by its ability to precipitate pneumococcal C polysaccharide (224,225). C polysaccharide is a teichoic acid, and it was found later that CRP recognized the phosphocholine, rather than the carbohydrate, moiety of the C polysaccharide (226,227). Two laboratories independently discovered that CRP could bind to phosphatidylcholine (219,220,228). Agglutination and precipitation of emulsions of soybean phosphatidylcholine, caused by binding of CRP, was used as a method for isolating CRP from serum (228). CRP also reacted with highly fluidized nonliposomal emulsions containing cholesterol and either phosphatidylcholine or sphingomyelin, resulting in activation of the classical complement pathway only when human Clq was present (219,220). These finding led us, in collaboration with H. Gewurz and his colleagues, to test if CRP would bind to, fix complement on, and induce complement-mediated damage to, liposomes (214). Interactions between CRP and liposomes resulted in both consumption of each of the components of the classical complement pathway and release of trapped liposomal glucose when positively charged liposomes containing DMPC and cholesterol were used (214). CRP-mediated complement consumption also occurred with liposomes containing egg phosphatidylcholine and either lysophosphatidylcholine or lysophosphatidylethanolamine (229). CRP-mediated complement consumption (215,229) and membrane damage (215) were enhanced when the liposomes contained certain ceramide lipids, including galactosyl ceramide, glucosyl ceramide, and ceramide, but not sphingomyelin. CRP-dependent glucose release and, to a lesser extent, consumption of hemolytic complement activity were inversely related to the fatty acyl chain length of the liposomal phospholipid (215). Phospholipid unsaturation and liposomal cholesterol concentration also influenced both complement consumption and membrane damage, suggesting that these phenomena require an optimum membrane fluidity (215). These effects of membrane composition on complement consumption and glucose release were related largely to requirements for binding of CRP to liposomes (230). Binding of

CRP to liposomes was dependent on the presence of stearylamine and did not require calcium (230). Although binding to the liposomes occurred mainly via stearylamine, it could be inhibited at least partially by soluble phosphocholine (230). The latter observations are consistent with the finding that CRP has a calcium-independent cation binding site that is probably distinct from the phosphocholine binding site (231). Increased cholesterol concentration in the liposomes caused decreased CRP binding (230) and noncomplement-mediated CRP-dependent glucose release (232), but it also caused increased complement consumption and complement-dependent glucose release (215). Although Hokama et al. (228) found that phospholipase C treatment of an emulsion of soybean lecithin greatly diminished CRP-dependent agglutination of lecithin, Volanakis and Wirtz (233) found that CRP could bind to phosphatidylcholine liposomes only when lysophosphatidylcholine was present as well. The binding to egg phosphatidylcholine/lysophosphatidylcholine liposomes was calcium-dependent, inhibited by phosphocholine and enhanced by the presence of galactosyl ceramide in the liposomes (229).

Although the binding properties were not well characterized, it has been claimed that CRP also binds to nonliposomal emulsions of a detergent (Span 60) and cholesterol, either with or without phosphatidylcholine (234,235). The binding of CRP to Span 60 was calcium-dependent and was partially inhibited by phosphocholine (235). In both of the studies cited above the Span 60 emulsions were added to whole serum that contained CRP, rather than to purified CRP. The Span 60 detergent emulsion might easily have solubilized phospholipids from the serum and further comment on this system must await studies with purified CRP.

Cholesterol-Dependent Activation

Human serum contains a factor that activates complement in the presence of liposomes containing a high concentration of cholesterol (greater than 50 mol%) (236). This activity was found in varying degrees in almost all human sera tested. Complement activation has also been observed in plasma exposed in vitro to crystalline cholesterol (237). Cholesterol-dependent complement activation may also operate in vivo. Hypercholesterolemia induced in rabbits was associated with deposition of complement on atherosclerotic plaques in the apparent absence of antibody (238).

Cholesterol-dependent complement damage to liposomes occurred in the absence of any recognizable antigen or antibody (236). Heating the serum (56°C, 30 min), or treatment with ethylenediaminetetracetic acid or ethylene glycol-bis-(β-aminoethyl ether)N,N'-tetraacetic acid or adsorption with insoluble immune complexes, inhibited this activity. The serum factor could be removed from serum by adsorption and was partially heat labile. The factor also could be transferred by adding heated high-reacting serum to unheated low-reacting human serum or guinea pig serum. When transfer experiments were performed using heated human serum containing anticholesterol factor and unheated C4-deficient guinea pig serum, there was no complement activation, thus indicating involvement of the classical complement pathway. The liposomal membrane composition influenced the cholesterol-dependent complement activation as well. When either galactosyl ceramide or ceramide was present, activities of about half of the human ser were enhanced. Activity increased with increasing phospholipid fatty acyl chain length, but no activity was observed when sphingomyelin was the sole phospholipid (236).

Activation by Lipid A

The cell wall lipopolysaccharide (LPS) of gram-negative bacteria has long been known to exhibit anticomplementary activity in the absence of antibodies (239). This anticomplementary activity has two components, one due to the lipid A portion of the LPS (240,241) and the other due to the polysaccharide portion of the molecule (242). Nonimmune activation of complement by lipid A occurs via the classical pathway (242-244). All of the latter studies used preparations of lipid A which probably had a heterogeneous chemical composition. Recently, preparations of lipid A, originally derived from *Shigella flexneri* LPS, have been purified further to yield a chloroform-soluble lipid A which can be separated into at least eight fractions by thin-layer chromatography (102). The chloroform-soluble lipid A caused much more complement activation than was obtained with the unpurified lipid A (Banerji, B. et al., unpublished observations, 1980). When the eight fractions obtained from the chloroform-soluble lipid A were tested for anticomplementary activity, two fractions completely lacked such activity, and of the remaining six fractions that were anticomplementary, none was as active as the original chloroform-soluble mixture (102). At the levels of lipid A tested, no anticomplementary activity was observed if the lipid A was incorporated into liposomes.

Alternative Pathway

Activation of the complement system can be initiated via C1 (classical pathway) or C3 (alternative pathway). The classical pathway typically is activated with immune complexes, whereas the alternative pathway involves activation of C3 directly on cell surfaces or cell surface components in the absence of antibody (reviewed in Refs. 201, 245, and 246).

A complex of cobra venom factor and purified alternative pathway factor B, after activation by trypsin treatment, reacted with human, rat, or guinea pig serum to generate a lytic factor. Human and sheep erythrocytes were not sensitive to this lytic factor, but liposomes prepared from the erythrocyte membrane lipids were lysed (247).

Positively charged liposomes containing galactosyl ceramide also can activate complement at C3 (214). This reaction resulted in consumption of components

C3 through C9, with no consumption of C1 or C4 (214,248). In this latter system, complement activation was independent of antibody, and alternative pathway factor D was required (248). Liposome-mediated activation of the alternative pathway was strongly influenced by the liposomal composition. C3 conversion required a positive charge and was influenced by liposomal cholesterol concentration and by phospholipid fatty acyl chain length and unsaturation (248). In addition, alternative pathway activation was observed only if a glycolipid such as galactosyl ceramide, glucosyl ceramide, lactosyl ceramide, digalactosyl diglyceride, or monogalactosyl diglyceride was present (248).

A serum factor was required for alternative pathway activation by certain liposomes. Preadsorption of the serum with liposomes abolished the capacity of fresh liposomes to activate the alternative pathway but did not affect activation by LPS or rabbit erythrocytes (249). The serum factor also could be adsorbed by phosphocholine-Sepharose beads. Liposomes lacking glycolipid or cholesterol could adsorb the serum factor even though they did not activate the alternative pathway. These experiments, and the finding that binding of the serum factor was inhibited by phosphocholine and glycerophosphocholine, suggested that the liposomal binding site for the uncharacterized serum factor was the phospholipid (DMPC). Liposomes containing dimyristoylphosphatidylethanolamine in place of DMPC did not require any adsorbable serum factor to activate the alternative pathway (249). Thus, membrane phospholipid can play an important role in the activation of the alternative pathway by liposomes.

Reactive Lysis

Cells can be lysed by the late-acting components of complement (C5-C9) without participation of antibody or the early complement components (C1-C3) by a process called "reactive lysis" (250-252). Reactive lysis has also been demonstrated with liposomes (207; reviewed in Ref. 158) and has been shown to require C567, C8, and C9 (209). Release of trapped liposomal marker occurred with C5b-8 and C5b-9 but not with C5b-7 (157). Studies of the effect of liposome composition on reactive lysis showed that marker release was the same from liposomes containing or lacking cholesterol (155). There was some effect of the phospholipid since marker release was less from liposomes containing saturated phosphatidylcholines than from liposomes containing unsaturated phosphatidylcholines or sphingomyelin (155). Release of trapped marker due to C5b-9 decreased with increasing acyl chain length in liposomes containing saturated phosphatidylcholine (157).

Mechanism of Complement Damage

The mechanism by which complement causes membrane damage and lysis has been studied extensively but has not yet been completely elucidated. Liposomes

are a useful model system for such studies since the membrane components can
be completely defined. Membrane damage to liposomes, measured as marker release, requires C$\overline{567}$, C8 and C9 (205,207,209), just as is found for erythrocytes.
Studies with liposomes have provided evidence that complement-dependent membrane damage does not occur by enzymatic attack on the membrane. Liposomes
prepared from lipids alone, containing no protein, are susceptible to complement
lysis (33,35) and thus no proteolytic activity is involved. No chemical changes
were detected in liposomal phospholipids after complement attack (253), and
membrane damage due to complement was observed with liposomes composed
of phosphinyl analogs of phosphatidylcholine, which are not susceptible to phospholipase activity (254), thus ruling out phospholipase activity as the cause of
complement lysis.

Lysis of erythrocytes by complement results in the formation of ringlike
lesions on the erythrocyte membrane which are detected by electron microscopy
(206). These lesions, when studied in freeze-etched preparations, appear only in
the outer half of the membranes (255,256). Such lesions also are produced in liposomes (208,257) and they occur only after addition of C$\overline{567}$, C8, and C9 (209).

The lytic activity of complement requires only the late-acting complement
components (C5b, C6, C7, C8, C9) (see "Reactive Lysis," p. 233). Until recently,
the mechanism by which the C5b-9 complex formed by these terminal components causes lysis was a matter of some controversy; two hypotheses were proposed to explain this activity. The "leaky patch" theory proposed that the C5b-9
complex had a detergent-like effect, causing a perturbation of the lipid bilayer structure in the vicinity of the C5b-9 complex (258-260). This disruption of the lipids
would result in altered membrane permeability in the region of the bilayer surrounding the C5b-9 complex. The "transmembrane channel" theory proposed that the
C5b-9 complex forms a discrete cylindrical pore, or channel, through the membrane which would allow free flow of molecules between inside and outside (158,
261). Recently, the latter hypothesis has gained support (182). The literature in
this general area is too extensive to be presented fully here; only the work relevant
to liposomes will be discussed.

The C5b-9 complex, as well as its precursor complexes C5b-7 and C5b-8, apparently insert into the lipid bilayer of membranes (262,263). The decrease in
membrane fluidity in sheep erythrocytes due to complement, as determined by
ESR spectroscopy of spin-labeled fatty acids, provides additional evidence for
penetration of complement protein into the membrane (264). Spin label studies
on flat lipid bilayers showed that C5b-6, C7, C8, or C9 alone had no effect on
the spectra, suggesting that they did not insert into the bilayer (258). Functional
C5b-7 interacted strongly with the bilayer surface but did not penetrate deeply,
whereas C5b-8 and C5b-9 caused marked changes in the spectra from spin label
probes located within the hydrocarbon phase of the bilayer, suggesting deep penetration into the bilayer (258). The spectral changes observed appeared to be the

result of reorientation of bilayer lipids due to strong binding of phospholipids to the complement proteins (258). The C7 subunit of C5b-7 complexes, whether bound to liposomes or in the fluid phase, was not susceptible to trypsin (265). The C5b-7 complex, when bound to liposomes, did not dissociate into its C5b, C6, and C7 subunits when treated with trypsin, although dissociation did occur in the absence of liposomes (265). These findings suggest a strong interaction between each component of the C5b-7 complex and the phospholipid bilayer but do not necessarily prove that the C5b-7 stage penetrates into the lipid bilayer.

The C5b-9 complex has been isolated from erythrocyte membranes following complement lysis and, by electron microscopy, it appeared to be identical in structure to the ringlike lesions observed on cells and liposomes (266-268). The isolated C5b-9 complex and some of its precusors (C5b-7 and C5b-8, but not C5b-6, C7, C8, or C9) bind phospholipid strongly (259,269); each C5b-9 complex bound over 1400 molecules of egg yolk phosphatidylcholine (259). Aqueous suspensions of phospholipid inhibited complement lysis of tumor cells (270), suggesting binding of phospholipid to fluid-phase complement components. Electron microscopy of isolated phospholipid-C5b-9 complexes showed no lipid bilayer structure, which is consistent with the C5b-9 complex forming phospholipid-protein micelles in membrane bilayers which could result in the formation of hydrophilic lipid channels surrounding the C5b-9 complex (259). Binding of C9 to C5b-8, which in turn had previously been bound to dioleoylphosphatidylcholine liposomes, induced the formation of typical ringlike complement lesions which appeared to be dimers of C5b-9 (271).

C5b-9 complexes isolated from sheep erythrocyte membranes also have been reconstituted in liposomes, resulting in hollow, cylindrical macromolecules possessing lipid-binding regions allowing one end of the cylinder to penetrate into the lipid bilayer (210). Liposomes containing C5b-9 complexes were stable; the C5b-9 complexes were not removed from the liposomes by density centrifugation or elution with salt solutions of high or low ionic strength or of high pH (211). It was proposed, based on these results, that the C5b-9 complexes were anchored in the membrane through apolar interactions similar to those deduced from elution studies on integral membrane proteins (211). Proteolysis of fluid-phase C5b-9 (SC5b-9) caused a transition of the molecule from an hydrophilic to an amphiphilic state which could bind lipid and was incorporated into lipid vesicles to yield a membrane-bound structure resembling the C5b-9 complement lesion (272). C5b-9 complexes which were incorporated into liposomes and then subjected to proteolysis in the presence of dithiothreitol lost the externally oriented annulus of the complex, leaving liposomes still retaining the membrane-bound, thin-walled cylindrical portion of the complex (273).

Interaction of complement and the isolated C5b-9 complexes with cells and artificial membranes strongly affects the membrane lipids. The electrical conductance of planar lipid bilayer membranes increased after addition of antigen, antibody, and complement (274). The increase in the electrical conductance of black lipid membranes after the sequential addition of C5b-6, C7, and C8 showed this phenomenon to be specific for the membrane attack complex of complement (275). Although the latter membranes were stable for 30-60 min, addition of C9 produced further increments of conductance increase, whereas low levels of C9 did not impair membrane stability (275). The conductance increase on addition of C8 depended on the voltage polarity, whereas this was not the case once C9 was added (275).

Spin label studies indicate that complement lysis induces reduced fluidity of membrane lipids (277,278). When a fatty acid spin label was used, the decrease in fluidity either first appeared on binding of C9 (279) or was enhanced by C9 binding (264). Treatment of erythrocytes with glutaraldehyde prior to complement lysis abolished the change in fluidity even though complement lesions were formed and lysis occurred (279). No changes in the ESR spectra were observed until after the addition of C9, when a phospholipid spin label was used (280). In this case, complement fluidized the phospholipid in the region of the lesion, whether or not the cells were lysed.

Complement lysis is also affected by, and affects, membrane fluidity. Increased phospholipid fatty acyl chain length (75,157) or increased phospholipid fatty acyl unsaturation (157) decreased complement lysis of liposomes, as measured by marker release. Above a molar ratio of cholesterol to phospholipid of 0.4, complement-dependent marker release decreased with increasing cholesterol concentration (157).

Erythrocyte membranes treated with antibody and complement had much less KSCN-dissociable cholesterol and phospholipid than did unlysed cells and this effect was related to the action of C8 (281). The authors concluded that complement lysis was accompanied by a major rearrangement of membrane lipids, even in cells containing one or a few complement lesions.

Complement damage is also associated with release of intact phospholipids from membranes. This phenomenon has been observed with antibody-sensitized sheep erythrocytes (281), antibody-sensitized tumor cells (282), antibody-sensitized *Escherichia coli* cells (283), and liposomes (157,284,285). Release of phospholipid from liposomes was dependent on the dose of complement (285), and the amount of phospholipid liberated from *E. coli* seemed to be proportional to the number of complement lesions (283). It was suggested by the latter authors that the release of phospholipid was due to displacement of membrane phospholipids by the C5b-9 complexes during their insertion into the membrane lipid bilayer. Part of the phospholipid removal from liposomes was mediated by C5b-7, and the remainder by C8 and/or C9 (285). Release of phospholipid from the liposomes was decreased with increasing liposomal phospholipid fatty acyl chain length (157). In contrast to release of trapped marker, release of phospholipid was directly related to liposomal cholesterol concentration and increased when phospholipid fatty acyl unsaturation was increased (157).

Recent studies suggest that complement lesions in membranes are transient (276,286). Stephens and Henkart (276) found that antibody-sensitized cultured neuroblastoma-glioma hybrid cells and primary rat and mouse muscle cells, when treated with small amounts of complement, initially showed membrane depolarization but recovered their original membrane properties within 1 hr. In contrast, complement-damaged erythrocytes were still releasing hemoglobin 24 hr after treatment with complement, suggesting very stable membrane lesions (261). When complement-dependent glucose release was measured from multilamellar liposomes separated into three size populations by a fluorescence activated cell sorter, the percentage of trapped glucose released was inversely related to liposomal size. The maximum release observed was 70% (286). It has been reported that small unilamellar liposomes release 100% of their trapped marker (169), but when large unilamellar liposomes were tested, the maximum glucose release was only 35% (286). These data suggested that the complement lesions were open for only a transient period since 100% release of trapped marker from the large unilamellar liposomes would be expected with permanently open lesions. Furthermore, it would be predicted that a transient lesion would result in a smaller percentage of marker release from large liposomes than from small ones, and this was observed (286).

CELL-MEDIATED IMMUNITY

Immunity Associated with Histocompatibility Antigens

Allogeneic cellular immunity—that is, T-lymphocyte-mediated immune reactions between different individuals (or strains) of the same species—is totally dependent on simultaneous participation of antigens derived from genes of the major histocompatibility complex (MHC) (see Ref. 287 for a comprehensive review of this topic).

The MHC antigens on target cells serve as receptors for cytotoxic lymphocytes (CTL). For example, if an animal is primed by immunization with a virus, T lymphocytes from that animal will recognize, and lyse, cells infected with the same virus *only* if the lymphocytes and the target cells have the same MHC antigens. The immune T cells recognize *both* the MHC antigens and the viral antigens expressed on the plasma membranes of the target cells. The MHC antigens have a close homology to immunoglobulins, both in primary and secondary structure (288,289) and the recognition phenomenon between compatible cells may be due to antibody-like activity.

In an exciting burst of activity, numerous laboratories have incorporated various MHC antigens into liposomes and the liposomes have been substituted for cells, both for primary and secondary induction of cell-mediated immunity—that is, for the initial priming antigenic stimulus and for restimulating lymphocytes already primed by a previous antigenic exposure. Liposomes containing appro-

priate antigens, including MHC antigens, also have been substituted for target cells
in cell-mediated immunity.

Incorporation of MHC Antigens into Liposomes

The MHC antigens that have been most widely studied are from human (HLA),
mouse (H-2 and Ia), or rat (AgB) cells. The HLA antigens have been the most
highly purified but, regardless of purity, H-2 (128,133,290-293) and AgB (131)
as well as HLA (129,130,132,135,137,294) antigens have been incorporated into
liposomes. In fact, the uptake of MHC antigens by liposomes is sufficiently great
that incorporation into liposomes has been proposed as a purification step for the
antigens (128).

    The MHC antigens are amphipathic proteins and are generally isolated in
dialyzable detergents, such as deoxycholate or octylglucoside. The presence of
membrane-active (and cytotoxic) detergent precludes the use of purified MHC
antigens in cell assays. This problem is solved by incorporating the MHC antigens
into liposomes and removing the detergent by dialysis. It is interesting that most
MHC antigens are incorporated into the liposomal membrane in the same "right-
side-out" orientation that is found in cells (128,129,132,135,137), although a
small fraction of liposomes might contain antigens in an inactive "inside-out"
orientation (128). Liposome-associated MHC antigens can bind to specific anti-
MHC antibodies (123,125,128-138), Semliki Forest virus (295), and even certain
gram-negative bacteria (129).

Induction of Cytotoxic Lymphocytes (CTL)

Numerous laboratories have obtained evidence that liposomes containing appropriate antigens, and always including appropriate histocompatibility antigens, can
induce a secondary stimulation of previously primed lymphocytes. Secondary
stimulation of immunity against viral antigens has been achieved by including
both the viral antigen and the appropriate MHC antigen in liposomes (125,290,
291,293,296-300). To obtain the MHC-restricted immune response, the viral
antigen and MHC antigen had to be in the same liposomes; the activity was lost
when the two antigens were in different liposomes (125,291,293). The latter
result suggested that in order to cooperate effectively with the lymphocyte to
induce an immune response, the viral and MHC antigens must be closer to each
other than 600-1000 Å (i.e., closer than the diameter of the liposomes) (291).

    Differences in MHC antigens alone (even xenogeneic differences, that is, be-
tween species) can be utilized in stimulating a secondary response with liposomes
(138,292,296,301) and the liposomal epitope density of MHC antigen can be important (296).

    Evidence that MHC antigens are required for stimulating CTL with liposomes
(in a murine system) include the requirement for the *appropriate* H-2 antigen, loss
of activity upon specific removal of the H-2 antigen with either anti-H-2 antibodies or on a protein A-Sepharose column, and retention of activity upon 20-
fold purification of the H-2 antigen (290).

Data have been presented that Ia-positive splenic adherent cells (i.e., macrophages) are involved in the processing of liposome-associated H-2 antigen, in generating a secondary CTL response in mice, and in the triggering of helper T lymphocytes (302-304). Apparently, macrophages are not always required, as removal of adherent cells did not drastically reduce the H-2-restricted immune response against a liposomally-associated viral antigen (125).

Although, as noted above, numerous examples of secondary induction of CTL have been described, there have also been two reports (but, to date, from only one laboratory) of induction of a low, but significant, *primary* MHC-restricted response (138,305). An analysis of the relatively great difficulty that has been experienced in producing a primary response by using liposomes led to the conclusion that the defect was due to absence of stimulation of T cell helper activity (305).

Liposomes have been used as carriers to introduce H-2 and viral antigens into cells by fusion and thus to render the cells sensitive to H-2-restricted immune cytolysis by cytotoxic T lymphocytes (123,306,307). Introduction of antigens into cells by fusion with liposomes is discussed further on p. 245.

Liposomes as Targets for Cellular Attack

Because of the success that liposomes have enjoyed as models for complement-mediated immune damage, they have also been tested as model targets for cell-mediated cytotoxicity. Numerous successful experiments have been reported, but this approach has generated some controversy because the systems are complex, and the results may be of small magnitude, erratic, difficult to reproduce, or may require special conditions.

Antibody-Dependent Cell-Mediated Cytotoxicity (ADCC)

In ADCC, target cells, such as erythrocytes that have been coated with antibodies, are lysed nonspecifically in a nonphagocytic process by cells that bind to the Fc regions of the antibodies (reviewed in Ref. 308). The antibodies typically are IgG but IgM has also been implicated (309). The effector cell may be a polymorphonuclear cell, a monocyte, or a still somewhat mysterious cell that has morphology similar to a small lymphocyte and is known as a killer, or K cell (308). Because the K cell has not been fully characterized, any cell population having Fc receptors, that induces ADCC against any target cell, is often referred to as having K-cell activity.

In 1975, Henkart and Blumenthal demonstrated that lipid bilayers (black lipid membranes) coated with specific antibodies could be attacked by lymphocytes (presumably K cells) in a process that seemed remarkably similar to ADCC (310). The effector cells caused increased electrical conductivity across the black lipid membranes and the process was interpreted as indicating that ion-conducting channels were formed across the lipid bilayer of the antibody-coated target membrane.

The above success with black lipid membranes stimulated analogous research with liposomes and several laboratories have now reported successful use of liposomes as targets in ADCC. Juy et al. (311) used liposomes containing cardiolipin as antigen. They showed release of ^{3}H-labeled liposomal glucose under conditions that were dependent on time, spleen cell concentration, and liposome concentration. The process worked with normal or nude (T-cell-depleted) mixed lymphocytes and with enriched T-lymphocyte populations but did not occur with lymphocytes depleted of Fc-binding cells. The experiments suffered from unexplained huge levels of liposomal glucose release in the presence of rabbit antiserum alone (in the absence of cells) and resultant small net levels of glucose release (10-15% of total trapped marker) that could be attributed to cells. Using liposomes containing Forssman antigen, Mayer et al. (158,159) confirmed the essential elements (including instability of control liposomes) reported by Juy et al. (311). The system was not always reliable and detailed results of 13 experiments, a few of which inexplicably did not work or worked poorly, were reported (159). Using liposomes having either DNP-PE or ganglioside (G_{M1} or G_{M2}) as antigen and a trapped fluorophore in the liposomes, Geiger and Schreiber (174) demonstrated binding of antibody-coated liposomes to lymphocytes, with no observed leakage of liposome contents over 90 min. After longer intervals (up to 11 hr), as much as 50% lysis of liposomes occurred and this lysis was attributed to ADCC (174). The effector cells had Fc receptors, were devoid of Thy-1 antigen, and were nylon nonadherent; all of which properties are compatible with K cells (174). In experiments by Ozato et al. (312), binding of K cells by antibody-coated liposomes was also demonstrated by showing inhibition of K-cell function in lysing ^{51}Cr-labeled Chang cells.

One of the technical difficulties frequently encountered in the studies on ADCC noted above was interference caused by agglutination of liposomes sensitized by antibodies. In addition, only small numbers of sensitized liposomes were bound by K cells. To solve these difficulties, Ozato et al. utilized antibody-coated liposomes immobilized on plastic surfaces (156,313). Such liposomes selectively depleted K cells from human lymphocyte populations and the liposomes were also lysed by the cells. However, caution was urged by the authors in interpreting these experiments and also in interpreting all of the experiments from other laboratories (156,313). This restraint was urged because of the *lack* of correlation between K cell function of cell populations (as determined by the ability to lyse antibody-coated Chang cells) and liposome lysis by the same cell populations. Furthermore, depletion of K cells by removal on immobilized immune complexes did not cause diminished liposome lysis. It was pointed out that genuine K cell lysis of liposomes may still have been operating but might have been masked by lysis induced by nonspecific liposome-cell interactions, such as endocytosis, fusion, or lipid exchange between membranes (156,313).

It appears to us that the criticisms and cautions noted above could have some validity and should be seriously considered. We might even suggest that one example of such a "nonspecific" interaction between liposomes, cells, and antibodies might be the reported lysis of liposomes by hydrophobic interactions with Fc regions of antibodies attached to bacteria (170). However, it should also be noted that more than one lytic mechanism might occur in ADCC. As in the complement system, ADCC-induced hemolysis of erythrocytes (generally acknowledged to occur by osmotic lysis) might be different from the killing of nucleated cells. Also, as with complement, liposomes might be a better universal model for the recognition and attack mechanisms per se than for particular cellular processes that occur after attack. Indeed, analogies between ADCC and complement have been discussed by McConnell and colleagues with regard to binding of neutrophils and macrophages to liposomes and subsequent triggered events such as a respiratory burst (314-316). This has also been discussed by Mayer et al. with regard to a possible attack mechanism (158). Furthermore, the possibility does exist that lymphocytes (K cells), polymorphonuclear cells, monocytes, and macrophages might have different cytolytic mechanisms.

In all of the systems cited above there was clearly at least some degree of immunological specificity probably relating to certain aspects of ADCC, and liposomes may yet turn out to be an excellent model for ADCC. For example, it is difficult not to be impressed by the photographs obtained by Lewis and McConnell in which ADCC induced by monocytes against erythrocytes was compared with ADCC-like monocyte-liposome interactions (171). Still, some uncertainties have been voiced (see 156, 313, and the audience discussion in Ref. 171) and a certain degree of controversy may be due in part to uncertainties among immunologists in defining all the biological aspects of ADCC per se.

With respect to the ADCC attack mechanism, we should like to point out a published phenomenon that may be quite nonspecific. It was reported that ADCC was inhibited by incubation of cells with synthetic phosphatidylcholine, or by an analog of phosphatidylcholine (317). This observation was cited as indicating that ADCC worked by an action of K-cell phospholipase A (317), or that ADCC membrane damage is due to transmembrane channels and that the hydrophobic peptides forming the channel were inhibited competitively by exogenous phospholipid or phospholipid analogs (158). In our opinion, the experimental results could also have been caused by an alternative mechanism, namely by changes of effector cell function secondary to transfer of cholesterol from the cells to liposomes. If the latter mechanism is correct, it might be a useful tool in studying certain aspects of ADCC function. Changes of lymphocyte function associated with liposome-induced alteration of membrane cholesterol concentration have been extensively studied (see "Alteration of Lymphocyte Lipids and Effects on Function," p. 245).

A particularly intriguing, and perplexing, liposome system has been proposed as a possible ADCC model relating to Graves ophthalmopathy (160). In the course of this disease, which is an eye disorder associated only with humans having Graves disease of the thyroid, autoantibodies to thyroid tissue and immune complexes (TgA) of thyroglobulin (Tg) and anti-Tg antibodies are often found circulating in serum. The Tg and TgA have high affinities for eye muscle membranes. Kriss and Mehdi (160) extracted an eye muscle protein (EMP) and incorporated it into liposomes. The EMP-liposomes bound TgA or Tg strongly. In the presence of normal human lymphocytes, EMP-liposomes having bound TgA but not those having bound Tg, were lysed. Lymphocytes from patients with Graves ophthalmopathy lysed EMP-liposomes that had bound either TgA or Tg. However, the patients' lymphocytes did not lyse liposomes that had Tg incorporated in place of EMP, or liposomes in which membrane proteins from other tissues were substituted for EMP. When the lymphocytes were fractionated, by rosetting, into those with characteristics of K cells and those lacking K cells, only the K-cell population was effective. The unique and obviously vital role of EMP was not fully understood.

Antibody-Independent Allogeneic Lysis of Liposomes

Despite the apparent successful use by several laboratories of liposomes as targets for ADCC, only one group has reported similar positive results for nonantibody-mediated lysis of liposomes (136,318). As in the system of Graves ophthalmopathy (see above), eye muscle protein (EMP) was a critical liposomal ingredient. The liposomes consisted of a mixture of DMPC, dipalmitoylphosphatidylcholine (DPPC), and cholesterol, and contained EMP and mouse H-2 histocompatibility antigen introduced into the lipid bilayer in a detergent solution. Release of trapped liposomal ^{51}Cr occurred when the liposomes were incubated with lymphocytes sensitized in mixed lymphocyte cultures against allogeneic cells. Lysis did not occur when the liposomes contained a different H-2 phenotype or contained the correct H-2 protein but lacked EMP. The cytotoxic cells responsible for lysis of liposomes were identified as thymocytes carrying Thy-1.2 antigen (136).

The foregoing system (136,318) had some unique and unusual characteristics: (a) EMP was required. As in the Graves ophthalmopathy system, the role of EMP was not fully understood and it was thought to serve some uncharacterized structural role. A "lumpy" membrane texture, due to EMP, was required, as visualized by freeze-etch electron microscopy (318). (b) A mixture of DMPC and DPPC was required. Either of the phospholipids alone did not work. (c) A very high protein/phospholipid ratio was required, 60%/40% (w/w). (d) The system was not uniformly reproducible and erratic results were thought to be due to variability of the human eye muscle preparations obtained at autopsy.

Liposomes as Carriers of Lymphokines

Supernatants from antigen- or mitogen-stimulated leukocytes contain soluble products (lymphokines) that profoundly influence immunological processes.

Since the first description of macrophage inhibitory factor (MIF) in 1966 (319, 320), nearly 100 lymphokine activities have been reported and named (reviewed by Waksman in Ref. 321). A major problem in analysis of lymphokine effects is the lack of precise chemical information about these molecules since most lymphokines are quantified by biological activity. Several lymphokines contained in supernatants that are rich in lymphokines due to prior stimulation can affect macrophage functions in vitro. Macrophages exposed to lymphokines in a variety of experimental situations will: (a) migrate toward increasing gradients of chemoattractants (chemotaxis), (b) cease random migration (migration inhibition), and (c) acquire enhanced nonspecific tumoricidal and microbicidal activities (activation). Liposome-macrophage interactions have been useful for studying lymphokine effects on macrophages. For example, the guinea pig macrophage receptor for MIF may be a glycolipid, possibly a fucosyl ganglioside (322-324). Liposomes containing mixed gangliosides adsorbed MIF activity from stimulated lymphocyte supernatants. Incubation of guinea pig macrophages with either mixed brain gangliosides or a soluble macrophage-derived glycolipid fraction, either alone or, much more strikingly, after integration of the glycolipid fraction into the liposomal lipid bilayer, enhanced macrophage activity. Presumably, the enhanced MIF activity occurred because of transfer of additional "receptor" glycolipids to the macrophages (322-324).

In an interesting and extended series of recent reports, Fidler and Poste and their colleagues investigated in vitro and in vivo effects of lymphokine solutions encapsulated in liposomes (325-334). The authors found that liposomes containing encapsulated lymphokines [particularly macrophage activating factor (MAF)] activated tumoricidal properties of macrophages. Liposome encapsulation both protected the lymphokines from extracellular inhibition or degradation and substantially enhanced the activities of the lymphokines. Populations of macrophages that were refractory to unencapsulated MAF were sensitive to liposome-encapsulated MAF. The liposome-encapsulated lymphokines were also quite active in vivo in mice, both in activating macrophages and in eradicating various experimental metastatic tumors.

Muramyl dipeptide (MDP) can be isolated from mycobacteria and is thought to be the moiety responsible for immunologic activity of complete Freund's adjuvant (335). Liposome-encapsulated MDP had enhanced effectiveness compared to unencapsulated MDP as an adjuvant for producing an immune response against ovalbumin (335). Macrophage nonspecific tumoricidal activity can also be induced by MDP (332,336,337). Activity of liposome-encapsulated MDP was synergistic with activity of encapsulated lymphokines (332,336).

The toxicities of lymphokines encapsulated in multilamellar liposomes were studied in mice and dogs (334). Slight elevations of serum alkaline phosphatase and glutamine oxaloacetic transaminase levels were observed, as well as occasional elevated bilirubins. Aside from these observations, there were no other histological, chemical, hematological, or immunological changes (334).

OTHER LIPOSOME-CELL INTERACTIONS

Immune Phagocytosis

The topic of phagocytosis is quite broad and includes numerous complicated
mechanisms and cell types (comprehensively reviewed in Refs. 338-340). It is
beyond the scope of the present review to discuss the detailed mechanisms of
phagocytosis per se, or even phagocytosis of liposomes. General aspects of interactions of liposomes with cells, including endocytosis, have been reviewed elsewhere (341,342).

Several laboratories have used immunological procedures to stimulate phagocytosis of liposomes. As discussed in "Targeting of Liposomes by Antibodies,"
p. 246, aggregated immunoglobulins have been coated on liposomes to cause enhanced phagocytosis of liposomes by macrophages or neutrophils (166,189,343-
351); reviewed in Ref. 352). Increased phagocytic uptake of liposomes containing
partially encapsulated antibodies has been reported (353,354). Opsonization of
liposomes by specific binding of antibodies to liposomal antigens, and resultant
binding and uptake of the liposomes by phagocytes have also been described
(171,175,315,316,355-358).

Presumably, most or all of the enhanced interactions of liposomes with cells
in the reports cited above were due to associated of liposome-bound immunoglobulins with Fc receptors on the phagocyte plasma membranes. $F(ab')_2$ did not
opsonize liposomes and opsonized liposomes were not bound to, or ingested by,
Fc receptor-negative cells (357,358). Indeed, it was even reported that IgG-
mediated phagocytosis of liposomes induced loss of the cell surface Fc receptor
but not the C3b receptor (175). Complement-mediated phagocytosis of liposomes has also been observed (359).

The binding to cells, and the triggering mechanism for endocytosis, by opsonized liposomes have been investigated in depth by McConnell and colleagues
(175,315,316,355). Antibody-dependent stimulation of the cellular respiratory
burst (315,355) and the kinetics of binding of opsonized liposomes to cells (316,
355) both were influenced by physical factors in the liposomal membrane, such
as fluidity.

As a matter of interest, it has been reported that the macrophage Fc receptor
for IgG2b is lipid-dependent. After incubation with phospholipase C, the Fc receptor was inactivated and activity was restored by reconstitution with liposomes (360).

Stimulation of macrophages following endocytosis of a liposome-encapsulated lymphokine (macrophage activating factor) is discussed in the section "Liposomes as Carriers of Lymphokines," p. 242. The requirement for macrophages in
processing liposomes for inducing cell-mediated immunity is discussed under
"Glycolipids," p. 212, and the macrophage-requirement for inducing humoral
immunity is discussed in the section "Immunization with Liposomal Lipids," p.
251.

Introduction of Antigenic Determinants into Cells

Liposomes have been employed as carriers to introduce lipid or protein antigens for various purposes into the plasma membranes of cells lacking the antigens. The antigens introduced into cells via liposomes include DNP or TNP derivatives of PE (312,361-365), liposome-linked horseradish peroxidase (366), liposome-linked dansyl-conjugated bovine serum albumin (145), viral glycoproteins (123,306,307, 365), and histocompatibility antigens (306,307).

The cells having newly incorporated antigens were tested as targets, either with antibodies (123,145,312,361-363,365) or with cytotoxic lymphocytes (123,145,306,307,312,313,363,365).

Alteration of Lymphocyte Lipids and Effects on Function

Incubation of lymphocytes with liposomes containing or lacking cholesterol can cause enrichment or depletion of lymphocyte plasma membrane cholesterol (367,368). This discovery has stimulated numerous investigations on the effects of changes in lymphocyte membrane cholesterol on cell function. The goal of many of these studies has been to correlate cholesterol concentration or lympho-cyte membrane fluidity ("microviscosity") with lymphocyte blast transformation induced by mitogenic lectins.

Enrichment of the lymphocyte plasma membrane with additional cholesterol causes suppression of mitogenic lymphocyte activation (368-370). Depletion of lymphocyte cholesterol by liposomes lacking cholesterol also suppresses mito-genesis (151,369-373). There is some disagreement as to whether depletion (369) or enrichment (370) is the stronger suppressive effect. These phenomena can be extraordinarily complex. For instance, Ip et al. (371) found that enrich-ment of lymphocytes with cholesterol *enhanced* mitogenesis, but in comparing their results with those of others they mention that enhancement depended on the use of rat serum or delipidated fetal calf serum in the culture medium. They stated (371) that use of nondelipidated fetal calf serum caused a "blunted" (or *decreased*) response by liposomes similar to the slight suppression observed by Chen and Keenan (369). The requirement for rat serum in order to see enhance-ment of mitogenesis was attributed to the low lipoprotein content of rat serum compared to fetal calf serum (371). Lipoproteins were also proposed by Rivnay et al. (370) as in vivo modulators of lymphocyte fluidity and function; indeed, high-density lipoproteins reversed the inhibition of mitogenesis of cholesterol-depleted lymphocytes (369).

In most experiments, incubation of lymphocytes with liposomes, with or without cholesterol, was performed over a prolonged period, typically 15-24 hr. However, the critical membrane events regulating appearance of mitogenesis in-duced by a lectin might occur much earlier, even within 30 min after adding the lectin (374). In another study (375), short-term incubation of cells with lipo-

somes did not influence cholesterol concentration but did cause marked changes
of mitogenesis that were dependent on liposomal *phospholipid* composition. Presumably, the liposomal phospholipids were transferred into the cell and altered
membrane fluidity, and this in turn affected mitogenesis (375). Although alterations of membrane cholesterol (or phospholipid) do change lymphocyte plasma
membrane fluidity, the mechanism of how this affects functional activity is not
entirely clear. It is possible that changes of ionic permeability play a role (376).

Incubation of liposomes with cells also affected other cellular activities, including lymphocyte cytotoxicity against allogeneic target cells and phytohemagglu-
tinin-dependent and antibody-dependent cell-mediated cytotoxicities (317,377),
capping of lymphocyte Ig (373,378), actively of responder cells (but not stimulator cells) in the mixed lymphocyte reaction (373), and phagocytic activity of
polymorphonuclear leukocytes (379).

Certain gangliosides, either from lymphocytes or brain, have been implicated
as mediators that can modulate (suppress) several aspects of murine lymphocyte
function, including T and B cell interactions and mitogenic stimulation by concanavalin A or LPS (380-385). Two laboratories have utilized liposomes to faci-
litate incorporation of the effector gangliosides into the lymphocyte plasma
membrane and thus influence lymphocyte function (380-382,385).

TARGETING OF LIPOSOMES BY ANTIBODIES

A major goal in the application of liposomes as drug carriers has been to formulate liposomes that are specifically targeted to certain sites in the body. Attachment of antibodies to liposomes for this purpose has been, and continues to be, a
focus of interest, and some interesting accomplishments, particularly recently,
have been achieved (earlier reviews in Refs. 354 and 386).

Three general methods of targeting of liposomes have been proposed: partial
encapsulation of antibody, with some active antibody remaining exposed on the
liposome surface; nonspecific attachment of heat-aggregated immunoglobulin on
the liposome surface; and specific binding, or covalent conjugation, of antibodies
to lipids in the liposome bilayer.

In 1972, Magee and Miller showed that antiviral antibodies that were "encapsulated" in the aqueous phase within liposomes still expressed antiviral activity at the liposome surface (387). When the liposomes were attached electro-
statically to tissue culture cells, the cells were protected against viral infection
(387). This observation raised the possibility that liposome-encapsulated antibodies might have therapeutic benefits and also suggested that antibodies might
be useful for certain types of targeting to cells. The targeting hypothesis was
later confirmed by Magee et al. in tissue culture. Liposomes having partially encapsulated antibodies against certain cell types were preferentially associated
with those cells (116,388,389). The liposomes carried either immune RNA that
proved toxic to tumor cells, actinomycin D, or poly(I):poly(C) (116,388,389).

Exposure of encapsulated antibodies at the liposome surface was later confirmed. A moderate amount of antibody was detected on the surface when tested with ^{125}I-labeled anti-Fab or anti-Fc antibodies (116). Antibodies that spontaneously became attached to liposomes during preparation of the liposomes by sonication also retained considerable antigen-binding capacity, and the antibodies were expressed on the surface (115,153).

Targeting of liposomes to certain cells, by means of partially encapsulated antibodies, was also shown by Gregoriadis et al. (390-393) and Margolis and Dorfman (153). Uptake of liposomes containing antibodies against tissue culture cells was greater with the cell used for immunization than with another cell type, and also, uptake was greater with specific IgG antibodies than with nonspecific IgG (153,390-392). Despite these promising in vitro experiments, an attempt to achieve selective in vivo targeting of liposomes was disappointing. Delivery of an encapsulated cytotoxic drug to a rodent tumor in vivo was only raised from 2.9% to 4.6% of the injected dose (392,393).

The most striking effect of partially encapsulated antibodies on in vivo behavior of liposomes was a pronounced increase in the uptake of liposomes by liver and spleen (392,393). The enhanced uptake by liver and spleen was abrogated by concurrent administration of either "empty" liposomes or a lipid emulsion (Intralipid) (392,393). Enhanced hepatic and splenic uptake of liposomes containing immunoglobulins was also observed by de Barsy et al. using morphologic and enzymatic techniques (353,354). Liposome-encapsulated antibodies against lysosomal α-glucosidase retained specific antibody activity in vivo, but the degree of activity was less than expected (353,354). Hepatic uptake of liposomes having encapsulated immunoglobulin was enhanced still further by including galactosyl ceramide or asialo mixed gangliosides in the liposomal bilayer or by conjugating galactose or mannose to the liposome surface (394). Presumably, the glycoconjugates recognized lectins that were present on hepatocyte or Kupffer cell plasma membranes, and this caused increased uptake of liposomes.

The enormous tendency of liposomes having encapsulated antibodies to be delivered to the liver, while being a difficult problem confronting liposomes as generalized drug carriers, was utilized as a therapeutic advantage in one study (395,396). Liposome-encapsulated antibodies against digoxin altered the distribution and excretion patterns of digoxin in rats. Whereas unencapsulated antibodies caused intravascular retention of digoxin, encapsulated antibodies caused increased hepatic and splenic uptake and an increased excretion rate of digoxin. This system was proposed as a possible treatment of drug overdose, not only for digoxin, but also for certain other drugs (395,396).

One significant theoretical limitation in the technique of targeting by the use of liposome-encapsulated antibodies is that such a procedure might preclude simultaneous encapsulation of certain other substances, such as drugs or enzymes. Therefore, nonspecific coating of *preformed* liposomes with aggregated immunoglobulins was the strategy underlying another method of liposome targeting by immunoglobulins proposed by Weissmann (166,189,343-351; reviewed in Ref. 352). In this system, liposomes are coated with nonspecific, heat-aggregated immunoglobulin (141). The purpose is to increase the lysosomotropic activity of liposome-encapsulated lysosomal enzymes and thus improve delivery of such enzymes in enzyme-deficient patients, particularly in lipidoses, such as Tay-Sachs disease (344,349). Liposomes coated with immunoglobulin might also have utility as a targeting mechanism in delivering enzymes in other types of disorders, such as chronic granulomatous disease (351). Although the in vitro experiments uniformly have looked promising for reconstituting deficient cells, to our knowledge only one in vivo animal study has been reported (346). During the latter presentation in 1977, Weissmann seemed pessimistic about the clinical future of utilizing this method to target liposomes to deliver enzymes (346). In all of these considerations it is important to keep in mind that enzyme delivery to deficient cells has proved to be a challenging clinical problem for various reasons in itself, irrespective of the enzyme carrier (397). Even idealized targeting of liposomes, or other carriers, may not provide all the conditions required for clinical success.

It has been presumed that increased uptake of liposomes coated with aggregated immunoglobulin is due to binding of the antibodies to Fc receptors on phagocytes (352). Indeed, receptor-mediated endocytosis of liposomes that had been coated with specific antibody to a liposomal antigen (DNP-Cap-PE) occurred via the Fc receptor (357,358).

One simple targeting method that has been employed is to use specific antibodies as a bridge between a hapten on cells and a hapten on liposomes (177, 398). A similar strategy was to incubate liposomes containing a hapten with myeloma tumor cells secreting an antibody against the hapten (399). These approaches did lead to uptake of liposomes by lymphocytes or tumor cells, but it may be significant that they did not lead to increased internal uptake of the trapped liposomal contents by the cells (177,398,399).

It has been reasoned that if antibodies could be covalently conjugated to liposomes, the antibodies might become a more effective targeting device. Three laboratories used different methods to conjugate monoclonal antibodies either to fatty acids or to a phospholipid (PE), and the complexes were subsequently introduced into liposomes (149-151). In two cases the derivatized complex was spontaneously inserted into preformed liposomes (149,151); in the other case the complex was added by the detergent dialysis method (150).

Three other laboratories covalently coupled specific antibodies directly to preformed liposomes (144,146-148). Heath et al. attached the antibodies, or F(ab')$_2$, by periodate oxidation of liposomal glycosphingolipids (146,147,400), while Torchilin et al. used glutaraldehyde (144), and Leserman et al. used a heterobifunctional cross-linking reagent (148) to conjugate antibodies to phosphatidylethanolamine in liposomes. In all cases, the antibodies retained activity. Under optimum conditions, up to 200 mg of IgG was bound per mmole of phospholipid and 80% of the targeted liposomes were associated with the target cells in vitro (146,147). The bound liposomes amounted to a lipid mass that was threefold greater than, and had a trapped aqueous volume that was one-third as much as, the cells to which they bound (146,147). To illustrate the advantage of using the covalently bound specific antibody targeting approach, Torchilin et al. demonstrated that they could visualize infarcted myocardium in dogs by a gamma scintigram obtained after injecting antimyosin-targeted liposomes that contained trapped ^{111}InCl$_3$ (144).

IMMUNIZATION WITH LIPOSOMES

The potential value of liposomes as carriers for immunization was recognized early (95,401,402; early reviews in Refs. 1, 402, and 403) and this has now emerged as one of the brightest prospects among projected possible clinical applications of liposomes (9). Initially, it was believed, based on results with diphtheria toxoid as antigen, that negatively charged liposomes caused increased antibody titers and that positively charged liposomes caused decreased antibody titers in animals (401). However, subsequent studies with other antigens did not support this generalization (402). From these modest beginnings, there have now appeared, by our count, more than 70 papers dealing with liposomes as adjuvants or as carriers of antigens for immunization. This section discusses mainly antibody production, while studies relating to cell-mediated immunity are handled elsewhere ("Immunity Associated with Histocompatibility Antigens," p. 237).

Liposomes as Adjuvants and Antigen Carriers

An adjuvant is a substance that potentiates the immunologic response against an antigen. Liposomes can fulfill adjuvant roles in many ways. Numerous lipids or hydrophobic proteins may be "solubilized" or presented in appropriate configurations in liposomes and thus increase their antigenicity (see Table 1). Substances having known adjuvant properties, such as LPS and lipid A (81,84,85,101,404, 405) or muramyl dipeptide (85,332,335,337,394,406-409), can be incorporated into liposomes. Liposomes may carry "recognition factors" such as histocompatibility antigens ("Immunity Associated with Histocompatibility Antigens," p. 237), "activation factors" such as lymphokines ("Liposomes as Carriers of Lymphokines," p. 242), or immune RNA (388,389). Many adjuvants exert their major effects in macrophages (410) and several studies have shown that macrophages readily ingest liposomes (175,315,316,333,341,355,411-413; see also "Immune Phagocytosis," p. 244).

An early argument in favor of the use of liposomes as adjuvants was that the liposomes themselves were nonimmunogenic (95). This is now known not to be completely true. Acidic phospholipids, such as cardiolipin and phosphatidylserine, are highly antigenic (13) and even lecithin is antigenic under certain conditions (29,81,82,84,85). Liposomes composed of phosphatidylcholine or sphingomyelin are not nonimmunogenic but they are certainly very poor antigens (see "Phospholipids," p. 219, for a complete discussion of antiliposome antibodies).

The existence of antiliposome antibodies, whether naturally occurring (74) or induced by immunization, does raise the theoretical problem that a protein antigen encapsulated within liposomes might be released from the liposomes by the action of complement-dependent immune damage prior to delivery to the appropriate immunologically competent cell. For two reasons we believe that this is unlikely to cause great practical difficulty. First, van Rooijen and van Nieuwmegen (414,415) have presented evidence suggesting that liposome-associated protein does not need to be encapsulated to be antigenic. They showed that liposomes served as adjuvants even for the small fraction of protein nonspecifically adsorbed on the liposome surface and that completely encapsulated proteins did not heighten the immune response. Second, certain proteins (perhaps many) have hydrophobic regions that cause them to be firmly embedded in the liposomal lipid bilayer rather than being only encapsulated (111; also see "Targeting of Liposomes by Antibodies," p. 246).

Kinsky and co-workers, using hapten-substituted phospholipid antigens, suggested that the magnitude of the anti-DNP immune response in mice was dependent on the transition temperature of the bulk phospholipids in the liposomes, with more solid liposomes giving a higher response (98,416,417). However, in our studies (B. G. Schuster and C. R. Alving, unpublished observations, 1979) we found no difference in the production of antiliposome antibodies in rabbits (as in Ref. 81) when the immunizing liposomes contained DMPC, DPPC, or DSPC, and Honnegger et al. (in the discussion section of Ref. 98) mentioned that certain hapten-phospholipid derivatives are immunogenic in fluid liposomes. Transition temperature may be an important consideration for certain immunogenic liposomal lipids but not for others. The effect of liposomal transition temperature on the production of antibodies against a liposomal protein antigen has not yet been reported.

Suppressive Effects of Liposomes

Liposomes can have potent immunosuppressive, as well as immunopotentiating, properties. A small amount of oxidized cholesterol, or purified 25-hydroxycholesterol, when present in liposomes had a potent suppressive effect on induction of plaque-forming anti-DNP responses in tissue culture. The antigens that were suppressed consisted either of DNP conjugated to Sepharose 4B (418) or DNP-Cap-PE in the liposomal bilayer (419). In another example of immunosuppression, in one experiment injection of liposomes (DMPC/cholesterol/dicetylphosphate) mixed with cholera toxin resulted in strong suppression of the anticholera toxin response in rabbits (420). The suppression occurred in eight rabbits but was not observed when repeated in older rabbits (420). The reason for the immunosuppression initially observed is not yet fully understood, but the possibility of a correlation between age of the animal and immunosuppression by liposomes is being investigated.

In another type of approach, Schwenk et al. found that reaginic (IgE) antibodies to DNP that had been induced in mice by a conjugate of DNP and ovalbumin were specifically suppressed by injecting liposomes containing DNP-Cap-PE. Antibodies against the ovalbumin were not suppressed (421). This interesting approach was proposed as a novel method to target liposomes to certain specific antibody-producing lymphocytes and perhaps even to certain neoplastic antibody-secreting cells.

Suppression or potentiation of IgM plaque-forming responses in tissue culture against liposomal DNP-Cap-PE was achieved by Humphries simply by altering the liposomal epitope density (419). Anti-sheep red blood cell plaque-forming response was not suppressed under conditions in which highly haptenated liposomes (i.e., suppressive liposomes) suppressed anti-DNP-Cap-PE responses against DNP-Cap-PE in separate liposomes that would normally be immunogenic (419). These results contrasted with an earlier in vivo study (422) in which suppression at high densities of liposomal DNP-Cap-PE was not observed. The reason for different results in these two reports is not yet clear. In another laboratory, when higher liposomal epitope densities were used, suppression was observed in vivo (423). Tadakuma et al. (424) found that pretreatment of cells in vitro with liposomes that were nonimmunogenic due to *low* epitope densities suppressed the immune response against liposomes that would normally be immunogenic.

Immunization with Liposomal Lipids

A series of interesting studies was made possible by the development, initially by Kinsky and colleagues, of various haptens conjugated to PE. A wide variety of haptens was employed, ranging from 2,4-dinitrophenyl-6-N-caproyl-PE (DNP-Cap-PE) (94) to spin label iodoacetamide PE derivatives (20,21,212) (see Table 1 for other examples). The initial approach was to immunize animals with liposomal lipids and examine either serum antibody activity, splenic plaque-forming cells, or delayed-type hypersensitivity reactions. Recently, in vitro plaque-forming cultures were employed with liposomal lipids as antigens.

The first studies utilizing liposomal hapten-conjugated phospholipids as immunogens (95,96,425,426) have been summarized and thoroughly reviewed elsewhere (1,13). Some of the important observations are summarized briefly as follows. The antihapten antibody response in rodents required the presence of antigen in the lipid bilayer and the response was dose dependent. Immunopotentiation by liposomes was not produced against nonliposomal antigen (95).

The length of the polar region of the hapten-lipid conjugate, and the extent to which it extended away from the surface of the liposomal bilayer, were correlated with the degree of antigenicity (427).

At least some lipid (i.e., fatty acyl constituent) was required for antigenicity of liposomal antigen. For delayed-type hypersensitivity (DTH), the lipid antigens did not require fatty acids (425). Exogenous adjuvant, such as complete Freund's adjuvant, had a marked stimulating effect on antigenicity (95) but adjuvant was not always required (96). These results when taken together suggested, at least with these hapten-conjugated phospholipid antigens, that the carrier function of liposomes was more important than other types of hypothetical adjuvant effects by liposomes.

Delayed-type hypersensitivity was not observed with liposomal DNP-Cap-PE (95) but DTH was found with ABA-Tyr-PE (425,428). The DTH reaction against ABA-Tyr-PE occurred regardless of whether ABA-Tyr-PE was or was not included in the liposomal bilayer. However, antibodies against ABA-Tyr-PE were greatly stimulated by inclusion of the hapten in liposomes (425). When DNP-Cap-PE and ABA-Tyr-PE were simultaneously present in liposomes, the antigenicity of DNP-Cap-PE was increased, and antigenicity of ABA-Tyr-PE was decreased, compared to results obtained with liposomes containing the antigens individually (426).

The importance of adjuvants in production of DTH against a tripeptide-enlarged antigen that was similar to ABA-Tyr-PE was examined by van Houte et al. (429). They found that DTH in mice was not produced against the liposomal hapten in the presence of complete Freund's adjuvant but that DTH did occur with a cationic detergent. In contrast, in guinea pigs, complete Freund's did stimulate DTH against the liposomal antigen (423).

Liposomes containing lipid antigens are primarily T cell-independent antigens. The immune response occurred in mice or in tissue culture populations genetically deficient in, or experimentally depleted of, T lymphocytes (419,422, 423,430). Antibodies obtained by immunizing with liposomes containing lipid haptens also had markedly restricted heterogeneity (96,98), and were mainly IgM, even after secondary immunization (419,422,423,430).

As mentioned above, antibodies produced against liposomal lipids, either in mice or in cultured mouse cells, are generally IgM, even after secondary immunization (419,422,423,430). However, it should be noted that IgG can be produced against liposomal LPS (404). In the absence of liposomes, the anti-LPS response was almost exclusively IgM, but the IgG response against LPS was increased 16-fold when LPS was injected into mice as a liposomal antigen (404).

In tissue culture, induction of plaque-forming cells secreting antibodies against liposomal lipid haptens generally requires the presence of adherent cells, presumably macrophages (430,431). One function of adherent macrophages in processing antigens might be to "present" the antigen in an appropriate fashion in the lipid bilayer on the macrophage plasma membrane. To test this hypothesis with a liposomal lipid antigen, Humphries (431) isolated a soluble factor derived from supernatants from concanavalin A-stimulated spleen cells. The soluble factor, which was not characterized, could replace adherent cells in stimulating immunity against liposomal lipid antigens, thus demonstrating that adherent macrophages are *not* necessarily required for presentation of liposomal lipid antigen (431). According to Humphries, the process of presenting the lipid antigen to B cells could be handled exclusively by the liposomal membrane itself, rather than by the macrophage, and the liposome might even serve as a model for this function as it occurs in macrophages. Using cultures depleted of adherent cells, but containing the soluble factor mentioned above, she showed that the antibody response against DNP-Cap-PE as a liposomal antigen could be either stimulated or suppressed by altering the liposomal epitope density. Epitope density was more important than total epitope dose in determining antigenic response (431). The role of epitope density was also examined by others in animals, and in cultures that contained adherent cells (423,424,427; see also "Suppressive Effects of Liposomes," p. 250).

The species, or strain, of animal can be a critical factor that can determine whether certain liposomal lipids are antigenic. We have found that lipid A, whether in liposomes or not, is highly immunogenic in rabbits (81,84) but is virtually nonimmunogenic in mice (432). Initially, we thought that antilipid A antibodies could also be produced in mice, but this turned out not to be true since the apparent antibody activity was caused by an effect of lipid A in changing the fluidity of liposomes and altering immune damage to liposomes in the assay procedure (86). Poor antigenicity of lipid A in mice has also been observed elsewhere (433,434). Antibodies against liposomal LPS have been produced in mice (404). Despite its lack of antigenicity, lipid A is still a strong adjuvant in mice and it has been used to induce mouse hybridoma antiliposome antibodies (86,90) (see "Phospholipids," p. 219). The mouse strain employed can also be important in determining antigenicity of a lipid antigen. The ability of certain inbred mouse strains to respond to DNP-Cap-PE did not parallel the ability to respond to a lipid antigen containing phosphocholine (98).

It has been reported that a glycolipid fraction, obtained from a virus (VSV) that had infected an SV40-transformed tumor cell, upon incorporation into liposomes induced antitumor immunity in hamsters (435).

Immunogenicity of Liposome-Associated Protein

Numerous protein antigens have been incorporated into or attached onto liposomes for immunization purposes. These include albumin, immunoglobulins, diphtheria toxoid, cholera toxin, malaria parasite antigens, viral proteins, various enzymes, myelin proteins, and histocompatibility antigens (references given in Table 1). Depending on the circumstances, liposomes have been used in the hope of either enhancing or inhibiting the antigenicity of proteins.

Information in this area is proliferating rapidly and some useful concepts have emerged. It appears that the antigenicity, either humoral or cellular, of nearly any protein can be enhanced by association with liposomes. There has been no single ideal formulation of lipids identified, although immunosuppressive lipids have been found (see "Suppressive Effects of Liposomes," p. 250). A given protein may do better with a certain type of liposome, such as a negatively charged one (401), whereas another protein may do equally well with neutral, positively or negatively charged liposomes (402,415). Probably certain proteins, whether due to the isoelectric points, particular sequences of charged amino acids, hydrophobicity, or other factors, are incorporated to a greater or lesser extent, or in different configurations, in liposomes having different lipid compositions, and the antigenicity is probably altered. As with lipid antigens, factors such as epitope density (296) may have importance.

To avoid the uncertainties of protein incorporation into liposomes, we devised an antigenic model in which the surface protein epitope density and the orientation of the protein were rigorously controlled by high-affinity binding to a liposomal receptor (84,85,420). In this system, the antigen consisted of cholera toxin bound to liposomes via its receptor (ganglioside G_{M1}). One advantage to this combination is that the toxicity of the cholera toxin was completely eliminated by high-affinity binding to a liposomal receptor (111). Liposomally bound toxin, given either intravenously or subcutaneously, induced a much higher humoral immune response in rabbits than did unbound toxin (84,420). When the liposomes also contained lipid A as an adjuvant, the antitoxin response was much higher, approximately equivalent to that induced by a combination of toxin plus complete Freund's adjuvant (84,420). Presumably, because liposomal G_{M1} was blocked by bound cholera toxin, there was no increased antibody titer against G_{M1} itself (85).

One of the earliest suggested clinical uses of liposomes was that lysosomal enzymes might be encapsulated into liposomes and delivered efficiently in disorders in which such enzymes were lacking (352,436,437). Interest in this approach began to wane, however, as the adjuvant properties of liposomes became more obvious (112). Indeed, investigators in one laboratory optimistically initially began to explore methods to encapsulate enzymes in liposomes (438). They then obtained evidence for adverse immunological effects against liposome-encapsulated enzymes (113) and because of these effects they now advocate encapsulation of enzymes in erythrocyte ghosts rather than in liposomes (112).

There are reasons for believing that pessimism about the future of liposome-encapsulated enzymes solely because of potential immunological problems is not fully warranted. Van Rooijen and van Nieuwmegen have proposed the interesting hypothesis that protein that is *fully* encapsulated within liposomes is *not* antigenic, but that protein adsorbed to the outer surface of liposomes is highly antigenic (414,415). This conclusion was based on indirect evidence, namely that the same immune response against albumin or immunoglobulin was obtained by injecting liposomes swollen in the protein solution (the liposomes having entrapped protein and nonspecifically adsorbed protein) or by injecting empty liposomes that were merely incubated in the protein solution after liposome formation (414,415). This concept of both nonantigenic and antigenic protein in liposomes might be important because it would suggest that enzymes that were truly encapsulated or "latent" by rigorous criteria (e.g., as proposed in Refs. 343 and 439) could be completely sequestered from the immune system. However, even though certain enzymes apparently can be completely encapsulated (105,109, 439), there is evidence that achieving complete latency of a protein in liposomes is not always an easy task. In our model system described above, although there was no detectable binding of cholera toxin to preformed "empty" liposomes lacking the G_{M1} receptor (110), when the liposomes lacking G_{M1} were swollen in cholera toxin during formation, only approximately two-thirds of the liposome-associated toxin was fully encapsulated (i.e., latent). It was found that one-third was associated with the lipid bilayer in some way, probably through hydrophobic bonds (111). Cholera toxin partially encapsulated in liposomes in this way had the same increased antigenicity observed with cholera toxin bound to liposomes containing G_{M1} (111). If the van Rooijen-van Nieuwmegen hypothesis is correct (and it still awaits a successful direct test), perhaps the most sensitive criterion for intraliposomal latency of a protein will be lack of antigenicity upon injection into an animal.

Liposomes can be used to prevent hypersensitivity reactions (Arthus reaction or serum sickness) against certain proteins in animals preimmunized against the protein (104,105,109). Such reactions are a major problem in the use of L-asparaginase in treatment of leukemia and they may be overcome by liposomes (105,109).

ABBREVIATIONS

ABA-Tyr, azobenzenearsonic acid-tyrosine

ABS-Tyr, azobenzenesulfonic acid-tyrosine

ADCC, antibody-dependent cell-mediated cytotoxicity

APPC-PPr, a phosphocholine-containing conjugate of PE

CDH, ceramide dihexoside

CMH, ceramide monohexoside

CRP, C-reactive protein

CTH, ceramide trihexoside

CTL, cytotoxic T lymphocyte

DMPC, dimyristoyl phosphatidyl-choline

DNP-Cap-PE, dinitrophenyl-amino-caproyl-phosphatidylethanolamine

DPPC, dipalmitoyl phosphatidyl-choline

DSPC, distearoyl phosphatidylcholine

DTH, delayed-type hypersensitivity

EMP, eye muscle protein

FITC, fluoroscein isothiocyanate

Gal, galactose

GalNAc, N-acetylgalactosamine
K cell, killer cell
MDP, muramyl dipeptide
MHC, major histocompatibility
  complex
MIF, macrophage inhibitory factor

LPS, lipopolysaccharide
PE, phosphatidylethanolamine
PIP, phosphatidylinositol phosphate
Tg, thyroglobulin
TgA, a complex of Tg bound to specific
  antibody

REFERENCES

1.   Kinsky, S. C. and R. A. Nicolotti. 1977. Immunological properties of
     model membranes. Annu. Rev. Biochem. 46:49.
2.   McConnell, H. M. 1978. Immunochemistry of model membranes contain-
     ing spin labels. Int. Rev. Biochem. 19:45.
3.   Taketomi, T. and K. Uemura. 1980. Immunological properties of glyco-
     lipids including some gangliosides in mammalian erythrocyte membranes.
     In *Structure and Function of Gangliosides*, L. Svennerholm, H. Dreyfus,
     and P.-F. Urban (Eds.). Plenum, New York, pp. 349-358.
4.   Yasuda, T. and T. Tadakuma. 1979. Immunological applications of lipo-
     somal model membranes. Taisha 16:1975.
5.   Humphries, G. M. K. 1980. The use of liposomes for studying membrane
     antigens as immunogens and as targets for immune attack. In *Liposomes
     in Biological Systems*, G. Gregoriadis and A. C. Allison (Eds.). Wiley, New
     York, pp. 345-376.
6.   Kinsky, S. C. 1980. Factors affecting liposomal model membrane immuno-
     genicity. In *Liposomes and Immunobiology*, B. H. Tom and H. R. Six
     (Eds.). Elsevier/North-Holland, New York, pp. 79-90.
7.   Gregoriadis, G. 1980. The liposome drug-carrier concept: its development
     and future. In *Liposomes in Biological Systems*, G. Gregoriadis and A. C.
     Allison (Eds.). Wiley, New York, pp. 25-86.
8.   Gregoriadis, G. 1980. Recent progress in liposome research. In *Liposomes
     in Biological Systems*, G. Gregoriadis and A. C. Allison (Eds.). Wiley, New
     York, pp. 377-398.
9.   Tom, B. H. 1980. An overview: liposomes and immunobiology—macro-
     phages, liposomes and tailored immunity. In *Liposomes and Immuno-
     biology*, B. H. Tom and H. R. Six (Eds.). Elsevier/North-Holland, New York,
     pp. 3-22.
10. Gregoriadis, G. 1981. Liposomes: a role in vaccines? Clin. Immunol.
     Newslett. 2:33.
11. van Rooijen, N. and R. van Nieuwmegen. 1982. Immunoadjuvant proper-
     ties of liposomes. In *Targeting of Drugs*, G. Gregoriadis, J. Senior and A.
     Trouet (Eds.). Plenum, New York, pp. 301-326.
12. Scherphof, G., J. Damen, and D. Hoekstra. 1981. Interactions of lipo-
     somes with plasma proteins and components of the immune system. In
     Liposomes: From Physical Structure to Therapeutic Applications, C. G.
     Knight (Ed.). Research Monographs in Cell and Tissue Physiology vol. 7.
     North-Holland, Amsterdam, pp. 299-322.

13.   Alving, C. R. 1977. Immune reactions of lipids and lipid model membranes. In *The Antigens,* Vol. 4, M. Sela (Ed.). Academic Press, New York, pp. 1–72.

14.   Tom, B. F. and H. R. Six (Eds.). 1980. *Liposomes and Immunobiology.* Elsevier/North-Holland, New York.

15.   Rapport, M. M. and L. Graf. 1969. Immunochemical reactions of lipids. Prog. Allergy 13:273.

16.   Kanemasa, Y. 1974. Electron microscopic observations of lipid antigens. With special reference to the Wasserman antigen. Jpn. J. Microbiol. 18:193.

17.   Alving, C. R., J. W. Fowble, and K. C. Joseph. 1974. Comparative properties of four galactosyl lipids as antigens in liposomes. Immunochemistry 11:475.

18.   Alving, C. R., D. H. Conrad, J. P. Gockerman, M. B. Gibbs, and G. H. Wirtz. 1975. Vitamin A in liposomes. Inhibition of complement binding and alteration of membrane structure. Biochim. Biophys. Acta 394:157.

19.   Humphries, G. M. K. and H. M. McConnell. 1975. Antigen mobility in membranes and complement-mediated immune attack. Proc. Natl. Acad. Sci. USA 72:2483.

20.   Brûlet, P. and H. M. McConnell. 1976. Lateral hapten mobility and immunochemistry of model membranes. Proc. Natl. Acad. Sci. USA 73:2977.

21.   Brûlet, P. and H. M. McConnell. 1977. Structural and dynamical aspects of membrane immunochemistry using model membranes. Biochemistry 16:1209.

22.   Humphries, G. M. K. and H. M. McConnell. 1977. Membrane-controlled depletion of complement activity by spin-label-specific IgM. Proc. Natl. Acad. Sci. USA 74:3537.

23.   Niedieck, B. 1975. On the function of lecithin and lecithin substitutes in the immune precipitation reaction of galactosyl lipids. Immunochemistry 12:807.

24.   Hruby, S., E. C. Alvord, Jr., and F. J. Seil. 1977. Synthetic galactocerebrosides evoke myelination-inhibiting antibodies. Science 195:173.

25.   Zalc, B., P. Dupouey, M. J. Coulon-Morelec, and N. A. Baumann. 1979. Immunogenic properties of glucosylceramide. Mol. Immunol. 16:297.

26.   Kundu, S. K., D. M. Marcus, and R. W. Veh. 1980. Preparation and properties of antibodies to G_{D3} and G_{M1} gangliosides. J. Neurochem. 34:184.

27.   Uchida, T. and Y. Nagai. 1980. Affinity chromatographic purification of antiglycolipid antibodies and their application to the membrane studies. Antigalactocerebroside antibodies. J. Biochem. 87:1829.

28.   Uchida, T. and Y. Nagai. 1980. Application of highly purified anti-galactocerebroside antibody to the analysis of lipid-lipid or lipid-protein interactions. J. Biochem. 87:1843.

29.   Niedieck, B., U. Kuck, and H. Gardemin. 1978. On the immune precipitation of phosphorylcholine lipids with TEPC 15 mouse myeloma protein and with anti-lecithin sera from guinea pigs. Immunochemistry 15:471.

30.   Greenberg, A. J., A. J. Trevor, D. A. Johnson, and H. H. Loh. 1979. Immunochemical studies of phospholipids: production of antibodies to triphosphoinositide. Mol. Immunol. 16:193.

31. Thudichum, J. L. W. 1884. *A Treatise on the Chemical Constitution of the Brain.* Reprinted by the Shoe String Press, Hamden, Conn., 1962.

32. Forssman, J. 1911. Die Herstellung hochwertiger spezifischer Schaf hämolysine ohne Verwendung von Schafblut. Ein Beitrag zur Lehre von heterologer Antikörperbildung. Biochem. Z. 37:78.

33. Haxby, J. A., C. B. Kinsky, and S. C. Kinsky. 1968. Immune response of a liposomal model membrane. Proc. Natl. Acad. Sci. USA 61:300.

34. Alving, C. R., S. C. Kinsky, J. A. Haxby, and C. B. Kinsky. 1969. Antibody binding and complement fixation by a liposomal model membrane. Biochemistry 8:1582.

35. Kinsky, S. C., J. A. Haxby, D. A. Zopf, C. R. Alving, and C. B. Kinsky. 1969. Complement-dependent damage to liposomes prepared from pure lipids and Forssman hapten. Biochemistry 8:4149. (Erratum in 10:1048.)

36. Inoue, K., T. Kataoka, and S. C. Kinsky. 1971. Comparative responses of liposomes prepared with different ceramide antigens to antibody and complement. Biochemistry 10:2574.

37. Alving, C. R., K. C. Joseph, and R. Wistar. 1974. Influence of membrane composition on the interaction of a human monoclonal "anti-Forssman" immunoglobulin with liposomes. Biochemistry 13:4818.

38. Sweeley, C. C. (Ed.). 1980. *Cell Surface Glycolipids.* ACS Symposium Series No. 128. American Chemical Society, Washington, D.C.

39. Esselman, W. J., R. A. Laine, and C. C. Sweeley. 1972. Isolation and characterization of glycosphingolipids. Methods Enzymol. 28:140.

40. Hakomori, S. 1972. Preparation of antisera against glycolipids. Methods Enzymol. 28:232.

41. Hakomori, S. 1973. Glycolipids of tumor cell membrane. Adv. Cancer Res. 18:265.

42. Hakomori, S. 1975. Structures and organization of cell surface glycolipids, dependency on cell growth and malignant transformation. Biochim. Biophys. Acta 417:55.

43. Critchley, D. R. 1979. Glycolipids as membrane receptors important in growth regulation. In *Surfaces of Normal and Malignant Cells*, R. O. Hynes (Ed.). Wiley, Chichester, England, pp. 63-101.

44. Hakomori, S. and A. Kobata. 1974. Blood group antigens. In *The Antigens*, Vol. 2, M. Sela (Ed.). Academic Press, New York, pp. 79-140.

45. Hakomori, S. 1981. Blood group ABH and Ii antigens of human erythrocytes: chemistry, polymorphism, and their developmental change. Semin. Hematol. 18:39.

46. Marcus, D. M., S. K. Kundu, and A. Suzuki. 1981. The P blood group system: recent progress in immunochemistry and genetics. Semin. Hematol. 18:63.

47. Rapport, M. M., S. E. Karpiak, and S. P. Mahadik. 1979. Biological activities of antibodies injected into the brain. Fed. Proc. 38:2391.

48. Rapport, M. M., S. E. Karpiak, and S. P. Mahadik. 1980. Perturbation of CNS functions by antibodies to gangliosides. Speculations on biological roles of ganglioside receptors. In *Structure and Function of Gangliosides*,

L. Svennerholm, H. Dreyfus, and P.-F. Urban (Eds.). Plenum, New York, pp. 335-338.

49. Rapport, M. M., S. E. Karpiak, and S. P. Mahadik. 1980. Perturbation of behavior and other CNS functions by antibodies to ganglioside. In *Cell Surface Glycolipids,* C. C. Sweeley (Ed.). ACS Symposium Series No. 128. American Chemical Society, Washington, D.C., pp. 407-417.

50. Nagai, Y., T. Momoi, M. Saito, E. Mitsuzawa, and S. Ohtani. 1976. Ganglioside syndrome, a new autoimmune neurologic disorder, experimentally induced with brain gangliosides. Neurosci. Lett. 2:107.

51. Niedieck, B. 1975. On a glycolipid hapten of myelin. Prog. Allergy 18: 353.

52. Marcus, D. M. and G. A. Schwarting. 1976. Immunochemical properties of glycolipids and phospholipids. Adv. Immunol. 23:203.

53. Siddiqui, B. and S. Hakomori. 1971. A revised structure for the Forssman glycolipid hapten. J. Biol. Chem. 246:5766.

54. Koscielak, J., S. Hakomori, and R. W. Jeanloz. 1968. Glycolipid antigen and its antibody. Immunochemistry 5:441.

55. Uemura, K., M. Yuzawa, and T. Taketomi. 1978. Preparation and properties of antisera to glycolipid of guinea pig erythrocyte membrane. J. Biochem. 83:1199.

56. Young, W. W., Jr., S. Hakomori, and P. Levine. 1979. Characterization of anti-Forssman (anti-Fs) antibodies in human sera: their specificity and possible changes in patients with cancer. J. Immunol. 123:92.

57. Alving, C. R. 1983. Antibodies against lipids, lipid bilayers and liposomes: A theory of aging. In *The Liposome Letters,* A. D. Bangham (Ed.). Academic Press, New York, in press.

58. Young, W. W., Jr., E. M. S. MacDonald, R. C. Nowinski, and S. Hakomori. 1979. Production of monoclonal antibodies specific for two distinct steric portions of the glycolipid ganglio-N-triosylceramide (asialo GM_2). J. Exp. Med. 150:1008.

59. Joseph, K. C., C. R. Alving, and R. Wistar. 1974. Forssman-containing liposomes: complement-dependent damage due to interaction with a monoclonal IgM. J. Immunol. 112:1949.

60. Naiki, M. and D. M. Marcus. 1977. Binding of N-acetylgalactosamine-containing compounds by a human IgM paraprotein. J. Immunol. 119:537.

61. Nowinski, R., C. Berglund, J. Lane, M. Lostrom, I. Bernstein, W. Young, S. Hakomori, L. Hill, and M. Cooney. 1980. Human monoclonal antibody against Forssman antigen. Science 210:537.

62. Race, R. R. and R. Sanger. 1975. The P blood groups. In *Blood Groups in Man,* R. R. Race and S. Sanger (Eds.). Blackwell, Oxford, pp. 139-177.

63. Schwarting, G. A., S. K. Kundu, and D. M. Marcus. 1979. Reaction of antibodies that cause paroxysmal cold hemoglobinuria (PCH) with globoside and Forssman glycosphingolipids. Blood 53:186.

64. Tsai, C., D. A. Zopf, and V. Ginsburg. 1978. The molecular basis for cold agglutination: effect of receptor density upon thermal amplitude of a cold agglutinin. Biochem. Biophys. Res. Commun. 80:905.

65. Alving, C. R. and R. L. Richards. 1977. Immune reactivities of antibodies against glycolipids. II. Comparative properties, using liposomes, of purified antibodies against mono-, di- and trihexosyl ceramide haptens. Immunochemistry 14:383.

66. Slovick, D. I., T. Saida, R. P. Lisak, and A. Schreiber. 1980. A new assay for lytic anti-galactocerebroside (GC) antibodies employing [56]rubidium release from GC-labelled liposomes. J. Immunol. Methods 39:31.

67. Hamers, M. N., W. E. Donker-Koopman, M.-J. Coulon-Morelec, P. Dupouey, and J. M. Tager. 1978. Characterization of antibodies against ceramidetrihexoside and globoside. Immunochemistry 15:353.

68. Clarke, J. T. R. and J. A. Embil. 1979. Preparation and characterization of antibody to galactosyl(α1→4)galactosyl(β1→4)glucosylceramide. Biochim. Biophys. Acta 582:283.

69. Dupouey, P., A. Billecocq, and M. Lefroit. 1976. Comparative study of the immunological properties of galactosyldiglyceride and galactosylceramide included within natural membranes. Immunochemistry 13:289.

70. Moss, J., P. H. Fishman, R. L. Richards, C. R. Alving, M. Vaughan, and R. O. Brady. 1976. Choleragen-mediated release of trapped glucose from liposomes containing ganglioside G_{M1}. Proc. Natl. Acad. Sci. USA 73:3480.

71. Hirsch, H. E. and M. E. Parks. 1976. Serological reactions against glycolipid-sensitised liposomes in multiple sclerosis. Nature 264:785.

72. Richards, R. L. and C. R. Alving. 1980. Immune reactivities of antibodies against glycolipids. Natural antibodies. In *Cell Surface Glycolipids*, C. C. Sweeley (Ed.). ACS Symposium Series 128. American Chemical Society, Washington, D. C., pp. 461–473.

73. Uemura, K., M. Yuzawa-Watanabe, N. Kitazawa, and T. Taketomi. 1980. Liposome agglutination and liposomal membrane immune-damage assays for the characterization of antibodies to glycosphingolipids. J. Biochem. 87:1641.

74. Strejan, G. H., P. M. Smith, C. W. Grant, and D. Surlan. 1979. Naturally occurring antibodies to liposomes. I. Rabbit antibodies to sphingomyelin-containing liposomes before and after immunization with unrelated antigens. J. Immunol. 123:370.

75. Alving, C. R., K. A. Urban, and R. L. Richards. 1980. Influence of temperature on complement-dependent immune damage to liposomes. Biochim. Biophys Acta 600:117.

76. Laine, R. A., G. Yogeeswaran, and S. Hakomori. 1974. Glycosphingolipids covalently linked to agarose gel or glass beads. Use of the compounds for purification of antibodies directed against globoside and hematoside. J. Biol. Chem. 249:4460.

77. Alving, C. R. and R. L. Richards. 1977. Immune reactivities of antibodies against glycolipids. I. Properties of anti-galactocerebroside antibodies purified by a novel technique of affinity binding to liposomes. Immunochemistry 14:373.

78. Uemura, K., M. Yuzawa-Watanabe, N. Kitazawa, and T. Taketomi. 1980. Immunochemical studies of lipids. VI. Reactions of anti-sulfatide antibodies with sulfatide in liposomal and myelin membranes. J. Biochem. 87:1221.

79. Curtain, C. C. 1979. Lymphocyte surface modulation and glycosphingo-
     lipids. Immunology 36:805.
80. Wasserman, A., A. Neisser, and C. Bruck. 1906. Eine serodiagnostische
     Reaktion bei Syphilis. Dtsch. Med. Wochenschr. 32:745.
81. Schuster, B. G., M. Neidig, B. M. Alving, and C. R. Alving. 1979. Produc-
     tion of antibodies against phosphocholine, phosphatidylcholine, sphingo-
     myelin, and lipid A by injection of liposomes containing lipid A. J. Immu-
     nol. 122:900.
82. Banerji, B. and C. R. Alving. 1981. Anti-liposome antibodies induced by
     lipid A. I. Influence of ceramide, glycosphingolipids, and phosphocholine
     on complement damage. J. Immunol. 126:1080.
83. Chan, S. W., C. T. Tan, and J. C. Hsia. 1977. Antiliposome antisera activity
     against negatively charged phosphate amphiphils. Biochem. Biophys. Res.
     Commun. 79:631.
84. Alving, C. R., B. Banerji, J. D. Clements, and R. L. Richards. 1980. Adjuvanti-
     city of lipid A and lipid A fractions in liposomes. In *Liposomes and Immu-
     nobiology*, B. T. Tom and H. R. Six (Eds.). Elsevier/North-Holland, New
     York, pp. 67–78.
85. Alving, C. R., B. Banerji, T. Shiba, S. Kotani, J. D. Clements, and R. L. Rich-
     ards. 1980. Liposomes as vehicles for vaccines. In *New Developments in
     Human and Veterinary Vaccines*, A. Mizrahi, I. Hertman, M. A. Klingberg,
     and A. Kohn (Eds.). Proceedings of the 25th Oholo Conference. (In the
     series: *Progress in Clinical and Biological Research*, Vol. 47.) Alan R. Liss,
     New York, pp. 339–355.
86. Banerji, B., J. A. Lyon, and C. R. Alving. 1982. Membrane lipid composi-
     tion modulates the binding specificity of a monoclonal antibody against
     liposomes. Biochim. Biophys. Acta 689:319.
87. Sato, J. and I. Hara. 1972. Anti-cholesterol activity in antisera against
     human serum lipoproteins. Immunochemistry 9:585.
88. Sato, J., T. Fukuda, K. Suzuki, and I. Hara. 1976. Antibody-like activities
     against cholesterol ester in anti-LDL antiserum. An observation on the sur-
     face of LDL. Biomedicine 24:385.
89. Sato, J., T. Fukuda, and I. Hara. 1976. Preparation of anti-etiocholenic
     acid antiserum and anti-cholesterol succinate antiserum and their cross
     reactivities. Jpn. J. Exp. Med. 46:213.
90. Roerdink, F., B. J. Berson, R. L. Richards, G. M. Swartz, Jr., J. A. Lyon,
     and C. R. Alving. 1981. Specificity of a hybridoma monoclonal antibody
     against liposomes containing phosphatidylinositol phosphate. Fed. Proc.
     40(3):996.
91. Friedman, R. L., B. H. Iglewski, F. Roerdink, and C. R. Alving. 1982.
     Suppression of cytotoxicity of diphtheria toxin by monoclonal antibodies
     against phosphatidylinositol phosphate. Biophys. J. 37:23.
92. Alving, C. R., B. H. Iglewski, K. A. Urban, J. Moss, R. L. Richards, and J.
     C. Sadoff. 1980. Binding of diphtheria toxin to phospholipids in lipo-
     somes. Proc. Natl. Acad. Sci. USA 77:1986.
93. Uemura, K. and S. C. Kinsky. 1972. Active vs. passive sensitization of
     liposomes toward antibody and complement by dinitrophenylated deriva-
     tives of phosphatidylethanolamine. Biochemistry 11:4085.

94.   Six, H. R., K. Uemura, and S. C. Kinsky. 1973. Effect of immunoglobulin class and affinity on the initiation of complement-dependent damage to liposomal model membranes sensitized with dinitrophenylated phospholipids. Biochemistry 12:4003.

95.   Uemura, K., R. A. Nicolotti, H. R. Six, and S. C. Kinsky. 1974. Antibody formation in response to liposomal model membranes sensitized with N-substituted phosphatidylethanolamine derivatives. Biochemistry 13:1572.

96.   Uemura, K., J. L. Claflin, J. M. Davie, and S. C. Kinsky. 1975. Immune response to liposomal model membranes: restricted IgM and IgG anti-dinitrophenyl antibodies produced in guinea pigs. J. Immunol. 114:958.

97.   Kinsky, S. C. 1978. Immunogenicity of liposomal model membranes. Ann. N.Y. Acad. Sci. 308:111.

98.   Honegger, J. L., P. C. Isakson, and S. C. Kinsky. 1980. Murine immunogenicity of N-substituted phosphatidylethanolamine derivatives in liposomes: response to the hapten phosphocholine. J. Immunol. 124:669.

99.   Kataoka, T., K. Inoue, O. Lüderitz, and S. C. Kinsky. 1971. Antibody- and complement-dependent damage to liposomes prepared with bacterial lipopolysaccharides. Eur. J. Biochem. 21:80.

100. Kataoka, T., K. Inour, C. Galanos, and S. C. Kinsky. 1971. Detection and specificity of lipid A antibodies using liposomes sensitized with lipid A and bacterial lipopolysaccharides. Eur. J. Biochem. 24:123.

101. Dancey, G. F., T. Yasuda, and S. C. Kinsky. 1977. Enhancement of liposomal model membrane immunogenicity by incorporation of lipid A. J. Immunol. 119:1868.

102. Banerji, B. and C. R. Alving. 1979. Lipid A from endotoxin: antigenic activities of purified fractions in liposomes. J. Immunol. 123:2558.

103. Galanos, C., O. Lüderitz, and O. Westphal. 1971. Preparation and properties of antisera against the lipid-A component of bacterial lipopolysaccharides. Eur. J. Biochem. 24:116.

104. Gregoriadis, G. and A. C. Allison. 1974. Entrapment of proteins in liposomes prevents allergic reactions in pre-immunised mice. FEBS Lett. 45:71.

105. Fishman, Y. and N. Citri. 1975. L-Asparaginase entrapped in liposomes: preparation and properties. FEBS Lett. 60:17.

106. Allison, A. C. and G. Gregoriadis. 1976. Liposomes as immunological adjuvants. Recent Results Cancer Res. 56:58.

107. Wilson, T., D. Papahadjopoulos, and R. Taber. 1977. Biological properties of poliovirus encapsulated in lipid vesicles: antibody resistance and infectivity in virus-resistant cells. Proc. Natl. Acad. Sci. USA 74:3471.

108. Taber, R., T. Wilson, and D. Papahadjopoulos. 1978. The encapsulation of picorna viruses by lipid vesicles: physical and biological properties. Ann. N.Y. Acad. Sci. 308:268.

109. Neerunjun, E. D. and G. Gregoriadis. 1976. Tumour regression with liposome-entrapped asparaginase: some immunological advantages. Biochem. Soc. Trans. 4:133.

110. Fishman, P. H., J. Moss, R. L. Richards, R. O. Brady, and C. R. Alving. 1979. Liposomes as model membranes for ligand-receptor interactions: Studies with choleragen and glycolipids. Biochemistry 18:2562.

111.   Alving, C. R., J. Moss, R. L. Richards, and L. I. Alving. 1981. Liposomes as vehicles for vaccines. Increased antigenicity and lack of toxicity of a toxin bound to liposomes. Clin. Res. 29:531A.

112.   Hudson, L. D. S., M. B. Fiddler, and R. J. Desnick. 1979. Enzyme therapy. X. Immune response induced by enzyme- and buffer-loaded liposomes in C3H/HeJ Gus[h] mice. J. Pharmacol. Exp. Ther. 208:507.

113.   Desnick, R. J., M. B. Fiddler, S. D. Douglas, and L. D. S. Hudson. 1978. Enzyme therapy. XI. Immunologic considerations for replacement therapy with untrapped, erythrocyte- and liposome-entrapped enzymes. Adv. Exp. Med. Biol. (Enzymes, Lipid Metab.) 101:753.

114.   Hudson, L. D. S., M. B. Fiddler, and R. J. Desnick. 1980. Immunologic aspects of enzyme replacement therapy. An evaluation of the immune response to unentrapped, erythrocyte- and liposome-entrapped enzyme in C3H/HeJ Gus[h] mice. In *Enzyme Therapy in Genetic Diseases: 2*, R. J. Desnick (Eds.). Birth Defects: Original Articles Series, Vol. 16, No. 1. Alan R. Liss, New York, pp. 163–178.

115.   Huang, L. and S. J. Kennel. 1979. Binding of immunoglobulin G to phospholipid vesicles by sonication. Biochemistry 18:1702.

116.   Magee, W. E., J. H. Cronenberger, D. E. Thor, and R. E. Paque. 1980. Modulation of the immune response by targeted liposomes containing nucleic acids and other agents. In *Liposomes and Immunobiology*, B. H. Tom and H. R. Six (Eds.). Elsevier/North-Holland, New York, pp. 133–149.

117.   Sharom, F. J., D. G. Barratt, and C. W. M. Grant. 1977. Glycophorin and the concanavalin A receptor of human erythrocytes: their receptor function in lipid bilayers. Proc. Natl. Acad. Sci. USA 74:2751.

118.   Kasahara, M. and P. C. Hinkle. 1977. Reconstitution and purification of the D-glucose transporter from human erythrocytes. J. Biol. Chem. 252:7384.

119.   Boggs, J. M., I. R. Clement, M. A. Moscarello, E. H. Eylar, and G. Hashim. 1981. Antibody precipitation of lipid vesicles containing myelin proteins: dependence on lipid composition. J. Immunol. 126:1207.

120.   Manesis, E. K., C. H. Cameron, and G. Gregoriadis. 1978. Incorporation of hepatitis-B surface antigen ($HB_s Ag$) into liposomes. Biochem. Soc. Trans. 6:925.

121.   Gerlier, D., F. Sakai, and J.-F. Doré. 1978. Inclusion d'un antigène de surface cellulaire associé au virus de Gross dans des liposomes. C. R. Hebd. Seances Acad. Sci. Ser. D Sci. Nat. 286:439.

122.   Manesis, E. K., C. H. Cameron, and G. Gregoriadis. 1979. Hepatitis B surface antigen-containing liposomes enhance humoral and cell-mediated immunity to the antigen. FEBS Lett. 102:107.

123.   Morein B., D. Barz, U. Koszinowski, and V. Schirrmacher. 1979. Integration of a virus membrane protein into the lipid bilayer of target cells as a prerequisite for immune cytolysis. Specific cytolysis after virosome-target cell fusion. J. Exp. Med. 150:1383.

124.   Gregoriadis, G. and E. K. Manesis. 1980. Liposomes as immunological adjuvants for hepatitis B surface antigens. In *Liposomes and Immuno-*

biology, B. H. Tom and H. R. Six (Eds.). Elsevier/North-Holland, New York, pp. 271–283.

125. Hale, A. H. and M. J. Ruebush. 1980. Minimal molecuaar and cellular requirements for elicitation of secondary anti-vesicular stomatitis virus cytotoxic T lymphocytes. J. Immunol. 125:1569.

126. Sakai, F., D. Gerlier, and J. F. Doré. 1980. Association of gross virus-associated cell-surface antigen with liposomes. Br. J. Cancer 41:227.

127. Trudel, M., M. Ravaoarinoro, and P. Payment. 1980. Reconstitution of rubella hemagglutinin on liposomes. Can. J. Microbiol. 26:899.

128. Curman, B., L. Östberg, and P. A. Peterson. 1978. Incorporation of murine MHC antigens into liposomes and their effect in the secondary mixed lymphocyte reaction. Nature 272:545.

129. Klareskog, L., G. Banck, A. Forsgren, and P. A. Peterson. 1978. Binding of HLA antigen-containing liposomes to bacteria. Proc. Natl. Acad. Sci. USA 75:6197.

130. Turner, M. J. and A. R. Sanderson. 1978. The preparation of liposomes bearing human (HLA) transplantation antigens. Biochem. J. 171:505.

131. Willoughby, E. M., M. J. Turner, and A. R. Sanderson. 1978. Incorporation of rat histocompatibility (AgB) antigens into liposomes, and their susceptibility to immune lysis. Eur. J. Immunol. 8:628.

132. Engelhard, V. H., B. C. Guild, A. Helenius, C. Terhorst, and J. L. Strominger. 1978. Reconstitution of purified detergent-soluble HLA-A and HLA-B antigens into phospholipid vesicles. Proc. Natl. Acad. Sci. USA 75:3230.

133. Littman, D. R., S. E. Cullen, and B. D. Schwartz. 1979. Insertion of Ia and H-2 alloantigens into model membranes. Proc. Natl. Acad. Sci. USA 76:902.

134. Whisnant, C. C. and D. B. Amos. 1979. Interaction of lipid vesicles containing H-2 antigens with alloantibody and with alloimmune cytotoxic T lymphocytes. In *T and B Lymphocytes: Recognition and Functions*, F. H. Bach, B. Bonavida, E. S. Vitetta, and C. F. Fox (Eds.). ICN-UCLA Symposia on Molecular and Cellular Biology, Vol. 16. Academic Press, New York, pp. 633–640.

135. Acuto, O., O. Pugliese, M. Müller, and R. Tosi. 1979. Preparation of liposomes incorporating membrane components from human lymphoid cells. Tissue Antigens 14:385.

136. Hollander, N., S. Q. Mehdi, I. L. Weissman, H. M. McConnell, and J. P. Kriss. 1979. Allogeneic cytolysis of reconstituted membrane vesicles. Proc. Natl. Acad. Sci. USA 76:4042.

137. Curman, B., L. Klareskog, and P. A. Peterson. 1980. On the mode of incorporation of human transplantation antigens into lipid vesicles. J. Biol. Chem. 255:7820.

138. Hale, A. H. 1980. H-2 antigens incorporated into phospholipid vesicles elicit specific allogeneic cytotoxic T lymphocytes. Cell. Immunol. 55:328.

139. Juliano, R. L. and G. Lin. 1980. The interaction of plasma proteins with liposomes: protein binding and effects on the clotting and complement systems. In *Liposomes and Immunobiology*, B. H. Tom and H. R. Six (Eds.). Elsevier/North Holland, New York, pp. 49–66.

140. Sweet, C. and J. E. Zull. 1969. Activation of glucose diffusion from egg lecithin liquid crystals by serum albumin. Biochim. Biophys. Acta 173:94.

141. Weissmann, G., A. Brand, and E. C. Franklin. 1974. Interaction of immunoglobulins with liposomes. J. Clin. Invest. 53:536.

142. Kimelberg, H. K. 1976. Protein-liposome interactions and their relevance to the structure and function of cell membranes. Mol. Cell. Biochem. 10: 171.

143. Torchilin, V. P., V. S. Goldmacher, and V. N. Smirnov. 1978. Comparative studies on covalent and noncovalent immobilization of protein molecules on the surface of liposomes. Biochem. Biophys. Res. Commun. 85:983.

144. Torchilin, V. P., B. A. Khaw, V. N. Smirnov, and E. Haber. 1979. Preservation of antimyosin antibody activity after covalent coupling to liposomes. Biochem. Biophys. Res. Commun. 89:1114.

145. Hashimoto, Y., B. Yamanoha, H. Endoh, and I. Ishizuka. 1979. Direct and liposome-mediated introduction of a fluorescent hapten and a hapten-conjugated protein into tumor cell membrane, and induction of tumor-specific transplantation immunity with the hapten-conjugated tumor cells. GANN Monogr. Cancer Res. 23:135.

146. Heath, T. D., R. T. Fraley, and D. Papahadjopoulos. 1980. Antibody targeting of liposomes: cell specificity obtained by conjugation of F(ab')$_2$ to vesicle surface. Science 210:539.

147. Heath, T. D., B. A. Macher, and D. Papahadjopoulos. 1981. Covalent attachment of immunoglobulins to liposomes via glycosphingolipids. Biochim. Biophys. Acta 640:66.

148. Leserman, L. D., J. Barbet, and F. Kourilsky. 1980. Targeting to cells of fluorescent liposomes covalently coupled with monoclonal antibody or protein A. Nature 288:602.

149. Sinha, D. and F. Karush. 1979. Attachment to membranes of exogenous immunoglobulin conjugated to a hydrophobic anchor. Biochem. Biophys. Res. Commun. 90:554.

150. Huang, A., L. Huang, and S. J. Kennel. 1980. Monoclonal antibody covalently coupled with fatty acid, a reagent for in vitro liposome targeting. J. Biol. Chem. 255:8015.

151. Jansons, V. K. and P. L. Mallett. 1981. Targeted liposomes: a method for preparation and analysis. Anal. Biochem. 111:54.

152. Moss, J., R. L. Richards, C. R. Alving, and P. H. Fishman. 1977. Effect of the A and B protomers of choleragen on release of trapped glucose from liposomes containing or lacking ganglioside G_{M1}. J. Biol. Chem. 252:797.

153. Margolis, L. B. and N. A. Dorfman. 1977. Preparation of liposomes with immunological specificity. Bull. Exp. Biol. Med. (USSR) 83:60.

154. Endoh, H., Y. Hashimoto, Y. Kawashima, and Y. Suzuki. 1980. Agglutination microassay of hapten- or protein-modified liposomes using a multiple cell culture harvester. J. Immunol. Methods 36:185.

155. Hesketh, T. R., S. N. Payne, and J. H. Humphrey. 1972. Complement and phospholipase C lysis of lipid membranes. Immunology 23:705.

156. Ozato, K., H. K. Ziegler, and C. S. Henney. 1978. Liposomes as model mem-

brane systems for immune attack. II. The interaction of complement and K cell populations with immobilized liposomes. J. Immunol. 121:1383.

157.   Shin, M. L., W. A. Paznekas, and M. M. Mayer. 1978. On the mechanism of membrane damage by complement: the effect of length and unsaturation of the acyl chains in liposomal bilayers and the effect of cholesterol concentration in sheep erythrocyte and liposomal membranes. J. Immunol. 120:1996.

158.   Mayer, M. M., C. H. Hammer, D. W. Michaels, and M. L. Shin. 1979. Immunologically mediated membrane damage: the mechanism of complement action and the similarity of lymphocyte-mediated cytotoxicity. Immunochemistry 15:813.

159.   Mayer, M. M., M. K. Gately, M. Okamoto, M. L. Shin, and J. B. Willoughby. 1979. Two mechanisms of cell-mediated cytotoxicity: Ca^{++} transport modulation by lymphotoxin and transmembrane channel formation by antibody and nonadherent spleen cells. Ann. N.Y. Acad. Sci. 332:395.

160.   Kriss, J. P. and S. Q. Mehdi. 1979. Cell-mediated lysis of lipid vesicles containing eye muscle protein: implications regarding pathogenesis of Graves ophthalmopathy. Proc. Natl. Acad. Sci. USA 76:2003.

161.   Kinsky, S. C. 1974. Preparation of liposomes and a spectrophotometric assay for release of trapped glucose marker. In *Methods in Enzymology*, Vol. 32, part B, S. Fleischer and L. Packer (Eds.). Academic Press, New York, pp. 501–513.

162.   Kataoka, T., J. R. Williamson, and S. C. Kinsky. 1973. Release of macromolecular markers (enzymes) from liposomes treated with antibody and complement. An attempt at correlation with electron microscopic observations. Biochim. Biophys. Acta 298:158.

163.   Knudson, K. C., D. H. Bing, and L. Kater. 1971. Quantitative measurement of guinea pig complement with liposomes. J. Immunol. 106:258.

164.   Knudson, K. C. and D. H. Bing. 1972. A simplified method for the centrifugation of liposomes. Immunochemistry 9:587.

165.   Ruysschaert, J. M., A. Tenenbaum, C. Berliner, and M. Delmelle. 1977. Correlation between lateral lipid phase separation and immunological recognition in sensitized liposomes. FEBS Lett. 81:406.

166.   Weissmann, G., T. Collins, A. Evers, and P. Dunham. 1976. Membrane perturbation: studies employing a calcium-sensitive dye, arsenazo III, in liposomes. Proc. Natl. Acad. Sci. USA 73:510.

167.   Uemura, K., M. Yuzawa, and T. Taketomi. 1979. Immunochemical studies of lipids. V. Effect of modified hydrophobic moiety on immunogenicity and immunologic reactivity of Forssman glycolipid. Jpn. J. Exp. Med. 49:1.

168.   Hamers, M. N., W. E. Donker-Koopman, D.-J. Reijngoud, A. W. Schram, and T. M. Tager. 1978. An optical method for the detection and quantitation of antibodies to glycosphingolipids and other antigens. Immunochemistry 15:97.

169.   Six, H. R., W. W. Young, Jr., K. Uemura, and S. C. Kinsky. 1974. Effect of antibody-complement on multiple vs. single compartment liposomes. Application of a fluorometric assay for following changes in liposomal permeability. Biochemistry 13:4050.

170. Tagesson, C., K.-E. Magnusson, and O. Stendahl. 1977. Physicochemical consequences of opsonization: perturbation of liposomal membranes by *Salmonella typhimurium* 395 MS opsonized with IgG antibodies. J. Immunol. 119:609.

171. Lewis, J. T. and H. M. McConnell. 1978. Model lipid bilayer membranes as targets for antibody-dependent cellular- and complement-mediated immune attack. Ann. N.Y. Acad. Sci. 308:124.

172. Smolarsky, M., D. Teitelbaum, M. Sela, and C. Gitler. 1977. A simple fluorescent method to determine complement-mediated liposome immune lysis. J. Immunol. Methods 15:255.

173. Geiger, B. and M. Smolarsky. 1977. Immunochemical determination of ganglioside G_{M2}, by inhibition of complement-dependent liposome lysis. J. Immunol. Methods 17:7.

174. Geiger, B. and A. D. Schreiber. 1979. The use of antibody-coated liposomes as a target cell model for antibody-dependent cell-mediated cytotoxicity. Clin. Exp. Immunol. 35:149.

175. Petty, H. R., D. G. Hafeman, and H. M. McConnell. 1980. Specific antibody-dependent phagocytosis of lipid vesicles by RAW264 macrophages results in the loss of cell surface Fc but not C3b receptor activity. J. Immunol. 125:2391.

176. Smith, B. A. and H. M. McConnell. 1978. Determination of molecular motion in membranes using periodic pattern photobleaching. Proc. Natl. Acad. Sci. USA 75:2759.

177. Leserman, L. D., J. N. Weinstein, R. Blumenthal, S. O. Sharrow, and W. D. Terry. 1979. Binding of antigen-bearing fluorescent liposomes to the murine myeloma tumor MOPC 315. J. Immunol. 122:585.

178. Smith, L. M., J. W. Parce, B. A. Smith, and H. M. McConnell. 1979. Antibodies bound to lipid haptens in model membranes diffuse as rapidly as the lipids themselves. Proc. Natl. Acad. Sci. USA 76:4177.

179. Alving, C. R. and S. C. Kinsky. 1971. The preparation and properties of liposomes in the LA and LAC states. Immunochemistry 8:325.

180. Fry, J. M., R. P. Lisak, M. C. Manning, and D. H. Silbergerg. 1976. Serological techniques for detection of antibody to galactocerebroside. J. Immunol. Methods 11:185.

181. Takashi, T., K. Inoue, and S. Nojima. 1980. Immune reactions of liposomes containing cardiolipin and their relation to membrane fluidity. J. Biochem. 87:679.

182. Podack, E. R., H. J. Müller-Eberhard, H. Horst, and W. Hoppe. 1982. Membrane attack complex of complement (MAC): Three dimensional analysis of MAC-phospholipid vesicle recombinants. J. Immunol. 128: 2353.

183. Humphries, G. K. and H. M. McConnell. 1974. Immune lysis of liposomes and erythrocyte ghosts loaded with spin label. Proc. Natl. Acad. Sci. USA 71:1691.

184. Wei, R., C. R. Alving, R. L. Richards, and E. S. Copeland. 1975. Liposome spin immunoassay: a new sensitive method for detecting lipid substances in aqueous media. J. Immunol. Methods 9:165.

185. Rosenqvist, E. and A. I. Vistnes. 1977. Immune lysis of spin label loaded

liposomes incorporating cardiolipin; a new sensitive method for detecting anticardiolipin antibodies in syphilis serology. J. Immunol. Methods 15:147.

186.   Hsia, J. C. and C. T. Tan. 1978. Membrane immunoassay: principle and applications of spin membrane immunoassay. Ann. N.Y. Acad. Sci. 308:139.

187.   Esser. A. T. 1980. Principles of electron spin resonance assays and immunologic applications. In *Immunoassays: Clinical Laboratory Techniques for the 1980s,* R. M. Nakamura, W. R. Dito, and E. S. Tucker, III (Eds.). Proceedings of the Second Annual Conference on Immunoassays in the Clinical Laboratory. Alan R. Liss, New York, pp. 213–233.

188.   Schiefer, H.-G., U. Schummer, D. Hegner, U. Gerhardt, and G. H. Schnepel. 1975. Electron spin resonance studies on the lipid-protein interaction between cardiolipin and anti-cardiolipin antibodies. Hoppe-Seylers Z. Physiol. Chem. 356:293.

189.   Schieren, H., G. Weissmann, M. Seligman, and P. Coleman. 1978. Interactions of immunoglobulins with liposomes: an ESR and diffusion study demonstrating protection by hydrocortisone. Biochem. Biophys. Res. Comun. 82:1160.

190.   Henry, N., J. W. Parce, and H. M. McConnell. 1978. Visualization of specific antibody and Clq binding to hapten-sensitized lipid vesicles. Proc. Natl. Acad. Sci. USA 75:3933.

191.   Haga, M., H. Itagaki, S. Sugawara, and T. Okano. 1980. Liposome immunosensor for theophylline. Biochem. Biophys. Res. Commun. 95:187.

192.   Shiba, K., T. Watanabe, Y. Umezawa, S. Fujiwara, and H. Momoi. 1980. Liposome immunoelectrode. Chem. Lett. 2:155.

193.   Shiba, K., Y. Umezawa, T. Watanabe, S. Ogawa, and S. Fujiwara. 1980. Thin-layer potentiometric analysis of lipid antigen-antibody reaction by tetrapentylammonium (TPA$^+$) ion loaded liposomes and TPA$^+$ ion selective electrode. Anal. Chem. 52:1610.

194.   Gray, G. M. 1974. Glycosphingolipids in biological membranes. In *Perspectives in Membrane Biology,* S. Estrada-O. and C. Gitler (Eds.). Academic Press, New York, pp. 85–106.

195.   Brûlet, P., G. M. K. Humphries, and H. M. McConnell. 1977. Immunochemistry of model membranes containing spin-labeled haptens. In *34th Nobel Symposium,* S. Abrahamsson and I. Pascher (Eds.). Plenum Press, New York, pp. 321–329.

196.   Parce, J. W., N. Henry, and H. M. McConnell. 1978. Specific antibody-dependent binding of complement component Clq to hapten-sensitized lipid vesicles. Proc. Natl. Acad. Sci. USA 75:1515.

197.   Parce, J. W., M. A. Schwartz, J. C. Owicki, and H. M. McConnell. 1979. Kinetics of antibody association with spin-label haptens on membrane surfaces. J. Phys. Chem. 83:3414.

198.   Kinsky, S. C. 1972. Immune damage to a lipid model membrane. Ann. N.Y. Acad. Sci. 195:429.

199.   Kinsky, S. C. 1975. Immune reactions of model membranes. In *Cell Membranes: Biochemistry, Cell Biology and Pathology,* G. Weissmann and R. Claiborne (Eds.). H. P. Publication, New York, pp. 231–238.

200.   Kinsky, S. C. and H. R. Six. 1975. A model for the lytic action of complement. In *Proteases and Biological Control,* E. Reich, D. B. Rifkin, and

E. Shaw (Eds.). Cold Spring Harbor Laboratory, Cold Spring Harbor, New York, pp. 243-253.

201. Stroud, R. M., J. E. Volanakis, S. Nagasawa, and T. F. Lint. 1979. Biochemistry and biological reactions of complement proteins. In *Immunochemistry of Proteins,* vol. 3, M. Z. Atassi (Ed.). Plenum, New York, pp. 167-222.

202. Mayer, M. M. 1961. Complement and complement fixation. In *Experimental Immunochemistry,* 2nd ed. Charles C Thomas, Springfield, Ill., pp. 133-240.

203. Klein, P. G. and H. J. Wellensiek. 1965. Complement: hemolytic function and chemical properties. Int. Rev. Exp. Pathol. 4:245.

204. Eisen, H. N. 1973. Complement. In *Microbiology Including Immunology and Molecular Genetics,* 2nd ed., B. D. Davis, R. Dulbecco, H. N. Eisen, H. S. Ginsberg, and W. B. Wood, Jr. (Eds.). Harper & Row, New York, pp. 511-525.

205. Haxby, J. A., O. Götze, H. J. Müller-Eberhard, and S. C. Kinsky. 1969. Release of trapped marker from liposomes by the action of purified complement components. Proc. Natl. Acad. Sci. USA 64:290.

206. Humphrey, J. H. and R. R. Dourmashkin. 1969. The lesions in cell membranes caused by complement. Adv. Immunol. 11:75.

207. Lachmann, P. J., E. A. Munn, and G. Weissmann. 1970. Complement-mediated lysis of liposomes produced by the reactive lysis procedure. Immunology 19:983.

208. Hesketh, T. R., R. R. Dourmashkin, S. N. Payne, J. H. Humphrey, and P. J. Lachmann. 1971. Lesions due to complement in lipid membranes. Nature 233:620.

209. Lachmann, P. J., D. E. Bowyer, P. Nicol, R. M. C. Dawson, and E. A. Munn. 1973. Studies on the terminal stages of complement lysis. Immunology 24:135.

210. Bhakdi, S. and J. Tranum-Jensen. 1978. Molecular nature of the complement lesion. Proc. Natl. Acad. Sci. USA 75:5655.

211. Bhakdi, S. and J. Tranum-Jensen. 1980. Re-incorporation of the terminal C5b-9 complement complex into lipid bilayers: formation and stability of reconstituted liposomes. Immunology 41:737.

212. Humphries, G. M. K. and H. M. McConnell. 1976. Antibodies against nitroxide spin labels. Biophys. J. 16:275.

213a. Rubenstein, J. L. R., J. C. Owicki, and H. M. McConnell. 1980. Dynamic properties of binary mixtures of phosphatidylcholines and cholesterol. Biochemistry 19:569.

213b. Conrad, D. H., C. R. Alving, and G. H. Wirtz. 1974. The influence of retinal on complement-dependent immune damage to liposomes. Biochim. Biophys. Acta 332:36.

214. Richards, R. L., H. Gewurz, A. P. Osmand, and C. R. Alving. 1977. Interactions of C-reactive protein and complement with liposomes. Proc. Natl. Acad. Sci. USA 74:5672.

215. Richards, R. L., H. Gewurz, J. Siegel, and C. R. Alving. 1979. Interactions of C-reactive protein and complement with liposomes. II. Influence of membrane composition. J. Immunol. 122:1185.

216. Kitagawa, T. and K. Inoue. 1975. Effect of temperature on immune damage of liposomes prepared in the presence and absence of cholesterol. Nature 254:254.

217. Esser, A. F., R. M. Bartholomew, J. W. Parce, and H. M. McConnell. 1979. The physical state of membrane lipids modulates the activation of the first component of complement. J. Biol. Chem. 254:1768.

218. Parce, J. W., H. M. McConnell, R. M. Bartholomew, and A. W. Esser. 1980. Kinetics of antibody-dependent activation of the first component of complement on lipid bilayer membranes. Biochem. Biophys. Res. Commun. 93:235 and erratum in 94:735.

219. Kaplan, M. H. and J. E. Volanakis. 1974. Interaction of C-reactive protein complexes with the complement system. I. Consumption of human complement associated with the reaction of C-reactive protein with pneumococcal C-polysaccharide and with the choline phosphatides, lecithin and sphingomyelin. J. Immunol. 112:2135.

220. Volanakis, J. E. and M. H. Kaplan. 1974. Interaction of C-reactive protein complexes with the complement system. II. Consumption of guinea pig complement by CRP complexes: Requirement for human Clq. J. Immunol. 113:9.

221. Siegel, J., R. Rent, and G. Gewurz. 1974. Interactions of C-reactive protein with the complement system. I. Protamine-induced consumption of complement in acute phase sera. J. Exp. Med. 140:631.

222. Siegel, J., A. P. Osmand, M. F. Wilson, and H. Gewurz. 1975. Interactions of C-reactive protein with the complement system. II. C-reactive protein-mediated consumption of complement by poly-L-lysine polymers and other polycations. J. Exp. Med. 142:709.

223. Osmand, A. P., R. F. Mortensen, J. Siegel, and H. Gewurz. 1975. Interactions of C-reactive protein with the complement system. III. Complement-dependent passive hemolysis initiated by CRP. J. Exp. Med. 142:1065.

224. Tillet, W. S. and T. Francis, Jr. 1930. Serological reactions in pneumonia with a nonprotein somatic fraction of pneumococcus. J. Exp. Med. 52:561.

225. MacLeod, C. M. and O. T. Avery. 1941. The occurrence during acute infections of a protein not normally present in the blood. III. Immunological properties of the C-reactive protein and its differentiation from normal blood proteins. J. Exp. Med. 73:191.

226. Gotschlich, E. C. and G. M. Edelman. 1967. Binding properties and specificity of C-reactive protein. Proc. Natl. Acad. Sci. USA 57:706.

227. Volanakis, J. E. and M. H. Kaplan. 1971. Specificity of C-reactive protein for choline phosphate residues of pneumococcal C-polysaccharide. Proc. Soc. Exp. Biol. Med. 136:612.

228. Hokama, Y., R. Tam. W. Hirano, and L. Kimura. 1974. Significance of C-reactive protein binding by lecithin: a simplified procedure for CRP isolation. Clin. Chim. Acta 50:53.

229. Volanakis, J. E. and A. J. Narkates. 1981. Interaction of C-reactive pro-

tein with artificial phosphatidylcholine bilayers and complement. J. Immunol. 126:1820.

230. Mold, C., C. P. Rodgers, R. L. Richards, C. R. Alving, and H. Gewurz. 1981. Interaction of C-reactive protein with liposomes. III. Membrane requirements for binding. J. Immunol. 126:856.

231. DiCamelli, R., L. A. Potempa, J. Siegel, L. Suyehira, K. Petras, and H. Gewurz. 1980. Binding reactivity of C-reactive protein for polycations. J. Immunol. 125:1933.

232. Ohsawa, T. 1980. Change in permeability of liposomal membranes mediated by C-reactive protein and its inhibition by cholesterol. Jpn. J. Exp. Med. 50:67.

233. Volanakis, J. E. and K. W. A. Wirtz. 1979. Interaction of C-reactive protein with artificial phosphatidylcholine bilayers. Nature 281:155.

234. Aho, K. 1969. Studies of syphilitic antibodies. IV. Evidence of reactant partner common for C-reactive protein and certain anti-lipoidal antibodies. Br. J. Vener. Dis. 45:13.

235. Tsujimoto, M., K. Inoue, and S. Nojima. 1980. C-reactive protein induced agglutination of lipid suspensions prepared in the presence and absence of phosphatidylcholine. J. Biochem. 87:1531.

236. Alving, C. R., R. L. Richards, and A. A. Guirguis. 1977. Cholesterol-dependent human complement activation resulting in damage to liposomal model membranes. J. Immunol. 118:342.

237. Greenberg, C. S., D. E. Hammerschmidt, P. R. Craddock, and H. S. Jacob. 1979. Atheroma cholesterol activates complement and aggregates granulocytes: possible role in ischemic manifestations of atherosclerosis. Trans. Assoc. Am. Physicians 92:130.

238. Pang, A. S. D., A. Katz, and J. O. Minta. 1979. C3 deposition in cholesterol-induced atherosclerosis in rabbits: a possible etiologic role for complement in atherogenesis. J. Immunol. 123:1117.

239. Pillemer, L., M. D. Schoenberg, L. Blum, and L. Wurz. 1955. Properdin system and immunity. II. Interaction of the properdin system with polysaccharides. Science 122:545.

240. Galanos, C., E. T. Rietschel, O. Lüderitz, and O. Westphal. 1971. Interaction of lipopolysaccharides and lipid A with complement. Eur. J. Biochem. 19:143.

241. Morrison, D. C. and P. Verroust. 1973. Anticomplementary activity of lipid A isolated from lipopolysaccharides. Proc. Soc. Exp. Biol. Med. 143:1025.

242. Morrison, D. C. and L. F. Kline. 1977. Activation of the classical and properdin pathways of complement by bacterial lipopolysaccharides (LPS). J. Immunol. 118:362.

243. Loos, M., D. Bitter-Suermann, and M. Dierich. 1974. Interaction of the first (C$\overline{1}$), the second (C2) and the fourth (C4) component of complement with different preparations of bacterial lipopolysaccharides and with lipid A. J. Immunol. 112:935.

244. Cooper, N. R. and D. C. Morrison. 1978. Binding and activation of the first component of human complement by the lipid A region of lipopolysaccharides. J. Immunol. 120:1862.

245. Fearon, D. T. 1979. Activation of the alternative complement pathway. Crit. Rev. Immunol. 1:1.

246. Müller-Eberhard, H. J. and R. D. Schreiber. 1980. Molecular biology and chemistry of the alternative pathway of complement. In *Advances in Immunology,* Vol. 29, H. G. Kunkel and F. J. Dixon (Eds.). Academic Press, New York, pp. 1–53.

247. Miyama, A., T. Kato, J. Yokoo, and S. Kashiba. 1975. Trypsin-activated complex of human factor B with cobra venom factor (CVF), cleaving C3 and C5 and generating a lytic factor for unsensitized guinea pig erythrocytes. II. Physicochemical characterization of the activated complex. Biken J. 18:205.

248. Cunningham, C. M., M. Kingzette, R. L. Richards, C. R. Alving, T. F. Lint, and H. Gewurz. 1979. Activation of human complement by liposomes: a model for membrane activation of the alternative pathway. J. Immunol. 122:1237.

249. Mold, C. and H. Gewurz. 1980. Activation of human complement by liposomes: serum factor requirement for alternative pathway activation. J. Immunol. 125:696.

250. Thompson, R. A. and D. S. Rowe. 1968. Reactive haemolysis—a distinctive form of red cell lysis. Immunology 14:745.

251. Thompson, R. A. and P. J. Lachmann. 1970. Reactive lysis: the complement-mediated lysis of unsensitized cells. I. The characterization of indicator factor and its identification as C7. J. Exp. Med. 131:629.

252. Lachmann, P. J. and R. A. Thompson. 1970. Reactive lysis: the complement-mediated lysis of unsensitized cells. II. The characterization of activated reactor as C$\overline{56}$ and the participation of C8 and C9. J. Exp. Med. 131:643.

253. Inoue, K. and S. C. Kinsky. 1970. Fate of phospholipids in liposomal model membranes damaged by antibody and complement. Biochemistry 9:4767.

254. Kinsky, S. C., P. P. M. Bonsen, C. B. Kinsky. L. L. M. van Deenen, and A. F. Rosenthal. 1971. Preparation of immunologically responsive liposomes with phosphonyl and phosphinyl analogs of lecithin. Biochim. Biophys. Acta 233:815.

255. Iles, G. H., P. Seeman, D. Naylor, and B. Cinader. 1973. Membrane lesions in immune lysis. Surface rings, globule aggregates, and transient openings. J. Cell Biol. 56:528.

256. Bhakdi, S., V. Speth, H. Knüfermann, D. F. H. Wallach, and H. Fischer. 1974. Complement-induced changes in the core structure of sheep erythrocyte membranes: a study by freeze-etch electron microscopy. Biochim. Biophys. Acta 356:300.

257. Humphrey, J. H. 1972. The nature of complement-induced lesions in membranes. Haematologia (Budapest) 6:319.

258. Esser, A. F., W. P. Kolb, E. R. Podack, and H. J. Müller-Eberhard. 1979.

Molecular reorganization of lipid bilayers by complement: a possible mechanism for membranolysis. Proc. Natl. Acad. Sci. USA 76:1410.

259. Podak, E. R., G. Biesecker, and H. J. Müller-Eberhard. 1979. Membrane attack complex of complement: generation of high-affinity phospholipid binding sites by fusion of five hydrophilic plasma proteins. Proc. Natl. Acad. Sci. USA 76:897.

260. Esser, A. F. 1982. Interactions between complement proteins and biological and model membranes. In *Biological Membranes*, Vol. 4, D. Chapman (Ed.). Academic Press, New York.

261. Ramm, L. E. and M. M. Mayer. 1980. Life-span and size of the transmembrane channel formed by large doses of complement. J. Immunol. 124:2281.

262. Hammer, C. H., A. Nicholson, and M. M. Mayer. 1975. On the mechanism of cytolysis by complement: evidence on insertion of C5b and C7 subunits of the C5b,6,7 complex into phospholipid bilayers of erythrocyte membranes. Proc. Natl. Acad. Sci. USA 72:5076.

263. Hammer, C. H., M. L. Shin, A. S. Abramovitz, and M. M. Mayer. 1977. On the mechanism of cell membrane damage by complement: evidence on insertion of polypeptide chains from C8 and C9 into the lipid bilayer of erythrocytes. J. Immunol. 119:1.

264. Dahl, C. E. and R. P. Levine. 1978. Electron spin resonance studies on interaction of complement proteins with erythrocyte membranes. Proc. Natl. Acad. Sci. USA 75:4930.

265. Yamamoto, K. 1980. Proteolysis of the C5b-7 complex: cleavage of the C5b and C6 subunits and its effect on the interaction of the complex with phospholipid bilayers. J. Immunol. 125:1745.

266. Bhakdi, S., P. Ey, and B. Bhakdi-Lehnen. 1976. Isolation of the terminal complement complex from target sheep erythrocyte membranes. Biochim. Biophys. Acta 419:445.

267. Tranum-Jensen, J., S. Bhakdi, B. Bhakdi-Lehnen, O. J. Bjerrum, and V. Speth. 1978. Complement lysis: the ultrastructure and orientation of the C5b-9 complex on target sheep erythrocyte membranes. Scand. J. Immunol. 7:45.

268. Biesecker, G., E. R. Podack, C. A. Halverson, and H. J. Müller-Eberhard. 1979. C5b-9 dimer: isolation from complement lysed cells and ultrastructural identification with complement-dependent membrane lesions. J. Exp. Med. 149:448.

269. Podack, E. R. and H. J. Müller-Eberhard. 1978. Binding of desoxycholate, phosphatidylcholine vesicles, lipoprotein and of the S-protein to complexes of terminal complement components. J. Immunol. 121:1025.

270. Ohanian, S. H., S. I. Schlager, M. Yamazaki, and B. Ishida. 1979. Effect of specific phospholipids on the antibody-complement-mediated killing of nucleated cells. J. Immunol. 123:1014.

271. Podack, E. R., A. E. Esser, G. Biesecker, and H. J. Müller-Eberhard. 1980. Membrane attack complex of complement. A structural analysis of its assembly. J. Exp. Med. 151:301.

272. Bhakdi, S., B. Bhakdi-Lehnen, and J. Tranum-Jensen. 1979. Proteolytic transformation of SC5b-9 into an amphiphilic macromolecule resembling the C5b-9 membrane attack complex of complement. Immunology 37: 901.

273. Bhakdi, S., J. Tranum-Jensen, and O. Klump. 1980. The terminal membrane C5b-9 complex of human complement. Evidence for the existence of multiple protease-resistant polypeptides that form the trans-membrane complement channel. J. Immunol. 124:2451.

274. Wobschall, D. and C. McKeon. 1975. Step conductance increase in bilayer membranes induced by antibody-antigen-complement action. Biochim. Biophys. Acta 413:317.

275. Michaels, D. W., A. S. Abramovitz, C. H. Hammer, and M. M. Mayer. 1976. Increased ion permeability of planar lipid bilayer membranes after treatment with the C5b-9 cytolytic attack mechanism of complement. Proc. Natl. Acad. Sci. USA 73:2852.

276. Stephens, C. L. and P. A. Henkart. 1979. Electrical measurements of complement-mediated membrane damage in cultured nerve and muscle cells. J. Immunol. 122:455.

277. Giavedoni, E. B., R. P. Mason, and A. P. Dalmasso. 1976. A spin probe study of complement induced changes in membrane lipids. J. Immunol. 116:1733.

278. Mason, R. P., E. B. Giavedoni, and A. P. Dalmasso. 1977. Complement-induced decrease in membrane mobility: introducing a more sensitive index of spin-label motion. Biochemistry 16:1196.

279. Giavedoni, E. B., R. P. Mason, and A. P. Dalmasso. 1978. Complement-induced modifications in membrane fluidity: studies with resealed and glutaraldehyde-treated erythrocyte membrane ghosts. J. Immunol. 120: 2003.

280. Nakamura, M., S. Ohnishi, H. Kitamura, and S. Inai. 1976. Membrane fluidity change in erythrocytes induced by complement system. Biochemistry 15:4838.

281. Giavedoni, E. B. and A. P. Dalmasso. 1976. The induction by complement of a change in KSCN-dissociable red cell membrane lipids. J. Immunol. 116:1163.

282. Schlager, S. I., S. H. Ohanian, and T. Borsos. 1978. Identification of lipids synthesized and released by tumor cells under attack by antibody and complement. J. Immunol. 120:1644.

283. Inoue, K., T. Kinoshita, M. Okada, and Y. Akiyama. 1977. Release of phospholipids from complement-mediated lesions on the surface structure of *Escherichia coli.* J. Immunol. 119:65.

284. Kinoshita, T., K. Inoue, M. Okada, and Y. Akiyama. 1977. Release of phospholipids from liposomal model membrane damaged by antibody and complement. J. Immunol. 119:73.

285. Shin, M. L., W. A. Paznekas, A. S. Abramovitz, and M. M. Mayer. 1977. On the mechanism of membrane damage by C: exposure of hydrophobic sites on activated C proteins. J. Immunol. 119:1358.

286. Richards, R. L., C. R. Alving, and I. Scher. 1979. Complement-dependent immune damage to large and small uni- and multilamellar liposomes. Fed. Proc. 38:1468.

287. Zinkernagel, R. M. and P. C. Doherty. 1979. MHC-Restricted cytotoxic T cells: studies on the biological role of polymorphic major transplantation antigens determining T-cell restriction-specificity, function, and responsiveness. Adv. Immunol. 27:51.

288. Peterson, P. A., L. Rask, K. Sege, L. Klareskog, H. Anundi, and L. Östberg. 1975. Evolutionary relationship between immunoglobulins and transplantation antigens. Proc. Natl. Acad. Sci. USA 72:1612.

289. Terhorst, C., R. Robb, C. Jones, and J. L. Strominger. 1977. Further structural studies of the heavy chain of HLA antigens and its similarity to immunoglobulins. Proc. Natl. Acad. Sci. USA 74:4002.

290. Finberg, R., M. Mescher, and S. J. Burakoff. 1978. The induction of virus specific cytotoxic T lymphocytes with solubilized viral and membrane proteins. J. Exp. Med. 148:1620.

291. Loh, D., A. H. Ross, A. H. Hale, D. Baltimore, and H. N. Eisen. 1979. Synthetic lipid vesicles containing a purified viral antigen and cell membrane proteins stimulate the development of cytotoxic T lymphocytes. J. Exp. Med. 150:1067.

292. Sherman, L., S. J. Burakoff, and M. F. Mescher. 1980. Induction of allogeneic cytotoxic T lymphocytes by partially purified membrane glycoproteins. Cell. Immunol. 51:141.

293. Hale, A. H., D. S. Lyles, and D. P. Fan. 1980. Elicitation of anti-Sendai virus cytotoxic T lymphocytes by viral and H-2 antigens incorporated into the same lipid bilayer by membrane fusion and by reconstitution into liposomes. J. Immunol. 124:724. (Erratum in 124:2524.)

294. Trägardh, L., L. Klareskog, B. Curman, L. Rask, and P. A. Peterson. 1979. Isolation and properties of detergent-solubilized HLA antigens obtained from platelets. Scand. J. Immunol. 9:303.

295. Helenius, A., B. Morein, E. Fries, K. Simons, P. Robinson, V. Schirrmacher, C. Terhorst, and J. L. Strominger. 1978. Human (HLA-A and HLA-B) and murine (H-2K and H-2D) histocompatibility antigens are cell surface receptors for Semliki Forest virus. Proc. Natl. Acad. Sci. USA 75:3846.

296. Engelhard, V. H., J. L. Strominger, M. Mescher, and S. Burakoff. 1978. Introduction of secondary cytotoxic T lymphocytes by purified HLA-A and HLA-B antigens reconstituted into phospholipid vesicles. Proc. Natl. Acad. Sci. USA 75:5688.

297. Mescher, M. F., R. Finberg, L. Sherman, and S. Burakoff. 1979. Induction of virus-specific H-2 restricted murine CTL by liposomes. In *T and B Lymphocytes: Recognition and Function*, F. H. Bach, B. Bonavida, E. S. Vitetta, and C. F. Fox (Eds.). ICN-UCLA Symposia on Molecular and Cellular Biology, Vol. 16. Academic Press, New York, pp. 623–632.

298. Hale, A. H., D. S. Lyles, L. K. Paulus, and M. J. Ruebush. 1980. Minimal molecular requirements for reactivity of tumor cells with T cells. J. Immunol. 124:2063.

299. Hale, A. H., M. J. Ruebush, and D. T. Harris. 1980. Elicitation of anti-viral cytotoxic T lymphocytes with purified viral and H-2 antigens. J. Immunol. 125:428.

300. Hale, A. H., M. J. Ruebush, and D. T. Harris. 1980. Study of the minimal molecular and cellular requirements for elicitation of anti-vesicular stomatitis virus cytotoxic T lymphocytes using purified viral and cellular antigens incorporated into phospholipid vesicles. In *Liposomes and Immunobiology*, B. H. Tom and H. R. Six (Eds.). Elsevier/North-Holland, New York, pp. 211-224.

301. Engelhard, V. H., J. F. Kaufman, J. L. Strominger, and S. J. Burakoff. 1980. Specificity of mouse cytotoxic T lymphocytes stimulated with either HLA-A and -B or HLA-DR antigens reconstituted into phospholipid vesicles. J. Exp. Med. 152:54s.

302. Weinberg, O., S. H. Herrmann, M. F. Mescher, B. Benacerraf, and S. J. Burakoff. 1980. Cellular interactions in the generation of cytolytic T lymphocyte responses: role of Ia-positive splenic adherent cells in presentation of H-2 antigen. Proc. Natl. Acad. Sci. USA 77:6091.

303. S. J. Burakoff, O. Weinberger, S. H. Herrmann, and M. F. Mescher. 1981. Cellular interactions in the regulation of the cytolytic T-lymphocyte response: The effect of suboptimal antigen dose. Transpl. Proc. 13:1039.

304. Weinberger, O., S. Herrmann, M. F. Mescher, B. Benacerraf, and S. J. Burakoff. 1981. Antigen-presenting cell function in induction of helper T cells for cytotoxic T-lymphocyte responses: evidence for antigen processing. Proc. Natl. Acad. Sci. USA 78:1796.

305. Hale, A. H. and M. P. McGee. 1981. A study of the inability of subcellular fractions to elicity primary anti-H-2 cytotoxic T lymphocytes. Cell. Immunol. 58:277.

306. Hale, A. H., M. J. Ruebush, D. S. Lyles, and D. T. Harris. 1980. Antigen-liposome modification of target cells as a method to alter their susceptibility to lysis by cytotoxic T lymphocytes. Proc. Natl. Acad. Sci. USA 77:6105.

307. Hale, A. H., M. J. Ruebush, D. T. Harris, and M. P. McGee. 1981. Elicitation of anti-H-2 cytotoxic T lymphocytes with antigen-modified H-2 negative stimulator cells. J. Immunol. 126:1485.

308. Roitt, I. M., L. Shen, and A. H. Greenberg. 1976. Antibody-dependent cell-mediated cytotoxicity. In *The Role of Immunological Factors in Infectious, Allergic, and Autoimmune Processes*, R. F. Beers, Jr. and E. G. Bassett (Eds.). Raven Press, New York, pp. 281-288.

309. Lamon, E. W., H. D. Whitten, H. M. Skurzak, B. Andersson, and B. Lidin. 1975. IgM antibody-dependent cell-mediated cytotoxicity in the Moloney sarcoma virus system: the involvement of T and B lymphocytes as effector cells. J. Immunol. 115:1288.

310. Henkart, P. and R. Blumenthal. 1975. Interaction of lymphocytes with lipid bilayer membranes: a model for lymphocyte-mediated lysis of target cells. Proc. Natl. Acad. Sci. USA 72:2789.

311. Juy, D., A. Billecocq, M. Faure, and C. Bona. 1977. Damage of liposomes in antibody-dependent cell-mediated cytotoxicity. Scand. J. Immunol. 6: 607.

312. Ozato, K., H. K. Ziegler, and C. S. Henney. 1978. Liposomes as model membrane systems for immune attack. I. Transfer of antigenic determinants to lymphocyte membranes after interactions with hapten-bearing liposomes. J. Immunol. 121:1376.

313. Henney, C. S. 1980. Liposomes as targets for cell-mediated immune attack. In *Liposomes and Immunobiology*, B. H. Tom and H. R. Six (Eds.). Elsevier/North-Holland, New York, pp. 167–178.

314. Hafeman, D. G., J. W. Parce, and H. M. McConnell. 1979. Specific antibody-dependent activation of neutrophils by liposomes containing spin-label lipid haptens. Biochem. Biophys. Res. Commun. 86:522.

315. Hafeman, D. G., J. T. Lewis, and H. M. McConnell. 1980. Triggering of the macrophage and neutrophil respiratory burst by antibody bound to a spin-label phospholipid hapten in model lipid bilayer membranes. Biochemistry 19:5387.

316. Lewis, J. T., D. G. Hafeman, and H. M. McConnell. 1980. Kinetics of antibody-dependent binding of haptenated phospholipid vesicles to a macrophage-related cell line. Biochemistry 19:5376.

317. Frye, L. D. and G. J. Friou. 1975. Inhibition of mammalian cytotoxic cells by phosphatidylcholine and its analogue. Nature 258:333.

318. Mehdi, S. Q., J. T. Lewis, B. R. Copeland, and H. M. McConnell. 1980. Freeze-fracture of reconstituted model membranes used as targets for cell-mediated cytotoxicity. Biochim. Biophys. Acta 600:590.

319. Bloom, B. R. and B. Bennett. 1966. Mechanism of a reaction in vitro associated with delayed-type hypersensitivity. Science 153:80.

320. David, J. R. 1966. Delayed hypersensitivity in vitro: its mediation by cell-free substances formed by lymphoid cell-antigen interaction. Proc. Natl. Acad. Sci. USA 56:72.

321. Waksman, B. H. 1979. Overview: biology of the lymphokines. In *Biology of the Lymphokines,* S. Cohen, E. Pick, and J. J. Oppenheim (Eds.). Academic Press, New York, pp. 585–616.

322. Higgins, T. J., A. P. Sabatino, H. G. Remold, and J. R. David. 1978. Possible role of macrophage glycolipids as receptors for migration inhibitory factor (MIF). J. Immunol. 121:880.

323. Liu, D. Y., K. D. Petschek, H. G. Remold, and J. R. David. 1980. Role of sialic acid in the macrophage glycolipid receptor for MIF. J. Immunol. 124:2042.

324. Poste, G., R. Kirsh, and I. J. Fidler. 1979. Cell surface receptors for lymphokines. I. The possible role of glycolipids as receptors for macrophage migration inhibitory factor (MIF) and macrophage activation factor (MAF). Cell. Immunol. 44:71.

325. Poste, G., R. Kirsh, W. E. Fogler, and I. J. Fidler. 1979. Activation of tumoricidal properties in mouse macrophages by lymphokines encapsulated in liposomes. Cancer Res. 39:881.

326. Poste, G. and R. Kirsh. 1979. Rapid decay of tumoricidal activity and loss of responsiveness to lymphokines in inflammatory macrophages. Cancer Res. 39:2582.

327. Sone, S., G. Poste, and I. J. Fidler. 1980. Rat alveolar macrophages are susceptible to activation by free and liposome-encapsulated lymphokines. J. Immunol. 124:2197.

328. Fidler, I. J. 1980. Therapy of spontaneous metastases by intravenous injection of liposomes containing lymphokines. Science 208:1469.

329. Fogler, W. E., A. Raz, and I. J. Fidler. 1980. In situ activation of murine macrophages by liposomes containing lymphokines. Cell. Immunol. 53: 214.

330. Poste, G., R. Kirsh, A. Raz, S. Sone, C. Bucana, W. Fogler, and I. J. Fidler. 1980. Activation of tumoricidal properties in macrophages by liposome-encapsulated lymphokines: in vitro studies. In *Liposomes and Immunobiology*, B. H. Tom and H. R. Six (Eds.). Elsevier/North-Holland, New York, pp. 93–107.

331. Fidler, I. J., I. R. Hart, A. Raz, W. E. Fogler, R. Kirsh, and G. Poste. 1980. Activation of tumoricidal properties in macrophages by liposome-encapsulated lymphokines: in vivo studies. In *Liposomes and Immunobiology*, B. H. Tom and H. R. Six (Eds.). Elsevier/North-Holland, New York, pp. 109–118.

332. Sone, S. and I. J. Fidler. 1980. Synergistic activation by lymphokines and muramyl dipeptide of tumoricidal properties in rat alveolar macrophages. J. Immunol. 125:2454.

333. Fidler, I. J., A. Raz, W. E. Fogler, L. C. Hoyer, and G. Poste. 1981. The role of plasma membrane receptors and the kinetics of macrophage activation by lymphokines encapsulated in liposomes. Cancer Res. 41:495.

334. Hart, I. R., W. E. Fogler, G. Poste, and I. J. Fidler. 1981. Toxicity studies of liposome-encapsulated immunomodulators administered intravenously to dogs and mice. Cancer Immunol. Immunother. 10:157.

335. Chedid, L., C. Carelli, and F. Audibert. 1979. Recent developments concerning muramyl dipeptide, a synthetic immunoregulating molecule. J. Reticuloendothel. Soc. 26(Dec. Suppl.):631.

336. Sone, S. and I. J. Fidler. 1981. In vitro activation of tumoricidal properties in rat alveolar macrophages by synthetic muramyl dipeptide encapsulated in liposomes. Cell. Immunol. 57:42.

337. Fidler, I. J., S. Sone, W. E. Fogler, and Z. L. Barnes. 1981. Eradication of spontaneous metastases and activation of alveolar macrophages by intravenous injection of liposomes containing muramyl dipeptide. Proc. Natl. Acad. Sci. USA 78:1680.

338. Stossel, T. P. 1975. Phagocytosis. Recognition and ingestion. Semin. Hematol. 12:83.

339. Stossel, T. P. 1977. Phagocytosis. Clinical disorders of recognition and ingestion. Am. J. Pathol. 88:741.

340. Walters, M. N.-I. and J. M. Papadimitriou. 1978. Phagocytosis: a review. CRC Crit. Rev. Toxicol. 5:377.

341. Tyrrell, D. A., T. D. Heath, C. M. Colley, and B. E. Ryman. 1976. New aspects of liposomes. Biochim. Biophys. Acta 457:259.

342. Pagano, R. E. and J. N. Weinstein. 1978. Interactions of liposomes with mammalian cells. Annu. Rev. Biophys. Bioeng. 7:435.

343. Weissmann, G., D. Bloomgarden, R. Kaplan, C. Cohen, S. Hoffstein, T. Collins, A. Gotlieb, and D. Nagle. 1975. A general method for the introduction of enzymes, by means of immunoglobulin-coated liposomes, into lysosomes of deficient cells. Proc. Natl. Acad. Sci. USA 72:88.

344. Cohen, C. M., G. Weissmann, S. Hoffstein, Y. C. Awasthi, and S. K. Srivastava. 1976. Introduction of purified hexosaminidase A into Tay-Sachs leukocytes by means of immunoglobulin-coated liposomes. Biochemistry 15:452.

345. Weissmann, G., C. Cohen, and S. Hoffstein. 1976. Introduction of missing enzymes into the cytoplasm of cultured mammalian cells by means of fusion-prone liposomes. Trans. Assoc. Am. Phys. 89:171.

346. Weissmann, G., H. Korchak, M. Finkelstein, J. Smolen, and S. Hoffstein. 1978. Uptake of enzyme-laden liposomes by animal cells in vitro and in vivo. Ann. N.Y. Acad. Sci. 308:235.

347. Finkelstein, M. C., S. H. Kuhn, H. Schieren, G. Weissmann, and S. Hoffstein. 1980. Selectivity in the uptake of liposomes by human leukocytes: a comparison of monocytes, lymphocytes and polymorphonuclear leukocytes. In *Liposomes and Immunobiology*, B. H. Tom and H. R. Six (Eds.). Elsevier/North-Holland, New York, pp. 255–270.

348. Finkelstein, M. C., S. H. Kuhn, H. Schieren, G. Weissmann, and S. Hoffstein. 1981. Liposome uptake by human leukocytes. Enhancement of entry mediated by human serum and aggregated immunoglobulins. Biochim. Biophys. Acta 673:286.

349. Wiktorowicz, J. E., P. S. Baur, and S. K. Srivastava. 1977. Introduction of liposome-sequestered human hexosaminidase A and ferritin into lysosomes of mouse macrophages. Cytobiologie 14:401.

350. Torchilin, V. P., V. R. Berdichevsky, A. A. Barsukov, and V. N. Smirnov. 1980. Coating liposomes with protein decreases their capture by macrophages. FEBS Lett. 111:184.

351. Ismail, G., L. A. Boxer, and R. L. Baehner. 1979. Utilization of liposomes for correction of the metabolic and bactericidal deficiencies in chronic granulomatous disease. Pediatr. Res. 13:769.

352. Finkelstein, M. and G. Weissmann. 1978. The introduction of enzymes into cells by means of liposomes. J. Lipid Res. 19:289.

353. de Barsy, T., P. Devos, and F. Van Hoof. 1976. A morphologic and biochemical study of the fate of antibody-bearing liposomes. Lab. Invest. 34:273.

354. de Barsy, T. and F. Van Hoof. 1980. Effect of antibodies administered in liposomes. In *Liposomes in Biological Systems*, G. Gregoriadis and A. C. Allison (Eds.). Wiley, New York, pp. 211–218.

355. Lewis, J. T., D. G. Hafeman, and H. M. McConnell. 1980. Specific antibody-dependent macrophage phagocytosis of lipid vesicles containing lipid

hapten. In *Liposomes and Immunobiology*, B. H. Tom and H. R. Six
(Eds.). Elsevier/North-Holland, New York, pp. 179-191.

356. Petty, H. R., D. G. Hafeman, and H. M. McConnell. 1981. Disappearance
of macrophage surface folds after antibody-dependent phagocytosis. J.
Cell Biol. 89:223.

357. Leserman, L. D., J. N. Weinstein, R. Blumenthal, and W. D. Terry. 1980.
Receptor-mediated endocytosis of antibody-opsonized liposomes by tumor
cells. Proc. Natl. Acad. Sci. USA 77:4089.

358. Leserman, L. D. and J. N. Weinstein. 1980. Receptor-mediated binding
and endocytosis of drug-containing liposomes by tumor cells. In *Liposomes
and Immunobiology*, B. H. Tom and H. R. Six (Eds.). Elsevier/North-
Holland, New York, pp. 241-251.

359. Alving, C. R., J. J. Mooney, and G. E. Olson. 1971. Use of liposomes as
a model for studying immune phagocytosis. Fed. Proc. 30:693.

360. Anderson, C. L. 1980. The murine macrophage Fc receptor for IgG2b is
lipid dependent. J. Immunol. 125:538.

361. Martin, F. J. and R. C. MacDonald. 1976. Lipid vesicle-cell interactions.
III. Introduction of a new antigenic determinant into erythrocyte membranes. J. Cell Biol. 70:515.

362. Schroit, A. J. and R. E. Pagano. 1978. Introduction of antigenic phospholipids into the plasma membrane of mammalian cells: organization
and antibody-induced lipid redistribution. Proc. Natl. Acad. Sci. USA
75:5529.

363. Ozato, K. and C. S. Henney. 1978. Studies on lymphocyte-mediated
cytolysis. XII. Hapten transferred to cell surfaces by interactions with
liposomes is recognized by antibody but not by hapten-specific H-2 restricted cytotoxic T cells. J. Immunol. 121:2405.

364. Maeda, T. and H. M. McConnell. 1979. Specificity of memory cells
raised against trinitrophenyl-conjugated syngeneic cells. Proc. Natl.
Acad. Sci. USA 76:1537.

365. Poste, G., N. C. Lyon, P. Macander, C. W. Porter, P. Reeve and H. Bach-
meyer. 1980. Liposome-mediated transfer of integral membrane glycoproteins into the plasma membrane of cultured cells. Exp. Cell Res. 129:
393.

366. Heath, T. D., D. Robertson, M. S. C. Birbeck, and A. J. S. Davies. 1979.
Liposomal fusion as a means of introducing surface antigens into living
cell membranes. GANN Monogr. on Cancer Res. 23:255.

367. Shinitzky, M. and M. Inbar. 1974. Difference in microviscosity induced
by different cholesterol levels in the surface membrane lipid layer of normal lymphocytes and malignant lymphoma cells. J. Mol. Biol. 85:603.

368. Alderson, J. C. E. and C. Green. 1975. Enrichment of lymphocytes with
cholesterol and its effect on lymphocyte activation. FEBS Lett. 52:208.

369. Chen, S. S.-H. and R. M. Keenan. 1977. Effect of phosphatidylcholine
liposomes on the mitogen-stimulated lymphocyte activation. Biochem.
Biophys. Res. Commun. 79:852.

370.   Rivnay, B., A. Globerson, and M. Shinitzky. 1978. Perturbation of
       lymphocyte response to concanavalin A by exogenous cholesterol and
       lecithin. Eur. J. Immunol. 8:185.
371.   Ip, S. H. C., J. Abraham, and R. A. Cooper. 1980. Enhancement of
       blastogenesis in cholesterol-enriched lymphocytes. J. Immunol. 124:87.
372.   Ng, M. H., W. S. Ng, W. K. K. Ho, K. P. Fung, and J. P. Lamelin. 1978.
       Modulation of phytohemagglutinin-mediated lymphocyte stimulation by
       egg lecithin. Exp. Cell Res. 116:387.
373.   Ostro, M. J., L. Welling, J. Summers, and S. Dray. 1980. Effect of lipo-
       somes on lymphocyte surface Ig, the con A receptor, and histocompati-
       bility-associated proteins. In *Liposomes and Immunobiology*, B. H. Tom
       and H. R. Six (Eds.). Elsevier/North-Holland, New York, pp. 225-239.
374.   Toyoshima, S. and T. Osawa. 1976. Cholesterol inhibition of the tem-
       porary increase of membrane fluidity of lymphocytes induced by mito-
       genic lectins. Exp. Cell Res. 102:438.
375.   Ozato, K., L. Huang, and R. E. Pagano. 1978. Interactions of phospho-
       lipid vesicles with murine lymphocytes. II. Correlation between altered
       surface properties and enhanced proliferative response. Membr. Biochem.
       1:27.
376.   Kramers, M. T. C., J. Patrick, J. M. Bottomley, P. J. Quinn, and D. Chap-
       man. 1980. Studies of liposome interactions with rat thymocytes. Eur. J.
       Biochem. 110:579.
377.   Dabrowski, M. P., W. E. Peel, and A. E. R. Thomson. 1980. Plasma mem-
       brane cholesterol regulates human lymphocyte cytotoxic function. Eur.
       J. Immunol. 10:821.
378.   Ostro, M. J., B. Bessinger, J. F. Summers, and S. Dray. 1980. Liposome
       modulation of surface immunoglobulins on rabbit spleen cells. J. Immunol.
       124:2956.
379.   Dahlgren, C., E. Kihlström, K.-E. Magnusson, O. Stendahl, and C. Tagesson.
       1977. Interaction of liposomes with polymorphonuclear leukocytes. II.
       Studies on the consequences of interaction. Exp. Cell Res. 108:175.
380.   Miller, H. C. and W. J. Esselman. 1975. Modulation of the immune re-
       sponse by antigen-reactive lymphocytes after cultivation with gangliosides.
       J. Immunol. 115:839.
381.   Esselman, W. J. and H. C. Miller. 1977. Modulation of B cell responses
       by glycolipid released from antigen-stimulated T cells. J. Immunol. 119:
       1994.
382.   Correa, M., H. C. Miller, and W. J. Esselman. 1980. Antigen induced
       modulation by shed lymphocyte membrane gangliosides. Immunol.
       Commun. 9:543.
383.   Lengle, E. E., R. Krishnaraj, and R. G. Kemp. 1979. Inhibition of the
       lectin-induced mitogenic response of thymocytes by glycolipids. Cancer
       Res. 39:817.
384.   Ryan, J. L. and M. Shinitzky. 1979. Possible role for glycosphingolipids
       in the control of immune responses. Eur. J. Immunol. 9:171.

385. Yates, A. J., C. L. Hitchcock, S. S. Stewart, and R. L. Whisler. 1980. Immunological properties of gangliosides. In *Cell Surface Glycolipids*, C. C. Sweeley (Ed.). ACS Symposium Series 128. American Chemical Society, Washington, D.C., pp. 419–433.

386. Gregoriadis, G. 1977. Targeting of drugs. Nature 265:407.

387. Magee, W. E. and O. V. Miller. 1972. Liposomes containing antiviral antibody can protect cells from virus infection. Nature 235:339.

388. Magee, W. E., J. H. Cronenberger, and D. E. Thor. 1978. Marked stimulation of lymphocyte-mediated attack on tumor cells by target-directed liposomes containing immune RNA. Cancer Res. 38:1173.

389. Magee, W. E. 1978. Potentiation of interferon production and stimulation of lymphocytes by polyribonucleotides entrapped in liposomes. Ann. N.Y. Acad. Sci. 308:308.

390. Gregoriadis, G. 1975. Homing of liposomes to target cells. Biochem. Soc. Trans. 3:613.

391. Gregoriadis, G. and E. D. Neerunjun. 1975. Homing of liposomes to target cells. Biochem. Biophys. Res. Commun. 65:537.

392. Gregoriadis, G., E. D. Neerunjun, and R. Hunt. 1977. Fate of a liposome-associated agent injected into normal and tumour-bearing rodents. Attempts to improve localization in tumour tissues. Life Sci. 21:357.

393. Neerunjun, E. D., R. Hunt, and G. Gregoriadis. 1977. Fate of a liposome-associated agent injected into normal and tumour-bearing rodents: attempts to improve localization in tumour tissues. Biochem. Soc. Trans. 5:1380.

394. Siddiqui, W. A., D. W. Taylor, S.-C. Kan, K. Kramer, S. M. Richmond-Crum, S. Kotani, T. Shiba, and S. Kusumoto. 1978. Vaccination of experimental monkeys against *Plasmodium falciparum:* a possible safe adjuvant. Science 201:1237.

395. Tyrrell, D. A., P. I. Campbell, N. G. L. Harding, A. Munro, and B. E. Ryman. 1978. Anti-digoxin antibody incorporation into liposomes: a potential therapy for digoxin toxicity. Biochem. Soc. Trans. 6:1239.

396. Campbell, P. I., N. G. L. Harding, B. E. Ryman, and D. A. Tyrrell, 1980. Redistribution and altered excretion of digoxin in rats receiving digoxin antibodies incorporated in liposomes. Eur. J. Biochem. 109:87.

397. Beutler, E. 1981. Enzyme replacement therapy. Trends Biochem. Sci. 6:95.

398. Weinstein, J. N., R. Blumenthal, S. O. Sharrow, and P. A. Henkart. 1978. Antibody-mediated targeting of liposomes. Binding to lymphocytes does not ensure incorporation of vesicle contents into the cells. Biochim. Biophys. Acta 509:272.

399. Leserman, L. D., J. N. Weinstein, J. J. Moore, and W. D. Terry. 1980. Specific interaction of myeloma tumor cells with hapten-bearing liposomes containing methotrexate and carboxyfluorescein. Cancer Res. 40:4768.

400. Papahadjopoulos, D., R. Fraley, and T. Heath. 1980. Optimization of liposomes as a carrier system for the intracellular delivery of drugs and macromolecules. In *Liposomes and Immunobiology*, B. H. Tom and H. R. Six (Eds.). Elsevier/North-Holland, New York, pp. 151–164.

401. Allison, A. C. and G. Gregoriadis. 1974. Liposomes as immunological adjuvants. Nature 252:252.

402. Heath, T. D., D. C. Edwards, and B. E. Ryman. 1976. The adjuvant properties of liposomes. Biochem. Soc. Trans. 4:129.

403. Gregoriadis, G. 1976. The carrier potential of liposomes in biology and medicine. N. Engl. J. Med. 295:704-710 and 765.

404. Ruttkowski, E. and K. Nixdorff. 1980. Qualitative and quantitative changes in the antibody-producing cell response to lipopolysaccharide induced after incorporation of the antigen into bacterial membrane phospholipid vesicles. J. Immunol. 124:2548.

405. van Rooijen, N. and R. van Nieuwmegen. 1980. Endotoxin enhanced adjuvant effect of liposomes, particularly when antigen and endotoxin are incorporated within the same liposome. Immunol. Commun. 9:747.

406. Kotani, S., F. Kinoshita, I. Morisaki, T. Shimono, T. Okunaga, H. Takada, M. Tsujimoto, Y. Watanabe, and K. Kato. 1977. Immunoadjuvant activities of synthetic 6-O-acyl-N-acetylmuramyl-L-alanyl-D-isoglutamine with special reference to the effect of its administration with liposomes. Biken J. 20:95.

407. Siddiqui, W. A., S.-C. Kan, K. Kramer, S. Case, K. Palmer, and J. F. Niblack. 1981. Use of a synthetic adjuvant in an effective vaccination of monkeys against malaria. Nature 289:64.

408. Mašek, K., M. Zaoral, J. Ježek, and R. Straka. 1978. Immunoadjuvant activity of synthetic N-acetyl muramyl dipeptide. Experientia 34:1363.

409. Mitchell, G. H., S. Cohen, C. D. V. Black, A. Voller, F. M. Dietrich, and P. Dukor. 1980. Novel adjuvants in malaria vaccination. In *The Host Invader Interplay,* H. van den Bossche (Ed.). Elsevier/North-Holland Biomedical Press, Amsterdam, pp. 629-632.

410. Allison, A. C. 1979. Mode of action of immunological adjuvants. J. Reticuloendothel. Soc. 26(Dec. Suppl.):619.

411. Segal, A. W., E. J. Wills, J. E. Richmond, G. Slavin, C. D. V. Black, and G. Gregoriadis. 1974. Morphological observations on the cellular and subcellular destinations of intravenously administered liposomes. Br. J. Exp. Pathol. 55:320.

412. Mattenberger-Kreber, L., G. Auderset, M. Schneider, A. Louis-Broillet, M. S. Benedetti, and A. Malnoë. 1976. Phagocytosis of liposomes by mouse peritoneal macrophages. Experientia 32:1522.

413. Raz, A., C. Bucana, W. E. Fogler, G. Poste, and I. J. Fidler. 1981. Biochemical, morphological, and ultrastructural studies on the uptake of liposomes by murine macrophages. Cancer Res. 41:487.

414. van Rooijen, N. and R. van Nieuwmegen. 1980. Liposomes in immunology: evidence that their adjuvant effect results from surface exposition of the antigens. Cell. Immunol. 49:402.

415. van Rooijen, N. and R. van Nieuwmegen. 1980. Liposomes in immunology: multilamellar phosphatidylcholine liposomes as a simple biodegradable and harmless adjuvant without any immunogenic activity of its own. Immunol. Commun. 9:243.

416. Yasuda, T., G. F. Dancey, and S. C. Kinsky. 1977. Immunogenicity of

liposomal model membranes in mice: dependence on phospholipid composition. Proc. Natl. Acad. Sci. USA 74:1234.

417. Dancey, G. F., T. Yasuda, and S. C. Kinsky. 1978. Effect of liposomal model membrane composition on immunogenicity. J. Immunol. 120: 1109.

418. Humphries, G. M. K. and H. M. McConnell. 1979. Potent immunosuppression by oxidized cholesterol. J. Immunol. 122:121.

419. Humphries, G. M. K. 1979. Specific stimulation and suppression of a primary in vitro plaque-forming cell response by monovalent lipid haptens in fluid liposomal membranes. J. Immunol. 123:2126.

420. Alving, C. R. 1982. Therapeutic potential of liposomes as carriers in leishmaniasis, malaria, and vaccines. In *Targeting of Drugs,* G. Gregoriadis, J. Senior, and A. Trouet (Eds.). Plenum, New York, pp. 337–353.

421. Schwenk, R., W. Y. Lee, and A. H. Sehon. 1978. Specific suppression of anti-hapten reaginic antibody titers with hapten-coated liposomes. J. Immunol. 120:1612.

422. Yasuda, T., G. F. Dancey, and S. C. Kinsky. 1977. Immunogenic properties of liposomal model membranes in mice. J. Immunol. 119:1863.

423. van Houte, A. J., H. Snippe, and J. M. N. Willers. 1979. Characterization of immunogenic properties of haptenated liposomal model membranes in mice. Immunology 37:505.

424. Tadakuma, T., T. Yasuda, S. C. Kinsky, and C. W. Pierce. 1980. The effect of epitope density on the in vitro immunogenicity of hapten-sensitized liposomal model membranes. J. Immunol. 124:2175.

425. Nicolotti, R. A. and S. C. Kinsky. 1975. Immunogenicity of liposomal model membranes sensitized with mono(p-azobenzenearsonic acid)tyrosylphosphatidylethanolamine derivatives. Antibody formation and delayed hypersensitivity reaction. Biochemistry 14:2331.

426. Kochibe, N., R. A. Nicolotti, J. M. Davie, and S. C. Kinsky. 1975. Stimulation and inhibition of anti-hapten responses in guinea pigs immunized with hybrid liposomes. Proc. Natl. Acad. Sci. USA 72:4582.

427. Dancey, G. F., P. C. Isakson, and S. C. Kinsky. 1979. Immunogenicity of liposomal model membranes sensitized with dinitrophenylated phosphtidylethanolamine derivatives containing different length spacers. J. Immunol. 122:638.

428. Nicolotti, R. A., N. Kochibe, and S. C. Kinsky. 1976. Comparative immunogenic properties of N-substituted phosphatidylethanolamine derivatives and liposomal model membranes. J. Immunol. 117:1898.

429. van Houte, A. J., H. Snippe, G. T. M. Peulen, and J. M. N. Willers. 1981. Characterization of immunogenic properties of haptenated liposomal model membranes in mice. Immunology 42:165.

430. Yasuda, T., T. Tadakuma, C. W. Pierce, and S. C. Kinsky. 1979. Primary in vitro immunogenicity of liposomal model membranes in mouse spleen cell cultures. J. Immunol. 123:1535.

431. Humphries, G. M. K. 1981. Evidence for direct control of an in vitro plaque-forming cell response by quantitative properties of intact, fluid,

haptenated liposomes: a potential model system for antigen presentation by macrophages. J. Immunol. 126:688.

432. Banerji, B., J. J. Kenney, I. Scher, and C. R. Alving. 1982. Antibodies against liposomes in normal and immune-defective mice. J. Immunol. 128:1603.

433. Mullan, N. A., P. M. Newsome, P. G. Cunnington, G. H. Palmer, and M. E. Wilson. 1974. Protection against gram-negative infections with antiserum to lipid A from *Salmonella minnesota* R595. Infect. Immun. 10:1195.

434. Kolb, C., R. Di Pauli, and E. Weiler. 1974. Induction of IgG by lipid A in the newborn mouse. J. Exp. Med. 139:467.

435. Huet, C. and S. Ansel. 1977. SV40 tumor rejection induced by vesicular stomatitis virus bearing SV40 tumor-specific transplantation antigen (SV40-TSTA). II. Association of SV40-TSTA activity with liposomes containing VSV glycolipids. Int. J. Cancer 20:61.

436. Gregoriadis, G., P. D. Leathwood, and B. E. Ryman. 1971. Enzyme entrapment in liposomes. FEBS Lett. 14:95.

437. Weissmann, G., H. Korchak, M. Finkelstein, J. Smolen, and S. Hoffstein. 1978. Uptake of enzyme-laden liposomes by animal cells in vitro and in vivo. Ann. N.Y. Acad. Sci. 308:235.

438. Steger, L. D. and R. J. Desnick. 1977. Enzyme therapy. VI. Comparative in vivo fates and effects on lysosomal integrity of enzyme entrapped in negatively and positively charged liposomes. Biochim. Biophys. Acta 464:530.

439. Sessa, G. and G. Weissmann. 1970. Incorporation of lysozyme into liposomes. A model for structure-linked latency. J. Biol. Chem. 245:3295.

440. Dorfman, S. H., J. M. Fry, D. H. Silberberg, C. Grose, and M. C. Manning. 1978. Cerebroside antibody titers in antisera capable of myelination inhibition and demyelination. Brain Res. 147:410.

441. Hakomori, S., S.-M. Wang, and W. W. Young, Jr. 1977. Isoantigenic expression of Forssman glycolipid in human gastric and colonic mucosa: its possible identity with "A-like antigen" in human cancer. Proc. Natl. Acad. Sci. USA 74:3023.

442. Feizi, T., R. A. Childs, S. Hakomori, and M. E. Powell. 1978. Blood-group-Ii-active gangliosides of human erythrocyte membranes. Biochem. J. 173:245.

443. Ehnholm, C. and D. B. Zilversmit. 1972. Use of Forssman antigen in the study of phosphatidylcholine exchange between liposomes. Biochim. Biophys. Acta 274:652.

444. Ehnholm, C. and D. B. Zilversmit. 1973. Exchange of various phospholipids and of cholesterol between liposomes in the presence of highly purified phospholipid exchange protein. J. Biol. Chem. 248:1719.

445. Nagai, Y. and T. Ohsawa. 1974. Production of high titer antisera against sialoglycosphingolipids and their characterization using sensitized liposome. Jpn. J. Exp. Med. 44:451.

446. Lingwood, C. A., R. K. Murray, and H. Schachter. 1980. The prepara-

tion of rabbit antiserum specific for mammalian testicular sulfogalacto-
glycerolipid. J. Immunol. 124:769.

447. Alving, C. R., K. C. Joseph, H. B. Lindsley, and M. J. Schoenbechler.
1974. Immune damage to liposomes containing lipids from *Schistosoma
mansoni* worms. Proc. Soc. Exp. Biol. Med. 146:458.

448. Ohsawa, T. and Y. Nagai. 1975. Immunological evidence for the localization of sialoglycosphingolipids at the cell surface of sea urchin spermato-
zoa. Biochim. Biophys. Acta 389:69.

449. van Houte, A. J., H. Snippe, G. T. M. Peulen, and J. M. N. Willers. 1981.
Characterization of immunogenic properties of haptenated liposomal
model membranes in mice. III. Specificity of delayed-type hypersensitivity and antibody formation. Immunology 42:233.

450. van Rooijen, N. and R. van Nieuwmegen. 1977. Liposomes in immunology: the immune response against antigen-containing liposomes. Immunol.
Commun. 6:489.

451. van Rooijen, N. and R. van Nieuwmegen. 1978. Liposomes in immunology: further evidence for the adjuvant activity of liposomes. Immunol.
Commun. 7:635.

452. van Rooijen, N. and R. van Nieuwmegen. 1979. Liposomes in immunology: impairment of the adjuvant effect of liposomes by incorporation of
the adjuvant lysolecithin and the role of macrophages. Immunol. Commun. 8:381.

453. van Rooijen, N. and R. van Nieuwmegen. 1979. Attempts to study the
localization of liposomes and liposome entrapped antigen in the spleen.
Acta Histochem. 65:41.

454. Ryman, B. E., R. F. Jewkes, K. Jeyasingh, M. P. Osborne, H. M. Patel,
V. J. Richardson, M. H. N. Tattersall, and D. A. Tyrrell. 1978. Potential
applications of liposomes to therapy. Ann. N.Y. Acad. Sci. 308:281.

455. Willenborg, D. O. and T. J. Higgins. 1981. Liposomes containing myelin
basic protein (BP) suppress but do not induce allergic encephalomyelitis
in Lewis rats. Aust. J. Exp. Biol. Med. Sci. 59:135.

456. Manesis, E. K., C. H. Cameron, and G. Gregoriadis. 1979. Hepatitis-B-
surface antigen-containing liposomes enhance humoral and cell-mediated
immunity to the antigen. Biochem. Soc. Trans. 7:678.

457. Kramp, W. J., H. R. Six, S. Drake, and J. A. Kasel. 1979. Liposomal enhancement of the immunogenicity of adenovirus type 5 hexon and fiber
vaccines. Infect. Immun. 25:771.

458. Six, H. R., W. J. Kramp, and J. A. Kasel. 1980. Effect of liposomes on
serological responses following immunization with adenovirus purified
type 5 subunit vaccines. In *Liposomes and Immunobiology*, B. H. Tom
and H. R. Six (Eds.). Elsevier/North-Holland, New York, pp. 119–131.

459. Sanchez, Y., I. Ionescu-Matiu, G. R. Dreesman, W. Kramp, H. R. Six, F. B.

Hollinger, and J. L. Melnick. 1980. Humoral and cellular immunity to hepatitis B virus-derived antigens: comparative activity of Freund complete adjuvant, alum, and liposomes. Infect. Immun. 30:728.

460. Lawman, M. J. P., P. T. Naylor, L. Huang, R. J. Courtney, and B. T. Rouse. 1981. Cell-mediated immunity to herpes simplex virus: induction of cytotoxic T lymphocyte responses by viral antigens incorporated into liposomes. J. Immunol. 126:304.

461. Ruebush, M. J., A. H. Hale, L. Lefrancois, D. T. Harris, and D. E. Burgess. 1981. Elicitation and specificity of anti-vesicular stomatitis virus-specific cytotoxic T lymphocytes. J. Immunol. 126:2053.

462. Gerlier, D., F. Sakai, and J. F. Doré. 1980. Induction of antibody response to liposome-associated Gross-virus cell-surface antigen (GCSAa). Br. J. Cancer 41:236.

463. Burakoff, S. J., V. H. Engelhard, J. Kaufman, and J. L. Strominger. 1980. Induction of secondary cytotoxic T lymphocytes by liposomes containing HLA-DR antigens. Nature 283:495.

7

Therapeutic Applications of Liposomes

Eric Mayhew / Roswell Park Memorial Institute, Buffalo, New York

Demetrios Papahadjopoulos / University of California, Medical School, San Francisco, California

INTRODUCTION

Drug-Carrier Systems

During the last few years, considerable interest has been raised concerning the potential of liposomes as an in vivo drug-carrier system. The literature on this subject is both voluminous and to a certain extent, conflicting. Several recent reviews have described the initial studies in considerable detail (1-9). In this discussion we attempt to analyze critically some of the conflicting results, point out the problems and difficulties, and emphasize the potential of liposomes in cancer chemotherapy.

The aim of all therapy is to selectively treat the pathological condition while minimizing nonspecific effects on the rest of the organism. Rational development of drugs for therapy has centered on exploiting known differences between the target diseased tissues and/or cells and other normal tissues. Often the differences in susceptibility to drugs between target and nontarget tissues are quantitative rather than qualitative. This is especially so in cancer chemotherapy, where there is still a lack of knowledge of truly cancer-related differences between the cancer and normal cells. This leads to the consequence that most therapies for cancer are toxic for normal cells as well as the tumor cells.

The use of carriers for drugs has been proposed to improve selectivity by:

1. Altering the pharmacokinetics of the drug, such as clearance, metabolism, and excretion in favor of the therapeutic as opposed to the toxic effects of the drug
2. Delivering the drug specifically to the target organ, tissue, or cellular or subcellular sites
3. Decreasing toxicity by control of drug and/or metabolite levels in blood and/ or at organ sites
4. Reducing the dosages needed to produce clinical effects at acceptable levels of toxicity
5. Controlling the site or rate of release of drug to enable more precision to be obtained in regulating the blood and organ levels of drug and/or its metabolites
6. At a more practical level by prolonging the intervals needed between administration of the drug

A number of different drug-carrier systems have been proposed (10-13) and are under investigation for their usefulness as delivery systems. However, for any drug carrier to be developed for clinical use a number of steps have to be undertaken:

1. The drug-carrier combination must be able to be prepared in a standardized way.
2. The preparation must be stable for a sufficient length of time, before administration, so that it can be distributed and used reproducibly.
3. There has to be knowledge of the mechanism by which the drug is released from the carrier.
4. There has to be knowledge as to where the drug is released (i.e., at the site of injection or implant or at tissue, cellular, or intracellular sites).
5. The pharmacokinetics in animals and in the clinical situation should be understood, thereby allowing a rational dosage regimen to be developed.
6. Ideally, there should be in vitro tests for the characteristics of the carrier which can be correlated with the results from in vivo experiments.

7. There should be clear evidence developed that the optimal dosing of the drug-
    carrier complex is at least as therapeutically effective as the optimal sched-
    uling of the "free" drug.

Liposomes as a Drug-Carrier System

The basic premises for using liposomes as a drug carrier are relatively simple:

1. Drugs and other macromolecules can be encapsulated within the interior of
    the relatively impermeable bilayer membranes, where they can be protected
    from the environment during transit to their target areas.
2. Liposomes can be taken up by cells without overt cytotoxic effects by any of
    a number of mechanisms and can therefore enhance the cellular uptake of the
    encapsulated material.
3. Encapsulation results in altered pharmacokinetics, which, coupled with the
    exciting possibility of producing liposomes with specificity for different cell
    types, may yield an enhanced therapeutic index.

It is not proposed in this discussion to compare in detail the properties of different carriers. However, liposomes can be considered to be one specific type of
drug carrier which has certain advantages and disadvantages compared to other
carriers.

Advantages of liposomes compared with other carriers include:

1. They are made from natural constituents and thus are biodegradable and (as
    will be discussed) quite nontoxic.
2. They can be made, relatively easily in many cases, from a wide variety of
    components. This leads to different biological properties for different types
    of liposomes. In principle, these properties can be modified in accordance
    with the goals of the specific therapeutic application.
3. For basic liposome-drug preparations no chemical bond formation is required,
    although covalent bonding, such as the coupling of protein, can be done to
    modify the basic liposome structure. In general, drugs are trapped without
    chemical modification either in solution in the internal aqueous space(s) of
    the liposome (water-soluble drugs) or in the lipid bilayer (lipid-soluble drugs).
    Thus, when the drug is released at the appropriate site, it is usually released
    in the form it was entrapped.
4. Liposomes can be made in a variety of sizes; the smallest liposomes are small
    compared with many other types of drug carrier. Small size has advantages
    in delivery of the liposome and its contents to specific cellular and intracellu-
    lar sites.
5. It is possible that intact liposomes may be able to deliver drugs intracellularly

by mechanisms other than endocytosis, which is the usual pathway for other drug carriers. It is possible that fusion between the liposome and cellular membrane systems could enable delivery of drugs into cellular compartments not able to be reached by other drug carriers. However, this may be difficult to achieve in the in vivo situation.

6. Depending on the liposome-drug combination used, liposomes can be made with a relatively high drug/carrier molecular ratio.

Some of the possible disadvantages of liposomes compared with other drug carriers include:

1. Fully chemically defined polymers can be made with a more precise and a broader range of in vivo drug-permeability properties than with current liposomes.
2. Once liposomes are administered, they cannot be removed. A number of types of implantable drug carriers are removable, can be reimplanted, and drug release controlled by external sources.
3. There is a great range of possible types of nonliposomal drug carriers. Although each type may not be as versatile as liposomes, in toto they may have greater versatility.

We shall discuss here the current status of liposomes as drug carriers for in vivo applications. We shall not discuss in detail the preparation of liposomes or their uses in immunologic applications, as these have been discussed in detail in other chapters in this book, except as they relate to the therapeutic applications.

PREPARATION OF LIPOSOMES FOR IN VIVO USE

Since liposomes were first described in detail by A. D. Bangham and colleagues (14) as a model membrane system, they have been intensely studied by a large number of biochemists and biophysicists as a simple experimental model for a variety of membrane phenomena (15-19). Because of this basic interest in liposomes as a model membrane system, we have at our disposal a large pool of information on the general properties of lipid bilayers and how they can be manipulated with predictable results. Although this aforesaid activity is not directed toward the use of liposomes as drug carriers, it has important consequences for the future development of liposomes in relation to specific applications.

Unfortunately, most of the work that has been done up to now with liposomes as a drug carrier has utilized methodologies developed more than 10 years ago; furthermore, these methods were developed fortuitously and were not optimized for the purpose of drug delivery. There is an obvious need, therefore, for optimization of the liposome system according to the requirements of a particular pharmacological application. This can be accomplished best by the proper understanding and methodical utilization of the physicochemical properties of the system. These properties are reasonably well understood and it is precisely the capacity of determining a variety of such properties that makes liposomes a promising and versatile system for drug delivery. The methodology for further optimization of the system exists and its proper utilization may lead to a variety of useful applications.

Methodologies for Different Liposome Preparations

Methodologies for liposome preparation have been discussed in Chapter 1 and by Szoka and Papahadjopoulos (18). Liposome properties vary widely depending on the preparation method used. Important parameters to be considered are the following: (a) the encapsulation (trapping) efficiency, expressed as the percentage of the initial drug that becomes associated with liposomes during preparations; (b) the captured volume, expressed as volume of internal aqueous space per milligram (or micromole) of lipid; (c) particle size, expressed as the average diameter of individual liposomes [particle size has been shown to be very important in determining clearance rates and tissue disposition of injected liposomes (20-22)]; and (d) size heterogeneity, expressed as size range for the majority of liposomes. The properties of each liposome membrane will vary widely depending on the chemical composition of the lipids.

Thus, liposomes can be made to possess a surface charge varying from positive to neutral to negative. In addition, liposomes can be made with phospholipids that vary in the fatty acyl chain configuration, such that they can be either "fluid" or "solid" at ambient temperature. Generally, liposomes that are "solid" or contain a high ratio of cholesterol are much more stable and retain their contents for extended periods of time both on storage and in the presence of plasma components following injection in vivo (18,23). Antitumor effects have been shown to vary drastically depending on the permeability of the injected liposomes to the encapsulated drug (23). Finally, liposomes can be made with various glycolipids which have been shown to alter the tissue specificity depending on the chemistry of the carbohydrate residues (24,25).

The original liposome preparation as described by Bangham et al. (14) consisted of multilamellar vesicles (MLVs) of widely varying diameter in the range 0.1-$10\,\mu$. It is a very simple procedure and has been used widely both as a model membrane system and as a drug carrier. Its main disadvantages as a drug delivery system stem from the very heterogeneous size distribution, the variable number of concentric lamellae per liposome, and the relatively low ratio of internal aqueous space per total lipid. In addition, the average size distribution can vary considerably depending on the type of lipid and degree of shaking during preparation (26). The very large liposomes have been found to be much more toxic in mice when injected intravenously in high amounts, probably due to occlusions in the microcirculation system (27). Brief sonication of the preparation, which has been used in many instances (1,6), would tend to increase the size heterogeneity by producing an additional population of very small vesicles. This would also tend to increase the variability between preparations. Extensive sonication produces a relatively homogeneous population of small unilamellar vesicles (SUVs) which can be separated from any remaining larger MLVs by gel filtration (28,29) or differential centrifugation (30). The SUVs are the best characterized liposomes (31) and their main advantage is the relative homogeneity in size distribution (200-300 Å in diameter), but their use as a drug carrier for water-soluble drug is limited by the very low efficiency of encapsulation and the extremely small ratio of encapsulated volume per lipid. However, for the same reasons, they can be efficient carriers of lipid-soluble drugs.

Small unilamellar vesicles have also been prepared by other methods, such as detergent removal and ethanol injection (18,32); although these methods are preferable for reconstitution of membrane proteins, they have limited applications for drug delivery because of the relatively low efficiency of encapsulation. Recent modifications of the detergent removal procedure have been reported to yield intermediate-size vesicles (1000 Å in diameter) with a reasonable efficiency of encapsulation (33).

Large unilamellar vesicles (LUVs) have been prepared by slow hydration at low ionic strength (34), by an ether infusion method (35), by calcium-induced fusion of small vesicles (36), and by a reverse-phase evaporation method (37). Of all these methods, the only one with proven advantages as a drug delivery system is the reverse-phase evaporation method because it combines the following characteristics: ease of preparation, encapsulation of a large percentage of the aqueous material, high ratio of entrapped aqueous volume per lipid, widely variable lipid composition and ionic conditions in aqueous phase, and a relatively narrow size distribution (0.2-0.4 μm). This procedure has already been used successfully for antitumor chemotherapy in animal systems (23).

The most recent methodological improvement involves the production of liposomes of well-defined size distribution by extrusion of MLV or LUV through nucleopore polycarbonate membranes (26). The pressure extrusion method allows the sequential passage of preformed liposomes through membranes of decreasing pore diameter. During the extrusion, all the lipid material passes through the membrane, but the size of liposomes is reduced to approximately the diameter of the membrane pores (26). This procedure can be applied to any type of liposome (38) and is rapid and simple. If the extrusion is performed in the presence of the mother liquid used for the initial encapsulation, it results in relatively good encapsulation efficiency.

Standardization and Reproducibility

For the successful use of any drug-carrier system, it is important that highly reproducible standard preparations can be made. This, in the case of liposomes, includes chemical and morphological reproducibility. It seems clear that under appropriate conditions, chemical purity could be achieved with sufficient care, and that possible long-term degradative changes in lipids and/or other components could be reduced or eliminated by storage in inert gas, at optimal temperatures, and perhaps by addition of antioxidants and other agents (39,40). It has also been suggested that liposomes and components can be reconstituted after freeze-drying (27,41). More of a problem may be the reproducibility of liposome size and morphology. Many investigations described so far have been made by using hand shaking or partially sonicated preparations. Moreover, particle size distribution in many of these preparations has not been described. From our experience, exact duplication of such preparations is impossible in this regard. Liposomal size distribution may greatly influence the tissue localization of liposomes (22,42); two similar but not identical preparations may show different tissue distribution and thus could have differential therapeutic effects. We have attempted to standardize size distribution of different liposome preparations by extrusion through different pore size membranes (26,38), but even here there were some size differences from preparation to preparation. Extensively sonicated liposome preparations are the most uniform preparations and therefore quite reproducible. This is a definite advantage in situations where this type of liposome is the best vesicle. Even so, it is important that the reproducibility of each preparation be characterized.

Another possible problem that has to be faced is that even though immediately after preparation the liposomes may be homogeneous, aggregation, fusion, or lysis of liposomes can occur during storage so that by the time the preparation is ready for injection, the size and contents of the particles could be altered. As yet, no in-depth study has been reported of investigations of the optimal long-term storage conditions of liposomes with respect to ionic strength, osmolarity, ionic composition, temperature, pH, and so on, of the suspending medium. If changes have occurred in the size distribution of the preparations by the time of injection, there will probably be important therapeutic consequences.

FATE OF INJECTED LIPOSOMES AND EFFECTS OF ROUTE OF INJECTION

If, up to the time of injection, the liposomes were in salt solution, their environment is drastically altered immediately upon injection. The presence of biological fluids can change the properties of liposomes very rapidly. After intravenous injection this can include destruction, aggregation, alteration of permeability, changes in size, and alteration of the liposome surface in terms of charge, adsorbed ions, and other substances (2,8,43).

The extent of these effects will depend on the liposome composition. Ideally, one would like minimal changes in liposome properties after injection, but as mentioned previously, one advantage of liposomes compared to some other types of carriers is that they are biodegradable. Thus, there has to be a trade-off between stability in biological fluids and their degradation. The primary concern is that liposomes should be stable enough in vivo so that either they can reach their site of action (if they can be targeted) or that they can release drugs over a sufficient period to exert their effects (if acting as a time-release system). There does not seem to be any reason why these conditions cannot be fulfilled if the appropriate composition is used. As has been shown, simple manipulation of cholesterol content and use of phospholipids with varying Tm (temperature of melting) can markedly alter the biological half-life for particular therapeutic applications (6,23,44-46).

The size distribution of liposomes can change rapidly after incubation with biological fluids. On the one hand, they can aggregate, and on the other, they can individually become smaller. Aggregation of liposomes could have particularly undesirable effects if the aggregated particles block blood vessels or capillaries. These effects would depend on the size of the aggregates. Presumably, if liposomes were injected at the appropriate dose rate, aggregation could be minimized. However, aggregation could also be caused by interaction between liposomes at their accumulation sites, perhaps with similar problems. In addition to problems caused by aggregation, other changes could result in altered distribution of liposomes or their contents, as will be discussed later.

Although there are a number of substances in blood which can interact with liposomes and alter their permeability, it is probable that the major interacting components are proteins and lipoproteins (47-49). A number of in vitro studies have been made showing that different proteins interact with liposomes (49-52), increasing their permeability to entrapped species. The effects on the liposomes are related to the ability of these proteins to interact directly and interdigitate into the lipid bilayer (53). Several studies have indicated that high- and low-density lipoproteins are the major interacting species in blood (54,55). This interaction probably involves exchange diffusion but also results in net transfer of lipid and/or cholesterol to the lipoproteins. If equal exchange of similar phospholipids occurred, the leakage of entrapped species would be minimal. However, studies have indicated that the characteristic liposome void volume peaks on Sephadex/Sepharose chromatography disappear in time and radioactive tracers appear in the lipoprotein peaks concurrent with loss of drugs or other low molecular weight species from the liposomes (54). High concentrations (up to 50

mol%) of cholesterol can significantly reduce the net destruction of liposomes
(56). In addition to these effects, liposome properties such as net surface charge
can be markedly altered by adsorption of serum factors (2,57). Thus, even intact liposomes could be coated with proteins and if the liposomes were initially
prepared with "targeting" groups at their surfaces, it is possible that these groups
could be masked in the blood environment.

Therefore, the possible alterations of liposome characteristics in blood should
be an important consideration in the design of liposomes for therapeutic applications. If liposomes are injected in vivo by routes and sites other than intravenous,
the rapid exchange and destruction may be decreased and thus the route of injection could have important consequences for the ultimate site of disposition of
liposomes and their contents.

Routes of Administration

1. *Intravenous.* In mice, the normal route is via the tail vein. Anatomically,
tail veins drain into the heart via the vena cava. Therefore, the first capillary bed
that liposomes would come in contact with is in the lung, and it is here that some
filtration of aggregated or large particles may be expected to occur (22). However, as cells the size of erythrocytes can pass through capillaries, it would be expected that smaller particles would pass through quite readily. After passing
through the lung, the liposomes would pass into the general circulatory network
and would be expected to pass through capillary beds unless the liposomes were
taken out of the circulation by interactions with capillary lining cells, in tissues
where they could penetrate into interstitial space, or by the kidney.

2. *Intraarterial.* Little seems to have been done with respect to intraarterial
injection of liposomes. If injected by this route, bypass of the first passage through
the lung would be achieved. If injected into the carotid artery (58) or in the appropriate heart chamber, the possibility of greater uptake into the brain is increased.

3. *Intraperitoneal.* This is a common route of injection in animal experiments
but usually not in humans. This route may be of use in diseases of the peritoneal
cavity and, for example, in ovarian cancer (59). Drainage of liposomes out of the
peritoneal cavity into the general circulation could occur both through capillary
lining of the cavity and by lymph drainage into the thoracic duct. It is not known
with certainty whether intact liposomes are able to reach the general circulation
after intraperitoneal injection, although it is clear that both bilayer markers and
entrapped species are able to do so, albeit significantly slower than by direct injection into the blood (8).

4. *Subcutaneous, intramuscular, footpad, lymphatic system.* These sites of
injection may be expected to give a slow release of liposomes and/or contents into
the general circulation and/or lymphatic system. Footpad injections have been

shown to increase the accumulation of liposome-associated materials at lymph nodes draining the area of injection (60,61). Some studies have been reported of the systemic accumulation of drugs and/or liposome components after subcutaneous or intramuscular injection (25,62).

5. *Intraarticular.* Injection directly into bone joints has been used in attempts to employ liposomes for treatment of joint diseases. Again, the injection into that area may be expected to confine the movement of liposomes within the joint resulting in a local release of drug (63,64).

6. *Intracerebrally.* Liposome-entrapped drugs can usually only enter the central nervous system very slowly after systemic injection, due to the blood-brain barrier (65), and therefore to increase the concentration of liposomes in this area, drug-containing liposomes have been injected intracerebrally (66,67). Although local uptake can be increased, relatively rapid release of liposomes into the systemic circulation occurs. The central nervous system could, in effect, act as a reservoir for the release of liposomes and their contents into the general circulation.

7. *Oral.* One of the more contentious areas concerning administration of liposomes has been the question of whether significant systemic effects can be observed after administration by the oral route. There have been some reports of increased blood levels (compared to free drug) of substances, notably insulin, entrapped in liposomes and augmented physiological effects have been observed (68-71). However, there have also been reports that liposomes are essentially completely degraded by the detergent action of bile salts (72). There is no convincing evidence that liposomes can be absorbed intact through the gut wall or that complexes resulting from the interaction of liposomes with gut fluids can penetrate the gut lining. However, it is quite possible that liposomes could be made which could be protected from destruction in the gut, be absorbed, and their contents eventually released into the circulation. To accomplish this, there is need for considerable progress not only in producing stable liposomes, but also in determining how liposomes and modifications thereof could pass intact through the gut wall.

8. *Topical.* Topical administration of liposomes has been reported to alter drug distribution and improve therapeutic effects in the rabbit (73).

9. *Intrabronchial.* Direct injection of liposomes into bronchi and alveoli has been attempted for possible use in treating respiratory distress syndrome (74) and for antitumor drugs (75,76). Although initial results have been reported, these studies are at an early stage.

In conclusion, we can say that for most possible routes of liposome administration, there have not been many reports of systematic studies on the fate of liposomes and entrapped drugs. Until there is better knowledge on the mechanisms of disposition of liposomes by different routes, it cannot be expected that liposomes can be exploited optimally for drug delivery.

TISSUE UPTAKE OF LIPOSOMES

General Tissue Disposition

After intravenous injection, intact liposomes are removed from the circulation by tissues and organs. Liposomes apparently are not bound significantly by circulating erythrocytes. Studies on binding by other cellular components in blood have not been reported. It may be expected that circulating leukocytes would pick up liposomes as tissue leukocytes have been shown to do (25). Platelets can interact with liposomes in vitro (77) and presumably would do so in vivo. Depending on the composition of the liposomes, platelet aggregation can be modified (77,78), but clotting time is not affected in short-term experiments (27). All tissues show some uptake of the labels initially associated with liposomes either as an internal (water-soluble) or bilayer (lipid-soluble) marker. Some tissues, such as brain and small intestine, show very low uptake. The organs that take up most of the liposomes on a percent injected dose per unit weight basis are liver, spleen, and lungs (see Refs. 2 and 8). In terms of absolute uptake, the liver may take up 50% or more of the injected dose by a few hours after injection. Some of the uptake may be accounted for by blood space within the organs, but in the major uptake sites, accumulation occurs by association of the liposomes with tissues. Assuming (this may be a crucial assumption) that intact liposomes are "bound" by tissues, the question arises as to how are they bound.

If liposomes are to reach tissue cells intact, they must pass through the lining of the capillaries into the interstitial spaces. In tissues where the capillary endothelium is fenestrated, such as the liver, this could be expected to occur, but in other tissues, unless liposomes can pass through endothelial cells and/or squeeze through spaces or pores between endothelial cells, it would be expected that they would remain associated with the endothelial cell surfaces. This is a major problem for any macromolecular drug carrier. Here they could remain until they were degraded, endocytosed, or cleared by cells specialized for removing such material—macrophages.

The most common means of determining the distribution of liposomes is by use of radiolabels (see Refs. 2 and 8), although spin labels (79,80), or other (25, 81,82) methods have been used. Determination of tissue labeling does not necessarily indicate the tissue location of the intact liposome, only that of the label. As discussed previously, on interaction with serum factors, molecular exchange, transfer of lipid, and leakage of entrapped contents can occur. Unless precautions are taken, the tissue uptake measured may not be that of intact liposomes and misleading conclusions may be drawn. Thus, careful consideration should be given to proving the intactness of liposomes in tissues. This can be quite difficult but it would seem that liposomes labeled both by a lipid bilayer marker and by an internal space marker should be used. However, the γ-ray perturbed angular

Table 1   Tissue Distribution of Free [^{14}C] Ara-C and [^{3}H] Ara-C Entrapped in DPPC-chol REV 1:1 Injected Simultaneously into the Same Mice (normal mice)

| | Percent dose per gram[a] | | Liposome/Free |
	[^{3}H] Lip-Ara-C	[^{14}C] Free Ara-C	Ara-C Ratio
Liver			
15 min	12.4	13.3	0.93
1 hr	17.2	4.1	4.2
4 hr	20.8	0.6	34.67
24 hr	21.6	0.2	108.0
Spleen			
15 min	144.4	21.0	6.88
1 hr	200.6	8.8	22.8
4 hr	265.1	2.7	98.19
24 hr	151.2	1.0	151.2
Lung			
15 min	103.2	17.3	5.97
1 hr	50.2	5.3	9.47
4 hr	35.7	0.7	51.0
24 hr	2.3	0.11	20.9
Brain			
15 min	1.3	3.1	0.42
1 hr	1.0	3.5	0.29
4 hr	0.8	0.64	1.19
24 hr	0.27	0.05	5.4
Small intestine			
15 min	1.6	18.0	0.09
1 hr	1.3	6.0	0.22
4 hr	1.3	0.8	1.63
24 hr	1.5	0.25	6.0
Kidney			
15 min	6.6	31.1	0.21
1 hr	4.1	6.9	0.59
4 hr	4.0	0.33	12.12
24 hr	1.3	0.12	10.83
Heart			
15 min	18.6	16.6	1.12
1 hr	7.3	4.9	1.49
4 hr	6.5	0.43	15.12
24 hr	1.0	0.12	8.33

(continued)

Table 1 (continued)

	Percent dose per gram[a]		Liposome/Free Ara-C Ratio
	^{3}H	^{14}C	
Whole blood			
15 min	287,540	159,640	1.8
1 hr	166,280	36,927	4.5
4 hr	139,813	3,280	42.63
24 hr	23,040	550	41.89
Dialyzed blood			
15 min	141,460	2,840	49.81
1 hr	121,760	2,850	42.72
4 hr	107,227	1,340	80.02
24 hr	15,610	340	45.91
Plasma			
15 min	50,080	140,110	0.36
1 hr	30,390	39,700	0.77
4 hr	75,960	1,373	55.32
24 hr	30,650	580	52.84

[a]Expressed as disintegrations per minute/ml. Each point represents the mean uptake for tissues from three mice.

correlation spectroscopy method (82) can reveal destruction of intact liposomes by perturbations in the correlation spectroscopy of labeled tissue samples. Determination of bilayer/internal space marker ratios should give an indication of the relative intactness of the liposomes, but even this system is not without drawbacks. Definitive proof of delivery of intact liposomes to tissues would require a consideration of the approach described above in combination with spectroscopic techniques. However, for the use of liposomes as a drug-delivery system, the key element is the delivery (bioavailability) of the drug, not the lipid, so that tissue determinations of drug content and positive identification of the metabolites, compared to distribution of nonentrapped drug, would at least show the alterations induced by liposome entrapment.

Table 1 shows the comparative distribution of cytosine-arabinoside in mice injected with [^{3}H] Ara-C (entrapped in liposomes) or [^{14}C] Ara-C (nonentrapped) in the same mice (27). This type of internally controlled experiment clearly shows the drastic alteration in distribution of the markers. Identification of the Ara-C metabolites in the various tissues has been made. The results show that, in general, the initially liposomal Ara-C remains mainly as Ara-C for prolonged periods, whereas the initially "free" Ara-C is rapidly degraded to inactive metabolites, mainly Ara-U. These types of studies need to be made on all specific liposome-drug combinations for the rational development of liposomes as a carrier for specific pathologic conditions.

Tumor Uptake of Liposomes

As noted previously, all tissues take up liposomes to some extent (8). Solid tumors do also, although the uptake is not particularly high and may be less than that of surrounding normal tissues (83,84). This may be related to the blood supply of tumors and the type of isotopic marker used. For example, the isotopically labeled phospholipid used in some studies (83,84) could exchange with lipoproteins in plasma (54), which could then be metabolized in the liver. Depending on the size and age of the tumor, there will be greater or lesser amounts of necrotic tissue present. There is no reason to suppose from the previous discussion that liposome-entrapped drug will be able to penetrate to the actively growing tumor cells more effectively than free drug.

Interaction of Liposomes with Cells and Uptake in Tissues

Although this communication is concerned with the possible uses of liposomes in therapy, unless we discuss the possible mechanisms of the cellular uptake of liposomes, any discussion of the therapeutic usefulness of liposomes will not be based on realistic predictions.

Most studies on the interaction of liposomes with cells have been made in vitro. It appears that with most types of liposomes the initial interaction is adsorption at the cell surface (85-89). This occurs rapidly and up to several percent of the liposomes initially present in the environment can bind.

What happens next is not clear. Some studies indicate that, independent of cell type, most liposomes remain attached at the cell surface for considerable periods, and very little internalization of liposome-associated material is seen (85,86,88,90). Other studies can be interpreted to show that after this initial binding stage, liposomes are internalized intact inside a typical endocytotic vacuole (85,91). The possibility that liposomes of the appropriate composition are able to fuse with membranes releasing their entrapped species (92,93) is still tenable, but it appears that in most systems in vitro (if the target cell surface is not deliberately modified to become more susceptible to fusion events) only a small proportion of the liposomes initially adsorbed can fuse with the cell membrane during the initial hours of incubation (88,94). Furthermore, it is possible that the same liposome types that are most likely to fuse (i.e., those with lipids that are above their transition temperatures at 37°C) are also most susceptible to degradation by plasma components. It would seem that in vivo delivery of liposomal contents to target cells by a fusion mechanism *using unmodified liposomes* may be unlikely. However, as indicated above, a small percentage of adsorbed liposomes may be able to enter cells by fusion. Although small quantitatively, this may be important qualitatively for bypassing permeability barriers. It is also possible that liposomes may fuse with phagosome membranes after endocytosis. Realistically, however, it is not easy to predict that the uptake process in vivo would be similar to that occurring in vitro. The liposomes initially adsorbed in tissues would be endocytosed if the cells were capable of endocytosis or if endocytosing cells were in the area. If the cells were incapable of endocytosis, the liposomes would gradually degrade, releasing their contents locally at relatively high concentration. This local release could result in higher uptake of drugs by target cells in the tissue and/or a prolonged bathing of the tissue by released drug (8).

Differential liposomal uptake by different organs could be explained by tissue trapping and/or uptake by endocytotic cells. Differential trapping could occur not only within spaces in the capillary lining but from alterations in local blood flow and by the presence of cells with particularly "sticky" surfaces. Sites of tissue damage may have altered binding properties for liposomes (95). Tissues differ greatly in the number of "macrophage" or macrophage-like cells, and tissues with great numbers of these cells would be expected to accumulate liposomes.

It is clear from the published data that many of the tissues that take up liposomes are rich in endocytotic cells that do clearly take up liposomes, supporting the general assumption that liposomes are cleared mainly by endocytotic cells (91,96,97). However, two pieces of data may perhaps suggest that endocytotic cell clearance of liposomes from the circulation is not the whole story. Some observations suggest first, that the cells accumulating liposomes are cells which may be at least partially recruited to the tissues *after* the liposomes have been already trapped in the tissues (27) and second, that if the reticuloendothelial system (RES) function is depressed by prior treatment with drugs or silica, only a proportion (<50%) of the tissue uptake is reduced in tissues with high RES clearance (27). This suggests that at least some of tissue-associated liposomes are not accumulated directly by endocytotic cells.

Unless the target cell is the endocytotic RES cell, the aim for chemotherapeutic applications should be to diminish RES cell uptake. Either the RES system could be blocked, perhaps by pretreatment with plain (empty) liposomes (42,98, 99), or liposomes could be made such that they avoid RES cells. The second alternative would be much preferable to the former since blocking RES cell function could have significant deleterious physiological effects. Furthermore, if liposomes are endocytosed by target cells, it may be expected that non-RES endocytotic blocking agents would block endocytotic uptake by the very cells that one wishes to attack.

Tissues vary markedly in the intactness of their endothelial linings. In liver, the endothelial linings are open and liposomes could be expected to reach the liver parenchymal cells. On the other hand, in the lung, the endothelia are tight and much penetration could not be expected. In tumors there may or may not be intact endothelia present. Endothelia are not impermeable to large molecules and even leukocytes can pass through endothelia either by passing directly through the cells or through pores. Thus, if conditions were correct, liposomes may even be able to pass through tight endothelia, but clearly, under the normal circumstances this would be a low-efficiency process. Use of drugs to open up endothelia may potentiate tissue cell uptake. Another means of increasing tissue cell uptake would be to use leukocytes as liposome carriers into the target tissues. These methods are, however, extremely speculative at present.

Metabolism and Excretion

Liposomal lipids are presumably susceptible to the same processes as natural circulatory and tissue-bound lipids. By various exchange and enzymatic processes the liposomal lipids are distributed into organ tissues and circulating lipoproteins. The low toxicity of plain liposomes (several grams per kilogram) (27,100) attests to the fact that formation of toxic metabolites does not occur to any significant extent. Radiolabel originally associated with liposomes can be detected in the urine within 1 hr of intravenous injection and most of the label is excreted over a period lasting from several hours to days. Although from some studies it appears possible that the excretion rate is slower for liposomes which are less susceptible to serum effects (27), no detailed studies on this have been published.

TARGETING OF LIPOSOMES IN VIVO

Specific Targeting

As mentioned previously, the aim in all chemotherapy is to deliver the active therapeutic agent specifically to its site of action. This aim is difficult to achieve in practice for many reasons, including lack of a specific basis for drug-target interaction, rapid clearance of the drug from the organism, deactivation of the drug and metabolism by tissues and blood, or by a need for the activation of the sequestered drug at a site distant from the specific targeted lesion. The drug may have greater or lesser specificity for its target site. Usually, drugs are not uniquely specific to the target cell and therefore have effects on other cells and tissues. These effects are particularly valid in cancer chemotherapy, where most differences between the behavior and properties of cancer cells and normal cells are quantitative rather than qualitative. These differences are usually related to properties such as growth rate, cell cycle distribution, expression of embryonic phenotype, and other effects which may be qualitatively similar in certain normal tissues in the same animal. The usual aim, therefore, is to attempt to exploit any quantitative differences that may exist.

How do liposomes as a drug carrier fit into this situation? Obviously, the aim in using any drug carrier is to be able to target the carrier to the specific site of action. There have been continuing attempts to link drugs directly to molecules which would "home" to the particular target cells (101-104). Targeting of liposomes or other carriers would be expected if there is a receptor on the surface membrane of the target cell which will specifically bind the targeting molecule sufficiently strongly, for a sufficient time, and in sufficient amounts for effects to occur. However, in most instances the drug with or without its carrier will still have to be internalized to reach the specific intracellular site of action.

As an example, a calculation (very approximate) can be made of the number of drug molecules that would be needed to be bound per target cell to cause cytotoxicity. For cytosine arabinoside, it has been shown that a tissue concentration of 0.1 μM Ara-CTP inhibits DNA synthesis 50% (105). Assuming that every Ara-C molecule arriving at the cell is converted to Ara-CTP, the necessary local concentration of Ara-C would also have to be 0.1 μM. Let us say that 0.1-1 μM intracellular Ara-C would inhibit DNA synthesis. L1210 cells have a volume of approximately 1000 μm^3 or 1×10^{-9} ml. From this value it can be calculated that 1×10^6 molecules per cell are needed to inhibit DNA synthesis. There are in the range of 10^4-10^5 hormone receptors per cell and for antigen there are approximately 10^5-10^6 receptor surface immunoglobulins per cell. Thus, it can be seen that for a one-to-one carrier-receptor interaction, some degree of amplification would be necessary. This could be possible for liposomes, where even if only one surface binding site interacted with a cell receptor, a large number of drug molecules could be delivered from one liposome. Under present preparation methods, an SUV liposome can trap 5×10^3 Ara-C molecules, and an LUV (0.4 μm in diameter), 5×10^6 molecules at a maximum. It can therefore be seen that about 10^3 SUV liposomes and their contents would need to be taken into the target cell, but only one LUV liposome. These numbers are quite important when considering the mechanisms of interaction of liposomes with cells. The number of liposomes that may adsorb optimally to cells can be of the order of 10^6 or more per cell. Thus, only a minute percent of bound liposomes and contents need to be taken in for cytotoxic effects to be observed. Thus, in spite of recent evidence suggesting that the fusion uptake mechanism may be occurring only infrequently, only a rare event is needed for cellular delivery of a cytotoxic dose of a drug.

There have been very few reports on the feasibility of targeting liposomes in vivo by inserting or binding proteins, including antibodies to cellular antigens, into or onto the surface of liposomes (103,106). Specific binding of liposomes coated with specific immunoglobulins compared with nonspecific immunoglobulins has been reported in vitro (101-104,106-110). Some specificity in the cytotoxic effects of immunoglobulin-labeled liposomes has been reported (111). Alterations in the tissue distribution patterns, including more rapid removal of coated liposomes from the blood into tissues such as liver and spleen, have been reported in vivo (102,107,112,113). Using antitumor antibodies bound to liposomes has been reported to stimulate some increase in the tumor uptake of sequestered drugs (112,113). However, it is not clear whether this increased tumor uptake represents specific antigen-antibody interactions or general increased tissue uptake.

Problems that have to be overcome in the use of antibody-antigen targeting systems include preparation of the specific antibody in large amounts and showing that the antibody can be specifically bound by the target cells in vivo. A new method for the efficient conjugation of IgG molecules to the surface of intact liposomes shows considerable promise in this respect (114,115) and may result in greatly increased specificity of vesicle-cell interaction. In the case of cancer chemotherapy, it has to be shown, particularly in the human situation, that there are truly tumor-specific antigens to which the antibodies can be made. It may be more feasible to prepare tissue-specific antibodies. However, in this case, if these antibodies are able to target drug carrying liposomes to specific tissues, it is quite likely that normal cells as well as the target cells will take up the antibody-liposome complex, probably causing nonspecific cytotoxic effects on these cells. Thus, it can be seen that a considerable number of problems will have to be overcome using this approach. Monoclonal antibodies will probably be of great value for future use in liposome targeting (114,115). The possibility of using other cellular receptors as specific binding sites for liposomes does not seem to have been explored extensively, but there could be possibilities in using lectin (116-118) and hormone receptors (119).

Alteration in tissue uptake may be achieved by attachment of molecules such as glycoproteins and glycolipids to liposomes (24,120-122). For example, it is possible that liver hepatocyte uptake of liposomes could be increased by coupling asialoglycoproteins to the vesicle bilayer. Attempts using crude preparations have not shown marked differences in the tissue uptake of liposomes with or without bound glycoproteins (123), although asialoglycoprotein can inhibit the uptake of asialoglycoprotein liposomes (124). There have been a number of studies where glycolipid-containing liposomes have been tested for their ability to interact with lectins (20,62,117,125), with the aim of using specific glycolipids to target liposomes to specific cell types. Some success has been reported where lactosylcerebroside-containing liposomes showed increased uptake by HeLa cells in culture (126). Moreover, significant therapeutic effects have been observed using glycolipid-liposomes against a malaria model (127).

Use of the γ-ray probe ^{111}In has shown that a 6-aminomannose residue derivative of cholesterol embedded in liposome membranes increases liposomal stability and also increases their binding to leucocytes (120). It has also been reported that the hydrophilic spacer arm distance in glycolipid liposomes can have marked effects on their agglutinability by lectins (128). It would appear that these attempts to target liposomes show great promise but are yet at an early stage for specific in vivo delivery of drugs to specific cellular sites.

Serendipitous Targeting

Advantage can be taken of the natural propensity of liposomes to be taken up by certain tissues and cells. Many studies of the distribution of liposomes administered by different routes have been made and it has been shown (see Refs. 2 and 8) that after intravenous injection, liposomes are cleared from the plasma into tissues throughout the organism but are found mainly in the liver, spleen, lungs, and kidney when assessed on a percent injected dose per gram of tissue basis.

Most of these studies have been made using radioactive markers for the lipid bilayer and/or the aqueous spaces and it is important to remember, as pointed out previously, that in many studies the determinations were of radioactivity only and thus may represent metabolic products or lipoproteins carrying lipid markers instead of intact liposomes. By manipulation of the liposome size and composition, quantitative changes can be made in the clearance rate from the plasma and in the relative distributions in different tissues. In general, larger liposomes are cleared more quickly than small ones, and positive and neutral liposomes (before injection) are cleared more slowly than negative liposomes (129, 130). Similar alterations in the clearance rate of various colloids of different size have been shown (131). It has also been demonstrated that large liposomes are retained for much longer periods in the lung than are smaller liposomes (22). One aim, therefore, is to use liposomes to carry drugs that would be active against pathological conditions of these liposomophilic organs. It has been shown that tumors, organs, and plasma retain liposome-entrapped drugs for considerably longer periods than when free drug is administered. However, in addition to determining whether the liposome-entrapped drug is retained by the organ of interest, one must also determine if the tissues and cells of interest in that organ also retain the drug. Thus, in cancer chemotherapy, it would appear reasonable to suggest that liposome-entrapped drugs should be active against liver tumors, since with some liposome types 50% of the total dose may be found in that organ. However, if liposomes are taken up by nontumor liver tissue, as indeed has been indicated in two human studies (83,84), perhaps the liposome-entrapped drug would be ineffective. If the major cell type that takes up liposomes is of reticuloendothelial origin, it is apparent that diseases associated with these cells should be accessible to treatment by liposome-entrapped drugs. This approach is being exploited in anti-leishmania therapy (see later). The implications in cancer chemotherapy are clear; many cancer cells may not be expected to take up liposomes more avidly than other sensitive normal tissues. Some relatively rare cancers of "macrophage" origin and other more common tumors associated with the reticuloendothelial system, such as lymphomas and leukemias, may be reasonable targets for chemotherapy using liposomes. Another consideration is that macrophages are often associated with tumors and it is possible that macrophages in tumors may nonspecifically endocytose liposomes and could, when they migrate to the tumor, allow leakage of the drug into or close to the target tumor cells.

Other Targeting Methods: Hyperthermia and pH

It is known that at their phase transition temperature liposomes become more leaky to entrapped species (15). Some initial attempts have been made to exploit these events in vitro (132,133) and in vivo (134-136). It has been shown that *Escherichia coli* can be killed by neomycin released from dipalmitoylphosphatidylcholine:distearoylphosphatidylcholine (DPPC:DSPC) liposomes at 42°C (132). More recently, evidence that methotrexate (135) and cis-platinum (136) can be locally released from liposomes and taken up by tumors under hyperthermic conditions has been published. However, local heating can also cause injury to the microvasculature, resulting in the generalized increase in macromolecular uptake (137). By use of local heating methods, this technique could have promise in cancer chemotherapy, although such methods require that the location of the tumor has to be known. Similarly, based on the premise that the pH in the vicinity of some tumors may be lower than that in the vicinity of normal tissues, it has been suggested that liposomes may be made such that they become more leaky at lower pH, thus releasing their contents in the area of the tumor, thereby increasing drug uptake and its cytotoxic effects (138, 139).

LIPOSOMES AS A DEPOT SYSTEM

Another characteristic of liposome-administered drugs that has been alluded to previously is the probability that the liposomes can act as a depot or time-release system. It is probable that targeted liposomes will have, in addition to their targeting capability, depot characteristics. Thus, it is important to consider the effects of sustained release of drugs from liposomes. In this regard, liposomes could be considered a special case of a variety of other time-release systems under investigation (140). The slow release of drugs from liposomes could occur while the liposome is still free in the circulation, after the liposome is bound to cells or tissues, or even after liposomes have been taken into cells intact as, for example, by endocytosis. The rate of release of the drug from the liposome will depend on both the location of the liposome and the liposome's composition and type. The range of effects that could be observed with different liposome types could vary. For example, unilamellar liposomes composed of phospholipids with a transition temperature above 37°C or liposomes stabilized with cholesterol will be degraded slowly and thereby release sequestered compounds over an extended time period. This may also be true for multilamellar liposomes, where sequestered molecules must traverse several lipid bilayers before being released to the extravesicular environment.

Compared with other polymeric slow-release systems, liposomes have advantages and disadvantages. One characteristic of liposomes is that they are biodegradable and quite nontoxic. Unfortunately, this biodegradability results in a relatively short in vivo liposome half-life, on the order of days or weeks compared to months for some polymers. Chemical modification and insertion of proteins and other substances into liposomes can either increase or decrease their stability, whereas many polymeric systems can be complexed to proteins without any effects on their leakage properties. The ability to regulate membrane permeability of liposomes easily gives the investigator tremendous flexibility in designing depot systems specifically suited to designated in vivo applications. Liposomes also have the advantage of a high drug/structural lipid mole ratio compared to other polymers. Furthermore, the drug can exist in the free state, whereas in some polymers the drug has to couple to the polymer itself, thus leading to chemical synthetic and drug release problems.

Liposomes differ from all the large implantable slow release polymeric delivery systems in that drugs are slowly released *in* the circulation and/or at the deep tissue sites where they are trapped without surgery. If the drugs are released while circulating in the blood, a very uniform distribution of the drug within the organism would be expected. If the drug is released within tissues or cells, locally high concentrations of drug could be expected to cause pharmacologic effects within the immediate vicinity of its release. If the slow release occurred in the vicinity of cells or tissues after specific or serendipitous targeting, greatly improved selectivity in drug action could be expected.

TOXICITY OF NONDRUG-CONTAINING LIPOSOMES

The in vivo and in vitro (141-147) toxicity of nondrug-containing liposomes is generally quite low and depends on vesicle size and composition. Positively charged vesicles containing stearylamine are the most toxic liposomes (27) while vesicles composed of acidic phospholipids are usually more toxic than neutral liposomes. Phosphatidylserine (PS)-containing liposomes are more toxic than, for example, phosphatidylglycerol (PG)-containing liposomes of otherwise similar composition and type (27), although at low PS/neutral phospholipid ratios these liposomes are completely nontoxic (27). Pure PS liposomes have been shown to inhibit brain glycolysis (142), possibly by the in vivo generation of lyso-PS (143). Similarly, dicetylphosphate:Chol:phosphatidylcholine (PC) and stearylamine:Chol:PC liposomes have been found to cause damage to the central nervous system in mice, whereas phosphatidic acid:Chol:PC and DPPC liposomes were found to be nontoxic (141). The in vivo toxicity of neutral and negative liposomes was not due to brain effects and was probably due to alterations in liver and/or RES function. We have observed marked initial gross necrotic effects on normal liver when mice were injected with 5 g/kg of DPPC:Chol liposomes. However, recovery was under way 3 days after treatment and by 7 days the liver appeared normal.

Cholesterol-containing liposomes have been reported to inhibit $(Na^+,K^+)/$ ATPase and adenyl cyclase activities in cultured cells (148). Whether the same effects occur in vivo is not known. It has also been reported that whereas neutral and negatively charged liposomes are without effect on the growth of Ehrlich ascites tumor cells, positively charged liposomes actually *enhance* the growth of that tumor (144).

Liposomes several microns in diameter are more toxic than those of less than 1 μm in vivo (27). However, the relative toxicities of sonicated liposomes (tens of nonometers in diameter) compared to those with diameters in the hundreds of nanometer range are not known.

TOXICITY OF DRUG-CONTAINING LIPOSOMES

The doses of lipid used in chemotherapy studies when liposomes have drugs entrapped in them is usually much lower than the toxic range of the lipids alone. Therefore, the toxicity of the drug-liposome combination usually results from the altered pharmacokinetics of the entrapped drug and not the lipid membrane. In the case of actinomycin D, it has been reported that the toxicity of encapsulated drug to gut and bone marrow is reduced and altered compared with the toxicity free drug (149), perhaps enabling altered dose regimens to be given to increase the therapeutic effects. Similarly, attempts to reduce the cardiotoxicity of anthracyclines (150-152) have been reported while maintaining their therapeutic effects. As yet no clinical studies have been reported.

THERAPEUTIC ACTIVITY AND TOXICITY

The relationship between the doses of drug required for toxicity and those needed to effect a desired therapeutic effect is a key consideration in clinical pharmacology. For many anticancer drugs the dose levels required for beneficial therapeutic effects and the doses that give significant toxicity are very close. Thus, the goal in chemotherapy, especially in cancer chemotherapy, is to increase the therapeutic effects and/or reduce the toxic effects in order to improve the therapeutic index.

THERAPEUTIC EFFECTS OF LIPOSOME-ENTRAPPED DRUGS

Anticancer Effects

A number of cancer chemotherapeutic agents have been trapped in liposomes and tested for activity in animal tumor systems. Table 2 lists various cancer chemotherapeutic agents entrapped in liposomes and their appropriate references. This table only includes reports from 1977 to 1981. Previous lists have been given in Refs. 1, 2, 8 and in Table 1 of Ref. 224. A number of the reported effects have been based on a small number of experiments and in many cases the initial reports have not been followed up, making it difficult to assess their significance. As can be seen from Table 2, the most extensively studied drugs have been cytosine arabinoside (Ara-C), methotrexate (MTX), adriamycin (ADR), and actinomycin D. All of these drugs have been used widely in different human cancers and there is considerable information about their mode of action in vivo and in vitro.

We feel that work done at Roswell Park Memorial Institute (RPMI) using Ara-C and MTX, and recent work using ADR, and actinomycin D can give some considerable insight into the mode of action of liposome-entrapped drugs against cancer and suggest the course of action for future work. The effects of the drugs are summarized in the following paragraphs.

Cytosine Arabinoside

Entrapment of this drug in liposomes markedly alters its effects against L1210 leukemia. Free Ara-C is effective only under a multiple dose schedule or by infusion (225). A single dose, even at 1000 mg/kg, does not significantly prolong the life of animals. However, after entrapment in liposomes, significant prolongation of survival time can be observed at single doses of 5-20 mg/kg, i.p. or i.v., against intraperitoneal or intravenous tumors at 10^5 or 10^6 cells initial tumor loads, respectively. If the doses are increased to 50 mg/kg, significant numbers of long-term survivors (>30 days) can be found. A significant increase in survival can be observed using the appropriate liposome compositions (e.g., DPPC:Chol 1:1) when mice, initially injected with 10^6 cells i.v., are treated intravenously 72 hr after tumor inoculation with a single dose of entrapped Ara-C (untreated animals die under these conditions at $\sim$120 hr after injection of tumor). Multiple doses of free Ara-C or infusion of drug does not have a significant effect if treatment is started at 72 hr. The therapeutic index has been calculated comparing a single dose of liposome Ara-C with a 50 day infusion of free Ara-C. The data indicate that the liposome-Ara-C has a therapeutic index approximately twice as great as the free drug (27,182). Thus, Ara-C entrapped in liposomes has increased therapeutic efficacy in treating murine L1210 leukemia compared to free drug. Entrapped Ara-C has been tested at RPMI against other tumors not normally susceptible to Ara-C. We have found increases in the survival of mice bearing intraperitoneal B16 melanoma from about 16 days for untreated mice to 22 days after treatment with 50 mg/kg of liposome-entrapped drug. No effects were observed against Lewis lung carcinoma or P1788 lymphoma. It would seem that some consideration could be given for future clinical trials of liposome-entrapped Ara-C against human leukemia.

Table 2 Cancer Chemotherapeutic Agents Trapped in Liposome[a]

Drug	Entrapment references	In vivo antitumor testing references[b]
Actinomycin D	129, 153–158	154(0), 155(0), 156(0), 158(0)
Adriamycin	150–152, 159–166	150(0), 151(0)
Amphotericin B	167	
L-Asparaginase	168	168(?)
8-Azaguanine	169, 170	169(0)
BCNU	171	171(+)
Bleomycin	104, 172, 173	173(+)
Carbazylquinone	174	174
cis-Platinum	61, 139, 175	61(+)
Cytosine arabinoside	22, 23, 45, 77, 100, 129, 153, 166, 176–186	23(+), 45(+), 100 (+), 177(+), 179 (+), 180(+), 182 (+), 183(+), 185 (+), 186(+)
Ara-C derivatives	187, 188	188(+)
Cyclophosphamide	166	
Daunorubicin/daunomycin	185, 189–191	
Estracyt	192	
5-Fluorouracil	185, 186	
Floxuridine	193	
Illudin-S	194, 195	194(+), 195(+)
Interferon	196	
Lymphokines	197–203	197(+)
Melphalan	174, 189	
6-Mercaptopurine	170	
Methotrexate	21, 67, 130, 134, 155, 189, 204–213	155(0), 206(?), 208(−), 211(+)
Methyl CCNU	174	174(+)
Mitomycin C	214	
Neocarzinostatin	215, 216	215(+), 216(+)
Neuraminidase	217	
Nitrogen mustard	218	

Table 2 (continued)

Drug	Entrapment references	In vivo antitumor testing references[b]
Prednisolone	219	
Protease inhibitors	220	
Thalicarpine	221	
Thioguanine	174	174(0)
Vinblastine	129, 153, 222, 223	220(0)

[a]Bibliography completed April 1981.

[b]0, no change in antitumor effects; ?, questionable effects; +, increased antitumor effects; –, decreased antitumor effects.

Methotrexate

MTX inhibits L1210 growth after a single injection at 5-10 mg/kg. Entrapment of MTX in liposomes shows some slight improvement in the antitumor activity in terms of survival time at equal doses but also shows signs of increased toxicity. There was no improvement in the therapeutic index. No effects were observed against Lewis lung carcinoma and B16 melanoma. Some reduction in the tumor size of P1788 lymphosarcoma was observed (226).

Adriamycin

No significant differences in the antitumor activities against Lewis lung carcinoma, B16 melanoma, or P1788 lymphosarcoma were found between free and entrapped ADR (27). However, significant reduction in toxicity of adriamycin entrapped in liposomes has been shown (150-152,227), probably related to reduction in its cardiotoxicity. This may be of considerable importance since the dose limitation for the clinical use of adriamycin is its cardiotoxicity. The reduction in cardiotoxicity may be related directly to reduced uptake of entrapped drug by the heart (152,227). Recent studies using well-defined ADR liposomes indicate that the therapeutic activity against L1210 tumor may be slightly increased (227).

Actinomycin D

Some early studies have suggested that actinomycin D entrapped in liposomes is more effective than the free drug in treating sensitive tumors (149,228). However, more recent work has indicated that liposome-entrapped drug is only as effective as free drug in treating a drug-resistant tumor system, both in vitro and in vivo (229). It has also been reported that actinomycin D trapped in lipo-

somes, is able to overcome membrane-related drug resistance problems in vitro
(230), whereas a more recent report (156) has shown that in vivo, actinomycin
D in liposomes is not effective against these same tumors. These conflicting and
confusing reports could reflect the different properties of the different systems
used in each study.

Thus, these four drugs show quite different effects when entrapped in liposomes. Ara-C, MTX, ADR, and actinomycin D can all readily penetrate cells
sensitive to the drugs. The primary mode of intracellular action of Ara-C, MTX,
and ADR is considered to be inhibition of DNA synthesis, although the three
drugs accomplish this inhibition in different ways. Actinomycin D inhibits primarily DNA-dependent RNA synthesis, which results in the cytotoxic effects.
Ara-C has to be activated to Ara-CTP, which is the active moiety causing inhibition of DNA synthesis, probably either by inhibiting DNA polymerase or by
causing premature termination of DNA chains. Ara-C acts only on cells during
DNA synthesis—S phase (231). MTX inhibits folic acid reductase, resulting in the
blocking of purine synthesis and therefore inhibition of DNA production (232).
The action of MTX differs from that of Ara-C in that RNA and protein synthesis is inhibited in addition to DNA synthesis so that cells are prevented from
reaching S phase. ADR is thought to bind strongly to DNA by intercalation between base pairs, resulting in uncoiling of the helix (233,234). Activity occurs
throughout the cell cycle but with a maximum in the S phase. Although proliferating cells are sensitive to actinomycin D, the effects are cell cycle nonspecific (235). Thus, Ara-C is the only drug which is truly S-phase specific and
is also the only one with consistent increased antitumor activity when encapsulated in liposomes. Why is this? From our previous discussion we have no reason to believe that liposomes are preferentially delivered to target tumor cells or
that liposome-entrapped drug that is delivered to the tumor cells is able to penetrate these cells more efficiently than free drug. If drugs sequestered within liposomes were able to enter tumor cells more efficiently than free drug in vivo, then
significant potentiation of the effects of MTX, ADR, and actinomycin D should
also be observed. Furthermore, we have shown that in the case of cells resistant
to Ara-C, addition of liposome-entrapped drug to cells in vitro or in vivo cannot
overcome the resistance (which is thought to be related to transport defects), suggesting that, in this case, liposomes do not significantly bypass normal permeability barriers (236a). However, if the resistance is only partially related to transport defects, this argument may be invalid. The most reasonable explanation for
the enhanced activity is that liposomes are acting as a type of depot system, pro-
tecting drugs like Ara-C from rapid deactivation while slowly releasing the drug
into the circulation. This mode of action would maintain a cytotoxic level of
the drug in the circulation for a prolonged period (8,23,100,179,181,183). Prolonged tissue retention of Ara-C which is entrapped in liposomes compared to
free drug has been shown (22). In order for Ara-C to be effective, its level

must be maintained in the tissue for a sufficient time to allow virtually all cells to proceed through the cell cycle to the DNA synthesis phase. This suggestion is strongly supported by the observation that the antitumor activity strongly correlates with the stability of the liposomes in biological fluids. Liposomes with increased cholesterol content are less susceptible to breadkown in the presence of serum, so that the rate of release of entrapped drug is reduced. The antitumor activity of relatively stable liposomes entrapping Ara-C is increased compared to less stable liposomes (23,45). Similarly, liposomes composed of lipids below their phase transition temperature show increased antitumor activity. The data also indicate that multilamellar liposomes are more active than are small unilamellar liposomes composed of the same phospholipids, although size and tissue-trapping differences are also involved (100).

Although it seems probable that in the examples above liposomes act by a "depot" mechanism, it is not clear what type of depot is operating. There may be a slow release of drug from circulating liposomes, or the slow release could come from tissue-bound liposomes or even from liposomes taken into RES cells. Thus, tissues like the liver could act as a reservoir for liposomes allowing for the slow release of the liposome-entrapped drug into the circulation. It is important to determine the type of depot in effect since it is possible to alter the vesicle tissue binding properties which would result in different levels of circulating drug. For example, it may be possible to cause pulsed release of liposomes bound by the liver or to inhibit liver binding resulting in a higher circulating amount of liposome-associated material (98). It must be emphasized that the postulated "depot" mechanism of activity applies only to nontargeted liposomes. If specific targeting is feasible in vivo, significant alterations in the activities of a range of anticancer agents sould be observed.

The use of entrapped lymphokines (197-203) and other natural products, such as interferon (196), has been studied to some extent. The evidence suggests that lymphokines can be entrapped efficiently in liposomes. Addition of the sequestered lymphokines to macrophages has been shown to enhance the macrophage tumoricidal response in vivo and in vitro. There has been a report (197) of antimetastatic activity of entrapped lymphokines against mouse B_{16} melanoma, possibly related to stimulation of lung macrophages. One of the difficulties of using impure natural products is that determination of the liposomal capture of the specific moieties responsible for the biological effects is difficult. There is the interesting possibility that liposomes may preferentially bind these agents and thus the macrophages are exposed to much higher concentration of membrane-bound lymphokines than of free lymphokine and for this reason their response is augmented. It has also been shown that the muramyl dipeptide can potentiate the tumoricidal activity of lung macrophages (236b,c). This is an area of considerable interest and should be investigated thoroughly.

Effects on Storage Diseases

The accumulation of substrates in the liver resulting from specific enzyme deficiencies (storage diseases) evoked the suggestion that perhaps by entrapping the appropriate enzymes in liposomes, they could be delivered to the liver, where they could act on their substrates (97). In vitro experiments where liposome-entrapped invertase was added to cultured cells preloaded with sucrose suggested that this approach was feasible. Following liposome-cell interaction, it was shown that glucose and fructose-sucrose hydrolysis products, were released into the medium (237). However, these experiments did not show whether liposome-entrapped invertase was more effective than "free" invertase under the same conditions. In vivo it was shown that both liposome-entrapped dextranase and free dextranase reduced liver dextran levels to approximately the same amount (238). Ryman has reported that injection of liposome-entrapped amyloglucosidase into a child suffering from Pompe's disease did have some objective beneficial effects (83) and cysteamine entrapped in liposomes has been used in attempts to reduce lysosomal cystine accumulation in cystinotic fibroblasts (239). However, it still remains to be proven that liposomes can be of real practical use in storage diseases.

A problem with in vivo tests with liposomes to treat enzyme storage diseases is that the liposomes not only have to "find" the correct organs and tissues, but they have to be taken up by the appropriate cells. For example, in Pompe's disease the liver parenchymal cells are involved in storing the excess glycogen, whereas it seems probable that the Kupffer cells may be the main site of uptake of large liposomes (96,97,123,240,241). Another problem with early experiments in this field may have been the low efficiency of encapsulation of the enzyme. These problems may be overcome by using smaller size vesicles, new methods that produce high encapsulation (18), and perhaps inserting galactose-containing glycolipids on the liposome membrane.

Antiparasitic Effects

Perhaps the area where liposomes show the greatest immediate promise is their use in carrying drugs against parasites which at certain stages in their life cycle invade and multiply in reticuloendothelial cells. There have been several reports (242-246) that entrapment of antimony compounds in liposomes and their subsequent injection into animals have very marked effects on experimental leishmania. The doses of liposome-entrapped drug required to inhibit parasitic growth and cure infected animals are approximately 1/100 to 1/800 of those required for the free drug. In addition, liposome-encapsulated drugs (not only those composed of antimony) have significant prophylactic activity against experimental leishmaniasis (247), which was not found for unentrapped drugs. Furthermore, differential interaction of "solid" and "fluid" liposomes with trypanosome membranes has been reported and thus different liposome types may have differential effects (248).

Other examples of treatment of parasitic diseases with liposomes have been reported. For example, liposomes, when prepared with neutral glycolipids bearing a terminal glucose or galactose residue, have been found to inhibit the development of malaria in mice (249), with obvious therapeutic possibilities. It has also been reported that quinacrine entrapped in liposomes has marked effects against a similar mouse malaria model (250). There are a number of parasitic diseases where liposomes carrying entrapped drugs should be tried, but again the question of the localization of the particular parasite in cells or tissue and the likelihood of accessibility of the liposome-entrapped drug to the parasite is important. Whenever the parasites pass through a liposome/drug-susceptible stage in cells of the RES, they will be most vulnerable to specific attack using liposomes.

Arthritis

Derivatives of cortisone have been trapped in liposomes and tested for effects on experimental arthritis in animals after direct injection into affected bone joints (251-256). Significant reduction in joint temperatures and reduction in inflammation was observed. Preliminary studies in humans have given similar results (257). The composition of the liposome may play a role in the effects observed (255) and it has been reported that the effects of the liposome-entrapped drugs are best for acute inflammation (256). However, it has been reported that when liposomes are compared with phospholipid emulsions as drug carriers, the effects were similar (258) and a cautionary note suggesting that the reported beneficial effects of liposome-entrapped cortisone are not strongly significant has been published (259). It is of interest to consider the possibility of using liposomes which could be made of the appropriate composition such that they were leaky at the slightly higher temperatures found in arthritic joints and be relatively impervious at normal body temperatures (see "Other Targeting Methods: Hyperthermia and pH").

Metal Chelation Therapy

This area of the application of liposomes has been recently reviewed (260). It seems clear that entrapped agents which chelate metals are often able to remove metals from organs more effectively than free drug at equal dose/schedules. These studies are of possible clinical significance for the removal of metals after poisoning or in diseases where metals, such as iron, may be stored in toxic amounts in some organs. For example, (ethylenedinitrilo)tetraacetic acid (EDTA) and diethylene triaminepentacetic acid encapsulated in liposomes were reported to be somewhat more effective in removing plutonium from organs of uptake and delivering it into the feces than the nonencapsulated drug (261). Clearance of the iron chelator deferrioxamine is considerably slowed by its entrapment in liposomes (262) and excretion of iron from preloaded mice was doubled after intravenous injection of deferrioxamine within liposomes (263). Specific targeted liposomes may be expected to improve this type of therapy.

Hemophilia

It has been reported that oral administration of factor VIII entrapped in liposomes results in elevated plasma levels of the factor for prolonged periods (264). Considering the problems associated with oral administration of liposomes, this study needs to be confirmed (265).

Myocardial Infarction

Liposomes tend to accumulate in regions of myocardial infarction (266,267) with the potential for delivery of drugs to such areas. As positive and neutral liposomes tended to accumulate more strongly than negative liposomes, it was suggested that these liposomes be used as carriers (266). However, as pointed out earlier, positive liposomes are themselves more toxic to cells and animals than neutral or negative liposomes, so that it would seem that neutral liposomes would be more appropriate for this application.

Antibacterial Effects

The antibacterial effects of liposome-entrapped antibiotics have not been studied extensively. However, there are some reports that entrapment of these agents may improve their activity. It has been reported that dihydrostreptomycin entrapped in liposomes was active against intracellular *Staphylococcus aureus* (268) and neomycin trapped in liposomes showed enhanced activity against *Escherichia coli* (269). Gentamycin entrapped in liposomes was distributed in mice in the same pattern observed for nondrug-containing liposomes (170). Much more work needs to be done in this area to draw any firm conclusions concerning the usefulness of liposomes as a carrier.

Epilepsy

There has been a report that γ-aminobutyric acid entrapped in liposomes was significantly more effective than free drug in reducing epilepsy in an experimental system (271).

Use of Liposomes as a Radiopharmaceutical Marker

Several studies have been reported using liposomes as a radiolabeled scanning tracer. This has involved labeling the liposomes with a suitable radioisotope and then injecting the liposomes into animals or man. A number of different isotopes have been used. ^{99m}Tc (272), ^{111}In (273), ^{131}I (274), and ^{67}Ga (275) have been coupled to different molecules and incorporated in liposomes. These studies have indicated that liposomes can be used for liver scanning (274) and for localization of lymph nodes (275) and possibly for the detection of tumors (276). Another possible use is to load liposomes with radiopaque materials to aid in selective delivery of radiation to areas where liposomes are absent (277).

HUMAN STUDIES USING LIPOSOMES

Table 3 gives a reasonably complete summary of published clinical investigations using liposomes. It can be seen that these few investigations have indicated that liposome distribution in humans is similar to that in animals and that only marginally significant biological effects have been noted for storage diseases. In the case of liver tumors, no increased uptake of radioactive label was observed. The interesting observation that in polycythemia vera, liver and spleen uptake was reduced (a disease characterized by loss of RES cell function) supports the idea that in humans, as in animals, cells of the reticuloendothelial system are important sites of liposome uptake. No studies have been reported on the therapeutic effects of anticancer agents in humans. The two studies of patients with enzyme storage diseases showed some marginal beneficial effects resulting from liposome treatment. The effects on arthritis and hemophilia have been commented on previously. It would seem inappropriate to pursue comprehensive clinical trials using liposomes until carefully controlled preclinical testing and optimization has been carried out for specific promising applications.

NUCLEIC ACIDS

The entrapment of nucleic acids in liposomes has been discussed previously in this book, and in Ref. 8 (283-285). Therapeutic applications would result from the ability to introduce undenatured functional nucleic acids into target cells, for example in genetic deficiency diseases, in order to alter permanently the genetic code of the cells. This is not feasible at the present time in vivo, but strategies can be visualized to accomplish this in the future. At present it would seem that the transplantation of cells, genetically altered in vitro, into animals would be technically possible. The introduction of genetic material into plant cells via liposomes for crop improvement, treatment, or prevention of plant diseases is also clearly a distinct possibility (283). Therapeutic applications related to the immunologic properties of liposomes and/or substances trapped in them have not been discussed in this chapter as they have been covered in detail in Chapter 6.

Table 3 Clinical Studies Using Liposomes

Reference	Number of patients	Disease	Treatment and/or label in liposome	Assessment reported
83	1	Pompe's disease (lack of lysosomal α gluco-sidase)	Amyloglucosidase	Liver size decrease, liver glycogen reduced, no overt toxicity
278	1	Gaucher's disease (lack of lysosomal gluco-cereboside: β-gluco-sidase)	Glucocereboside: beta-glucosidase (^{111}In)	Liver size decrease, RES function improved
279	1	Hepatoma	^{99m}Tc	No increased uptake by tumor
280	2	Hepatoma	^{99m}Tc	No increased uptake in tumor, except for one hepatoma; strong uptake in liver and spleen in general
280	4	Choriocarcinoma	^{99m}Tc	
280	1	Acute myeloblastic leukemia	^{99m}Tc	
280	1	Teratoma of testis	^{99m}Tc	
280	1	Hypernephoma	^{99m}Tc	
280	1	Heamangiopericytoma	^{99m}Tc	
280	1	Carcinoma of bronchus	^{99m}Tc	
280	1	Kaposi sarcoma	^{99m}Tc	

280	1	Polycythemia vera rubra	^{99m}Tc	Reduced liver and spleen uptake, increased bone marrow uptake
281	1	Hepatoma	^{99m}Tc	No localization in tumor
284	1	Hepatoma	[^{111}In] bleomycin	No increased uptake in tumor
284	1	Adenocarcinoma	[^{111}In] bleomycin	
282	1	Carcinoma of bladder	[^{131}I] albumin	Some increase in tumor uptake compared to normal liver tissue
282	1	Colon adenocarcinoma	[^{131}I] albumin	
282	1	Breast carcinoma	[^{131}I] albumin	
148	1	Hemophilia	Factor F VIII	Prolonged plasma levels
71	6	Arthritis	^{99m}Tc, cortisol palmitate	Reduced inflammation

FUTURE PROSPECTS

The three most prominent features of systems involving in vivo administration of liposomes are: (a) liposomes act as a depot system; (b) agents can be protected from degradation by entrapment in liposomes; and (c) liposomes are probably bound most effectively in vivo by cells of the reticuloendothelial system.

In terms of cancer chemotherapy, this suggests some experiments that should be attempted, for example, use of cell cycle stage specific but rapidly inactivated drugs against tumors of reticuloendothelial origin or of high endocytotoxic capacity. As the liver and spleen are organs of high liposome accumulation, tumors of these tissues may be more susceptible to locally released drug than sites where liposomes do not accumulate. At the same time, reduction in toxicity of some drugs may be obtained by virtue of the relatively reduced initial uptake of liposomes in the gut (e.g., actinomycin D) and perhaps heart (e.g., ADR).

The most obvious drawback of liposomes and indeed other macromolecular delivery systems is their lack of specificity in site of delivery. As discussed previously, if this can be solved, great strides in their use would be made. Specificity will probably result from the membrane insertion of glycolipids, (glyco)-proteins, or antibodies capable of interaction with cell surface sites in the tissue. Just as for cell-cell interactions, specificity will probably result from protein anc/or carbohydrate interactions but probably not lipid-lipid interactions. If liposomes are to be used to deliver drugs intracellularly by a fusion mechanism, it will be necessary to bring the cell and liposome surfaces sufficiently close so that fusion can occur. This is presumably the role of polyethylene glycol and agglutins in fusion between erythrocytes and target cells in vitro (286-288), and the role of viral proteins in vitro and in vivo. Irrespective of any possible targeting to tumor cells, liposomes may be valuable for the encapsulation of "rescue factors" or "radioprotective" molecules by increasing their uptake into the reticuloendothelial system. Other potentially "targetable" cells include macrophages and lymphocytes, with possible applications in specific immunosuppression or activation of immune surveillance mechanisms. Liposomes can also be used for coencapsulation of multiple drugs, which may otherwise show very different pharmacokinetic clearance patterns. In any case, it appears that the most useful aspect of liposomes as a drug carrier system is that their properties can be modified in various ways to affect their size, surface charge, permeability, presence of surface ligand groups, and so on, so that their effectiveness can be optimized for various applications as needed by the specific system under consideration. The highly active ongoing research on the physicochemical properties of liposomes in various laboratories will no doubt lead to successful use in some of these areas. It would seem to us unlikely that any single defined liposome type is going to be a "magic bullet" for all applications. Rather, it is more likely that specific liposomes will have to be designed for different applications.

To conclude, we would like to state that in many ways we think that the stage has been set, by the studies discussed in this chapter, for the practical application of liposomes. We hope that in the next few years this prediction will be fulfilled.

REFERENCES

1. Gregoriadis, G. 1976. The carrier potential of liposomes in biology and medicine. N. Engl. J. Med. 295:765.

2. Tyrrell, D. A., T. D. Heath, C. M. Colley, and B. E. Ryman. 1976. New aspects of liposomes. Biochim. Biophys. Acta 457:259.

3. Fendler, J. H. and A. Romero. 1977. Liposomes as drug carriers. Life Sci. 20:1109.

4. Poste, G., D. Papahadjopoulos, and W. J. Vail. 1976. Lipid vesicles as carriers for introducing biologically active materials into cells. Methods Cell Biol. 14:33.

5. Pagano, R. E. and J. N. Weinstein. 1978. Interactions of liposomes with mammalian cells. Annu. Rev. Biophys. Bioeng. 7:435.

6. Finkelstein, M. and G. Weissmann. 1978. The introduction of enzymes with cells by means of liposomes. J. Lipid Res. 19:289.

7. Juliano, R. L. 1978. Drug delivery systems. A brief review. Can. J. Physiol. Pharmacol. 56:683.

8. Kimelberg, H. K. and E. G. Mayhew. 1978. Properties and biological effects of liposomes and their uses in pharmacology and toxicology. CRC Crit. Rev. Toxicol. 6:25.

9. Papahadjopoulos, D. 1979. Liposomes as drug carriers. Annu. Rep. Med. Chem. 14:250.

10. Baker, R. (Ed.). 1980. *Controlled Release of Bioactive Materials.* Academic Press, New York.

11. Gurny, R., N. A. Peppas, D. I. Harrington, and G. S. Banke. 1981. Development of biodegradable and injectable latices for controlled release of potent drugs. Drug Dev. Ind. Pharm. 7:1.

12. Gregoriadis, G. (Ed.). 1979. *Drug Carriers in Biology and Medicine.* Academic Press, London.

13. Kostelnik, R. J. (Ed.). 1978. *Polymeric Delivery Systems.* Midland Macromolecular Monographs 5. Gordon and Breach, New York.

14. Bangham, A. D., M. M. Standish, and J. C. Watkins. 1965. Diffusion of univalent ions across the lamellae of swollen phospholipids. J. Mol. Biol. 13:238.

15. Papahadjopoulos, D. and H. K. Kimelberg. 1973. Phospholipid vesicles (liposomes) as models for biological membranes: their properties and interactions with cholesterol and proteins. In *Progress Surface Science,* Vol. 4, S. G. Davison (Ed.). Pergamon Press, Oxford, pp. 141–232.

16. Bangham, A. D., M. W. Hill, and N. G. A. Miller. 1974. Preparation and use of liposomes as models of biological membranes. In *Methods in Membrane Biology,* Vol. 1, E. D. Korn (Ed.). Plenum Press, New York, pp. 1–68.

17. Papahadjopoulos, D., G. Poste, and W. J. Vail. 1979. Studies on membrane fusion with natural and model membranes. In *Methods in Membrane Biology,* Vol. 10, E. D. Korn (Ed.). Plen Press, New York, pp. 1–21

18. Szoka, F. C. and D. Papahadjopoulos. 1980. Comparative properties and methods of preparation of lipid vesicles (liposomes). Annu. Rev. Biophys. Bioeng. 9:467.

19. Szoka, F. C. and D. Papahadjopoulos. 1981. Liposomes. *Preparation and Characterization in Liposomes: From Physical Structure to Therapeutic Applications,* G. Knight (Ed.). Elsevier/North-Holland, pp. 51–82.

20. Juliano, R. L. and D. Stamp. 1975. The effect of particle size and charge on the clearance rates of liposomes and liposome encapsulated drugs. Biochem. Biophys. Res. Commun. 63:651.

21. Kimelberg, H. K., T. F. Tracy, S. M. Biddlecome, and R. S. Bourke. 1976. The effect of entrapment in liposomes on the in vivo distribution of (^{3}H) methotrexate in a primate. Cancer Res. 36:2949.

22. Hunt, C. A., Y. M. Rustum, E. Mayhew, and D. Papahadjopoulos. 1979. Retention of cytosine arabinoside in mouse lung following administration in liposomes of different size. Drug Metab. Dispos. 7:124.

23. Mayhew, E., Y. M. Rustum, F. Szoka, and D. Papahadjopoulos. 1979. Role of cholesterol in enhancing the antitumor activity of cytosine arabinoside entrapped in liposomes. Cancer Treat. Rep. 63:1923.

24. Jonah, M. M., E. A. Cerny, and Y. E. Rahman. 1978. Tissue distribution of EDTA encapsulated within liposomes containing glyco lipids or brain phospholipids. Biochim. Biophys. Acta 541:321.

25. Mauk, M. R., R. C. Gamble, and J. D. Baldeschwieler. 1980. Vesicle targeting: timed release and specificity for leucocytes in mice by subcutaneous injection. Science 207:309.

26. Olson, F., C. A. Hunt, F. Szoka, W. J. Vail, and D. Papahadjopoulos. 1979. Preparation of liposomes of defined size distribution by extrusion through polycarbonate membranes. Biochim. Biophys. Acta 557:9.

27. Mayhew, E., D. Papahadjopoulos, Y. Rustum, and F. Szoka. 1978–1980. Report on Contract CM-77118 from National Institutes of Health.

28. Huang, C. H. 1969. Studies on phosphatidylcholine vesicles. Formation and physical characteristics. Biochemistry 8:344.

29. Papahadjopoulos, D., S. Nir and S. Ohki. 1972. Permeability properties of phospholipid membranes: effect of cholesterol and temperature. Biochim. Biophys. Acta 266:561.

30. Barenholz, Y., D. Gibbes, B. J. Litman, J. Goll, T. E. Thompson, and I. D. Carlson. 1977. A simple method for the preparation of homogeneous phospholipid vesicles. Biochemistry 16:2806.

31. Mason, J. T. and C. Huang. 1978. Hydrodynamic analysis of egg phosphatidylcholine vesicles. Ann. NY Acad. Sci. 308:29.

32. Batzri, S. and E. D. Korn. 1973. Single bilayer liposomes prepared without sonication. Biochim. Biophys. Acta 298:1015.

33. Enoch, H. G. and P. Strittmatter. 1979. Formation and properties of 1000

A diameter, single-bilayer phospholipid vesicles. Proc. Natl. Acad. Sci. USA 76:145.

34. Reeves, J. P. and R. M. Dowben. 1969. Formation and properties of thin-walled phospholipid vesicles. J. Cell. Physiol. 73:49.

35. Deamer, D. and A. D. Bangham. 1976. Large volume liposomes by an ether vaporization method. Biochim. Biophys. Acta 443:629.

36. Papahadjopoulos, D., W. J. Vail, K. Jacobson, and G. Poste. 1975. Cochleate lipid cylinders: formation by fusion of unilamellar lipid vesicles. Biochim. Biophys. Acta 394:483.

37. Szoka, F., Jr. and D. Papahadjopoulos. 1978. Procedure for preparation of liposomes with large internal aqueous space and high capture by reverse-phase evaporation. Proc. Natl. Acad. Sci. USA 75:4194.

38. Szoka, F., F. Olson, T. Heath, W. Vail, E. Mayhew, and D. Papahadjopoulos. 1980. Preparation of unilamellar liposomes of intermediate size (0.1-0.2 microns) by a combination of reverse phase evaporation and extrusion through polycarbonate membranes. Biochim. Biophys. Acta 601:559.

39. Konings, A. W., J. Damen, and W. B. Triehing. 1979. Protection of liposomal lipids against radiation induced oxidative damage. Int. J. Radiat. Biol. 35:343.

40. Schutte, H. H. and A. W. Konings. 1978. Oxidative damage to membrane lipids by ionizing radiation and protection by vitamin E. Int. J. Radiat. Biol. 34:551.

41. Evans, J. R., F. J. Fildes, and J. E. Oliver. 1978. Liposomes. Ger. Offen. Patent No. 2818655, Nov. 23 (Imperial Chemical Industries Ltd.).

42. Abra, R. M., M. E. Bosworth, and C. A. Hunt. 1980. Liposome disposition in vivo: effects of predosing with liposomes. Res. Commun. Chem. Pathol. Pharmacol. 29:349.

43. Kimelberg, H. K. and D. Papahadjopoulos. 1974. Effects of phospholipid and chain fluidity, phase transitions and cholesterol in ($Na^+ + K^+$)-stimulated adenosine triphosphatase. J. Biol. Chem. 249:1071.

44. Gregoriadis, G. and J. Senior. 1980. The phospholipid component of small unilamellar liposomes controls the rate of clearance of entrapped solutes from the circulation. FEBS Lett. 119:43.

45. Garapathi, R., A. Krishan, I. Wodinsky, C. G. Zubrod, and L. Lesko. 1980. Effect of cholesterol content on anti-tumor activity of liposome-encapsulated 1-β-D-arabino furanosylcytosine in vivo. Cancer Res. 40:630.

46. Gregoriadis, G. and C. Davis. 1979. Stability of liposomes in vivo and in vitro is promoted by their cholesterol content and the presence of blood cells. Biochem. Biophys. Res. Commun. 89:1287.

47. Hoekstra, D. and G. Scherphof. 1979. Effect of fetal calf serum and serum protein fractions on the uptake of liposomal phosphatidylcholine by rat hepatocytes in primary monolayer culture. Biochim. Biophys. Acta 551:109.

48. Jonas, A. 1979. Interaction of bovine serum high density lipoprotein with mixed vesicle of phosphatidylcholine and cholesterol. J. Lipid Res. 20:817.

49. Zborowski, J., F. H. Roerdink, and G. Scherphof. 1977. Leakage of su-

crose from phosphatidylcholine liposomes induced by interaction with serum albumin. Biochim. Biophys. Acta 497:183.

50. Gerlier, D., F. Sakai, and J. F. Dore. 1978. Isolation of a cell-surface antigen associated with gross virus in liposomes. C. R. Acad. Sci. (Paris) 286: 439.

51. Petri, W. A. and R. R. Wagner. 1979. Reconstitution into liposomes of the glycoprotein of vesicular stomatitis virus by detergent dialysis. J. Biol. Chem. 254:4313.

52. Segrest, J. P. 1976. Molecular packing of high density lipoproteins: A postulated functional role. FEBS Lett. 69:111.

53. Kimelberg, H. K. and D. Papahadjopoulos. 1971. Phospholipid-protein interactions: membrane permeability correlated with monolayer penetration. Biochim. Biophys. Acta 233:805.

54. Scherphof, G., F. Roerdink, M. Waite, and J. Parks. 1978. Disintegration of phosphatidylcholine liposomes in plasma as a result of interaction with high density lipoproteins. Biochim. Biophys. Acta 542:296.

55. Tall, A. R., V. Hogan, L. Askinazi, and D. Small. 1978. Interaction of plasma high density lipo proteins with dimyristoyl lecithin multilamellar liposomes. Biochemistry 17:322.

56. Tall, A. R. 1980. Studies on the transfer of phosphatidylcholine from unilamellar vesicles into plasma high density lipoproteins in the rat. J. Lipid Res. 21:354.

57. Black, C. D. V. and G. Gregoriadis. 1976. Interaction of liposomes with blood plasma proteins. Biochem. Soc. Trans. 4:253.

58. Jackson, A. J. 1980. The effect of route of administration on the disposition of insulin encapsulated in multi lamellar vesicles of defined particle size. Res. Commun. Chem. Pathol. Pharmacol. 27(2):293.

59. Dedrick, R. L., C. E. Myers, P. M. Bungay, and V. J. DeVita, Jr. 1978. Pharmacokinetic rationale for peritoneal drug administration in the treatment of ovarian cancer. Cancer Treat. Rep. 62:1.

60. Richardson, V. J., K. Jeyasingh, R. F. Jewkes, S. B. Kaye, E. S. Newlands, B. E. Ryman, and M. H. Tattersall. 1978. Distribution of [^{99m}Tc] technetium-labelled liposomes in patients with cancer. Br. J. Cancer 38:195.

61. Kaledin, V. I., N. A. Matienko, V. G. Budker, V. P. Nikolin, and E. V. Gruntenko. 1978. Inhibitory effect of cis-diammino-dichloroplatinum incorporated into liposomes on the lymphogenic metastases of a transplanted murine tumor. Dokl. Akad. Nauk. SSSR 242:473.

62. Mauk, M. R. and R. C. Gamble. 1979. Stability of lipid vesicles in tissues of the mouse: a gamma-ray perturbed angular correlation study. Proc. Natl. Acad. Sci. USA 76:765.

63. Shaw, I. H., J. T. Dingle, N. C. Phillips, D. P. Page-Thomas, and C. G. Knight. 1978. Liposomes in the treatment of experimental arthritis. Ann. N.Y. Acad. Sci. 308:435.

64. Shaw, I. H., C. G. Knight, and J. T. Dingle. 1976. Liposomal retention of a modified anti-inflammatory steroid. Biochem. J. 158:473.

65. Rapoport, S. I. 1976. *Blood-Brain Barrier in Physiology and Medicine.*
Raven Press, New York.

66. Adams, D. M., G. Joyce, V. J. Richardson, B. E. Ryman, and H. M. Wisniewski. 1977. Liposome toxicity in the mouse central nervous system.
J. Neurol. Sci. 31:173.

67. Kimelberg, H. K., T. F. Tracy, R. E. Watson, D. Kung, F. L. Reiss, and
R. S. Bourke. 1978. Distribution of free and liposome-entrapped [3]HMTX
in the central nervous system after intracerebroventricular injection in a
primate. Cancer Res. 38:706.

68. Dapergolas, G., E. D. Nurunjun, and G. Gregoriadis. 1976. Penetration of
target areas in the rat by liposome-associated bleomycin, glucose oxidase
and insulin. FEBS Lett. 63:235.

69. Gregoriadis, G., G. Dapergolas, and E. D. Nurunjun. 1976. Penetration of
target areas in the rat by liposome-associated agents administered parenter-
ally and intragastrically. Biochem. Soc. Trans. 4:256.

70. Patel, H. M. and B. E. Ryman. 1976. Oral administration of insulin by encapsulation within liposomes. FEBS Lett. 62:60.

71. Patel, H. M. and B. E. Ryman. 1977. The gastrointestinal absorption of
liposomally entrapped insulin in normal rats. Biochem. Soc. Trans. 5:1054.

72. Richards, M. H. and C. R. Gardner. 1978. Effects of bile salts on the structural integrity of liposomes. Biochim. Biophys. Acta 543:508.

73. Mezei, M. and V. Gulasekharam. 1980. Liposomes, a selective drug delivery system for the topical route of administration. 1. Lotion dosage form.
Life Sci. 26:1473.

74. Ivey, H. H., S. Roth, and J. Kattwinkel. 1976. Use of nebulized surfac-
tants in treatment of the respiratory distress syndrome (RDS) of infancy.
Pediatr. Res. 10:462.

75. Juliano, R. L., D. Stamp, and N. McCullough. 1978. Pharmacodynamics
of liposome encapsulated anti-tumor drugs and implications for therapy.
Ann. N.Y. Acad. Sci. 308:411.

76. McCullough, H. N. and R. L. Juliano. 1979. Organ selective studies of an
anti-tumor drug: pharmacologic studies of cytosine arabinoside administered via the respiratory system of the rat. J. Natl. Cancer Inst. 63:727.

77. Shattil, S. J., R. Anaya-Galindo, J. Bennett, R. W. Colman, and R. A.
Cooper. 1975. Platelet hypersensitivity induced by cholesterol incorporation. J. Clin. Invest. 55:636.

78. Shattil, S. J. and R. A. Cooper. 1976. Membrane microviscosity and human platelet function. Biochemistry 15:4832.

79. McDougall, I. R., J. K. Dunnick, M. G. McNamee, and J. P. Kriss. 1974.
Distribution and fate of synthetic lipid vesicles in the mouse: a combined
radionuclide and spin label study. Proc. Natl. Acad. Sci. USA 71:3487.

80. Yeagle, P. L., W. C. Hutton, R. B. Martin, B. Sears, and C.-H. Huang. 1976.
Transmembrane asymmetry of vesicle lipids. J. Biol. Chem. 251:2110.

81. Hwang, K. J. and M. R. Mauk. 1977. Fate of lipids in vivo: a gamma-ray
perturbed angular correlation study. Proc. Natl. Acad. Sci. USA 74:4991.

82. Mauk, M. R. and R. C. Gamble. 1979. Stability of lipid vesicles in tissues of the mouse: a gamma-ray perturbed angular correlation study. Proc. Natl. Acad. Sci. USA 76:765.

83. Ryman, B. E., R. S. Jewkes, K. Jeyasingh, M. P. Osborne, H. M. Patel, V. J. Richardson, M. H. N. Tattersall, and D. A. Tyrrell. 1978. Potential application of liposomes to therapy. Ann. N.Y. Acad. Sci. 308:281.

84. Segal, A. W., G. Gregoriadis, J. P. Lavender, D. Tarin, and T. J. Peters. 1976. Tissue and hepatic subcellular distribution of liposomes containing bleomycin after intravenous administration to patients with neoplasms. Clin. Sci. Mol. Med. 51:421.

85. Magee, W. E., C. W. Goff, J. Schoknecht, D. Smith, and K. Cherian. 1974. The interaction of cationic liposomes containing entrapped horseradish peroxidase with cells in culture. J. Cell Biol. 63:492.

86. Pagano, R. E. and M. Takeichi. 1977. Adhesion of phospholipid vesicles to Chinese hamster fibroblasts. J. Cell Biol. 74:531.

87. Tokes, Z. A., C. J. Der, and J. A. Todd. 1977. Cationic liposome interaction with normal and transformed rat liver epithelial cells in culture. Proc. Am. Assoc. Cancer Res. 18:247.

88. Mayhew, E., C. Gotfredsen, Y. J. Schneider, and A. Trouet. 1980. Interaction of liposomes with cultured cells: effect of serum. Biochem. Pharmacol. 29:877.

89. Szoka, F., K. Jacobson, Z. Derzko, and D. Papahadjopoulos. 1980. Fluorescence studies on the mechanism of liposome-cell interactions in vitro. Biochim. Biophys. Acta 600:1.

90. Martin, F. J. and R. C. MacDonald. 1976. Lipid vesicle-cell interactions. 11. Induction of cell fusion. J. Cell Biol. 70:506.

91. Wisse, E., G. Gregoriadis, and W. Th. Daems. 1976. Electron microscopic cytochemical localization of intravenously injected liposome-encapsulated horseradish peroxidase in rat liver cells. Adv. Exp. Med. Biol. 73:237.

92. Pagano, R. E. and L. Huang. 1975. Interaction of phospholipid vesicles with cultured mammalian cells. J. Cell Biol. 67:49.

93. Papahadjopoulos, D., E. Mayhew, G. Poste, and S. Smith. 1974. Incorporation of lipid vesicles by mammalian cells provides a potential method for modifying cell behaviours. Nature 252:163.

94. Szoka, F., K. E. Magnusson, J. Wojcieszyn, Y. Hou, Z. Derzko, K. Jacobson. 1981. Use of lectins and polyethylene glycol for fusion of glycolipid-containing liposomes with eukaryotic cells. Proc. Natl. Acad. Sci. USA 78: 1685.

95. Caride, V. J. and B. L. Zaret. 1978. Liposomal accumulation in regions of acute myocardial infarction: effect of surface charge. Ann. N.Y. Acad. Sci. 308:435.

96. Gregoriadis, G. and B. E. Ryman. 1972. Fate of protein-containing liposomes injected into rats. An approach to the treatment of storage diseases. Eur. J. Biochem. 24:485.

97. Gregoriadis, G. and B. E. Ryman. 1972. Lysosomal localization of β-fructo-

furanosidase-containing liposomes injected into rats: some implications in the treatment of genetic disorders. Biochem. J. 129:123.

98. Haynes, D. M. and C. H. Kang. 1978. Saturation of uptake of liposomes by the reticulo-endothelial system. Possible use for increased tumor specificity. Ann. N.Y. Acad. Sci. 308:440.

99. Mayhew, E., Y. Rustum, and F. Szoka. 1982. Therapeutic efficacy of cytosine arabinoside trapped in liposomes. In *Targeting of Drugs*, G. Gregoriadis, J. Senior, and A. Trouet (Eds.). Plenum Press, New York, pp. 249–260.

100. Rustum, Y. M., C. Dave, E. Mayhew, and D. Papahadjopoulos. 1979. Role of liposome type and route of administration in the antitumor activity of liposome-entrapped 1-β-D-arabinofuranosylcytosine against mouse L1210 leukemia. Cancer Res. 39:1390.

101. Gregoriadis, G. 1974. Structural requirements for the specific uptake of macromolecules and liposomes by target tissues. In *Enzyme Therapy in Lysosomal Storage Disease*, J. M. Tager, G. J. W. Hooghwinkel and W. Th. Daems (Eds.). North-Holland, Amsterdam, pp. 131–148.

102. Gregoriadis, G. 1977. Targeting of drugs. Nature 265:407.

103. Heath, T. D., R. Fraley, and D. Papahadjopoulos. 1980. Antidoby targeting liposomes: specific interaction of vesicles conjugated to anti-erythrocyte F(ab') with erythrocytes. Science 210:539.

104. Martin, F., W. Hubbell, and D. Papahadjopoulos. 1982. Immunospecific targeting of liposomes to cells: a novel and efficient method for covalent attachment of FAB' fragments via disulphide bonds. Biochemistry 20:4229.

105. Chou, T.-C., D. J. Hutchinson, and F. A. Schmid. 1975. Metabolism and selective effect of 1-β-D-arabinofuranosylcytosine in L1210 and host tissues in vivo. Cancer Res. 35:225.

106. Sharkey, R. M., D. M. Doldenberg, and F. J. Primus. 1979. Targeting of antibody coated liposomes to tumor cells producing carcinoembryonic antigen. Fed. Proc. 38:1089.

107. Gregoriadis, G. and E. D. Neerunjun. 1975. Homing of liposomes to target cells. Biochem. Biophys. Res. Commun. 65:537.

108. Leserman, L. D., J. N. Weinstein, R. Blumenthal, S. O. Sharrow, and W. D. Terry. 1979. Binding of antigen-bearing fluorescent liposomes to the murine myeloma tumor MOPC 315. J. Immunol. 122:585.

109. Weissmann, G., D. Bloomgarden, R. Kaplan, C. Cohen, S. Hoffstein, T. Collins, A. Gotlieb, and D. Nagle. 1975. A general method for the introduction of enzyme by means of immunoglobulin-coated liposomes into liposomes of deficient cells. Proc. Natl. Acad. Sci. USA 72:88.

110. Weissmann, G., A. Brand, and E. C. Franklin. 1974. Interaction of immunoglobulins with liposomes. J. Clin. Invest. 53:536.

111. Heath, T. D., J. A. Montgomery, J. R. Piper, and D. Papahadjopoulos. 1983. Antibody-targeted liposomes: Increase in specific toxicity of methotrexate-γ-aspartate. Proc. Nat. Acad. Sci. USA 80.

112. Dunnick, J. K., I. R. McDougall, S. Aragon, M. L. Goris, and J. P. Kriss.

1975. Vesicle interactions with polyamino acids and antibody: in vivo and in vitro studies. J. Neurol. Med. 16:483.

113. Dunnick, J. K., J. D. Rooke, S. Aragon, and J. P. Kriss. 1976. Alteration of mammalian cells by interaction with artificial lipid vesicles. Cancer Res. 36:2385.

114. Papahadjopoulos, D., R. Fraley, and T. Heath. 1982. Optimization of liposomes as a macromolecular carrier system: new methodology for cell targeting and intracellular delivery of DNA. In *Liposomes in the Study of Drug Activity and Immunocompetent Cell Functions,* C. Nicolau and A. Paraf (Eds.). Proc. Int. Symp., Grignon, France (in press).

115. Martin, F., and D. Papahadjopoulos. 1982. Irreversible coupling of immunoglobulin fragments to preformed vesicles: an improved method for liposome-targeting. J. Biol. Chem. 257:286.

116. Alderson, J. C. and C. Green. 1978. Lectin-induced cell agglutination and membrane cholesterol level. Exp. Cell Res. 114:475.

117. Orr, G. A., R. R. Rando, and F. W. Bangerte. 1979. Synthetic glycolipids and the lectin mediated aggregation of liposomes. J. Biol. Chem. 254: 4721.

118. Redwood, W. R., V. K. Jonsons, and B. C. Patel. 1975. Lectin-receptor interactions in liposomes. Biochim. Biophys. Acta 406:347.

119. Bhattacharya, A. and B. K. Vonderhaar. 1978. Reconstitution of prolactin receptors in artificial lipid vesicles. J. Cell Biol. 79:64a.

120. Mauk, M. R., R. C. Gamble, and J. D. Baldeschwieler. 1980. Targeting of lipid vesicles: specificity of carbohydrate receptor analogues for leukocytes in mice. Proc. Natl. Acad. Sci. USA 77:430.

121. Rahman, Y.-E., E. H. Lau, and K. R. Patel. 1979. In vivo cell targeting by liposomes containing glycolipids. J. Cell Biol. 83:286a.

122. Robbins, J. C. and M. S. Wu. 1980. Incorporation of synthetic glycolipids alters in vivo behavior of liposomes. Fed. Proc. 39:Abstr. 1976.

123. Gregoriadis, G. and E. D. Neerunjun. 1974. Control of the rate of hepatic uptake and catabolism of liposome-entrapped proteins injected into rats. Possible therapeutic applications. Eur. J. Biochem. 47:179.

124. Hildenbrandt, G. R. and N. N. Aronson, Jr. 1980. Uptake of asialoglycophorin-liposomes by the perfused rat liver. Biochim. Biophys. Acta 631: 499.

125. Surolia, A. and B. K. Bachhawat. 1977. Monosialoganglioside liposome-entrapped enzyme uptake by hepatic cells. Biochim. Biophys. Acta 497: 760.

126. Bussian, R. W. and J. C. Wriston, Jr. 1977. Influence of incorporated cerebrosides on the interaction of liposomes with HeLa cells. Biochim. Biophys. Acta. 471:336.

127. Alving, C. R., I. Schneider, G. M. Schwartz, and E. A. Steck. 1979. Sporozoite-induced malaria. Therapeutic effects of glycolipids in liposomes. Science 205:1142.

128. Slama, J. S. and R. R. Rando. 1980. Lectin-mediated aggregation of lipo-

somes containing glycolipids with variable hydrophilic spacer arms. Biochemistry 19:4595.

129. Juliano, R. L. and Stamp, D. 1978. Pharmacokinetics of liposome-encapsulated tumor drugs. Studies with vinblastine, actinomycin D, cytosine arabinoside and daunomycin. Biochem. Pharmacol. 27:21.

130. Kimelberg, H. K. 1976. Differential distribution of liposome-entrapped (^{3}H) methotrexate and labelled lipids after intravenous injection in a primate. Biochim. Biophys. Acta 448:531.

131. Saba, T. M. 1970. Physiology and physiopathology of the reticuloendothelial system. Arch. Intern. Med. 126:1031.

132. Yatvin, M. B., J. N. Weinstein, W. H. Dennis, and R. Blumenthal. 1978. Design of liposomes for enhanced local release of drugs by hyperthermia. Science 202:1290.

133. Yatvin, M. B., J. N. Weinstein, W. H. Dennis, and R. Blumenthal. 1978. Use of hyperthermia to promote selective local release of drugs from liposomes. *Proc. Conf. Clin. Prospects for Hypoxic Cell Sensitizers and Hyperthermia.*

134. Weinstein, J. N., R. L. Magin, M. B. Yatvin, and D. S. Zaharko. 1979. Liposomes and local hyperthermia: selective delivery of methotrexate to heated tumors. Science 204:188.

135. Weinstein, J. N., R. L. Magin, R. L. Cysyk, and D. S. Zaharko. 1980. Treatment of solid L1210 murine tumors with local hyperthermia and temperature sensitive liposomes containing methotrexate. Cancer Res. 40: 1388.

136. Yatvin, M. B., H. Muhlensiepen, W. Porschen, and L. E. Feinendegen. 1980. Selective delivery by hyperthermia of liposome encapsulated cis-dichlorodiammine platinum(II) and tumor growth delay. Proc. Am. Assoc. Cancer Res. 21:281.

137. Arturson, G. 1979. Microvascular permeability to macromolecules in thermal injury. Acta Physiol. Scand. Suppl. 463:111.

138. Yatvin, M. B., W. Kreutz, B. Horwitz, and M. Shinitzky. 1980. Induced drug release from lipid vesicles in serum by pH change. Biophys. Struct. Mech. 6:233.

139. Yatvin, M. B., W. Kreutz, B. A. Horwitz, and M. Shinitzky. 1980. pH-sensitive liposomes: possible clinical implications. Science 210:1253.

140. Blackshear, P. J. 1979. Implantable drug-delivery systems. Sci. Am. 241:66.

141. Adams, D. H., G. Joyce, V. J. Richardson, B. E. Ryman, and H. Wisniewski. 1977. Liposome toxicity in the mouse central nervous system. J. Neurol. Sci. 31:173.

142. Bigon, E., E. Boarato, A. Bruni, A. Leon, and G. Toffano. 1979. Pharmacological effects of phosphatidylserine liposomes: regulation of glycolysis and energy level in brain. Br. J. Pharmacol. 66:167.

143. Bigon, E., E. Boarato, A. Bruni, A. Leon, and G. Toffano. 1979. Pharmacological effects of phosphatidylserine liposomes: the role of lysophosphatidylserine. Br. J. Pharmacol. 67:611.

144. Sur, P., B. Hazra, and D. K. Roy. 1979. Effect of lipid vesicles on the growth of Ehrlich ascites carcinoma in mice. IRCS Med. Sci: Cancer 7: 453.

145. Behrens, B. C., N. Yamamoto, R. R. Brown, and G. T. Bryan. 1979. Cytotoxicity of liposomes towards cultured human and rat bladder cancer cell lines. Proc. Am. Assoc. Cancer Res. 20:160.

146. Chawla, R. S., I. W. Kellaway, I. M. Hunneyball, and J. Stevens. 1979. The effect of liposomal charge on drug toxicity and efflux. J. Pharm. Pharmacol. 31:86.

147. Panzner, E. A. and V. K. Jansons. 1979. Control of in vitro cytotoxicity of positively charged liposomes. J. Cancer Res. Clin. Oncol. 95:29.

148. Klein, I., L. Moore, and I. Pastan. 1978. Effect of liposomes containing cholesterol on adenylate cyclase activity of cultured mammalian fibroblasts. Biochim. Biophys. Acta 506:42.

149. Rahman, Y. E., E. A. Cerny, S. L. Tollaksen, B. J. Wright, S. L. Nance, and J. F. Thomson. 1974. Liposome-encapsulated actinomycin D; potential in cancer chemotherapy. Proc. Soc. Exp. Biol. Med. 146:1173.

150. Rahman, A., A. Kessler, N. More, B. Sikic, G. Rowden, P. Woolley, and P. S. Schein. 1980. Liposomal protection of adriamycin induced cardiotoxicity in mice. Cancer Res. 40:1532.

151. Maslow, D. E., E. Mayhew, F. Olson, and Y. Rustum. 1980. Reduction of inhibitory effect of adriamycin on myocardial contraction in vitro by entrapment in liposomes. Proc. Am. Assoc. Cancer Res. 21:281.

152. Forssen, E. A. and Z. A. Tokes. 1979. In vitro and in vivo studies with adriamycin liposomes. Biochem. Biophys. Res. Commun. 91:1295.

153. Juliano, R. L. and D. Stamp. 1979. Interactions of drugs with lipid membranes characteristics of liposomes containing polar or nonpolar antitumor drugs. Biochim. Biophys. Acta 586:137.

154. Kaye, S. B., J. A. Boden, K. D. Bagshawe, and B. E. Ryman. 1979. Effect of liposome encapsulation of actinomycin D on its therapeutic efficacy. Br. J. Cancer 40:818.

155. Kaye, S. B., J. A. Boden, and B. E. Ryman. 1980. Application of liposome entrapped cytotoxic drugs to the treatment in vivo of drug-resistant solid murine tumors. Proc. Am. Assoc. Cancer Res. 21:254.

156. Kaye, S. B. and B. E. Ryman. 1980. The fate of liposome-entrapped actinomycin D in vivo and its therapeutic effect in a solid murine tumor. Biochem. Soc. Trans. 8:107.

157. Macek, C., Y. Rahman, and B. Tom. 1977. Uptake of liposome encapsulated actinomycin D in human tumor cells. Fed. Proc. 36:1053.

158. Rahman, Y. E., W. R. Hanson, J. Bharucha, E. J. Ainsworth, and B. N. Jaroslow. 1978. Mechanisms of reduction of anti-tumor drug toxicity by liposome encapsulation. In *Annals of the New York Academy of Sciences,* Vol. 308, D. Papahadjopoulos (Ed.). New York Academy of Sciences, New York, pp. 325–342.

159. Forssen, E. and Z. Tokes. 1979. Adriamycin entrapment in liposomes: in vitro and in vivo studies. Proc. Am. Assoc. Cancer Res. 20:188.

160. Murphree, S. A., T. R. Tritton, and A. C. Sartorelli. 1977. Differential effects of adriamycin on fluidity and fusion characteristics of cardiolipin containing liposomes. Fed. Proc. 36:303.

161. Parker, R. J., R. Dixon, and S. M. Sieber. 1979. Fate of liposome-entrapped ^{14}C-adriamycin (^{14}C-ADR) given by IP injection to rats. Proc. Am. Assoc. Cancer Res. 20:238.

162. Parker, R. J. and S. M. Sieber. 1980. Effect of the route of administration on the tissue distribution of liposome-encapsulated adriamycin in rats. Proc. Am. Assoc. Cancer Res. 21:280.

163. Parker, R. J., K. D. Hartman, and S. M. Sieber. 1981. Lymphatic absorption and tissue disposition of liposome-entrapped (^{14}C)adriamycin following intraperitoneal administration to rats. Cancer Res. 41:1311.

164. Rahman, A., A. Kessler, J. Macdonald, V. Waravdekar, and P. Schein. 1979. Liposomal delivery of adriamycin. Proc. Am. Assoc. Cancer. Res. 20:288.

165. Tritton, R., S. A. Murphree, and A. C. Sartorelli. 1978. Adriamycin liposome interactions. In *Annals of the New York Academy of Sciences*, Vol. 308, D. Papahadjopoulos (Ed.). New York Academy of Sciences, New York, p. 438.

166. Yamanaka, N., K. Koizumi, T. Kato, K. Ota, and S. Shimizu. 1978. Encapsulation of anti-cancer drugs in large-volume liposome and distributions in vivo. Proc. 37th Annu. Meet. Jpn. Cancer Assoc. p. 143.

167. Schiffman, F. J. and I. Klein. 1977. Rapid induction of amphotericin B sensitivity in L-1210 leukemia cells by liposomes containing ergosterol. Nature 269:65.

168. Neerunjun, E. D. and G. Gregoriadis. 1976. Tumour regression with liposome-entrapped asparaginase: some immunological advantages. Biochem. Soc. Trans. 4:133.

169. Fendler, J. H. and A. Romero. 1976. Encapsulation of 8-azaguanine in single multiple compartment liposomes. Life Sci. 18:1453.

170. Kano, K. and J. H. Fendler. 1977. Enhanced uptake of drugs in liposomes: use of labile vitamin B12 complexes of 6-mercaptopurine and 8-azaguanine. Life Sci. 20:1729.

171. Ritter, C. and R. J. Rutman. 1980. Relative enhancement by various liposomes of BCNU effectiveness against L-1210 leukemia in vivo. Res. Commun. Chem. Pathol. Pharmacol. 30:123.

172. Neerunjun, E. D., R. Hunt, and G. Gregoriadis. 1977. Fate of a liposome associated agent injected into normal and tumor bearing rodents attempts to improve localization in tumor tissues. Biochem. Soc. Trans. 5:1380.

173. Sur, P. and D. K. Roy. 1979. Enhancement of the action of bleomycin using liposome with Ehrlich ascites carcinoma in mice. Indian J. Exp. Biol. 17:952.

174. Inaba, M., S. Tsukagoshi, and N. Yoshida. 1979. Effect of lipophilic antitumor agents entrapped in the lipid layer of liposomes. *Proc. 38th Annu. Meet. Jpn. Cancer Assoc.*, p. 194.

175. Robertson, D. 1977. Incorporation of cis-dichlorobiscyclopentylamine-

platinum(II) into liposomes enhances its uptake by ADJ/PC6A tumours implanted subcutaneously into mice. Biochem. Soc. Trans. 5:1326.

176. Griffin, M. J. and S. Lucid. 1977. Liposomal enhanced cytosine arabinoside chemotherapy of L-1210 in BDF-1 mice. Proc. Am. Assoc. Cancer Res. 18:183.

177. Griffin, M. J. and J. J. Killion. 1977. Liposomal encapsulated cytosine arabinoside in vivo and in vitro toxicity to L-1210 cells. In Vitro 13:152.

178. Juliano, R. L. and H. N. McCullough. 1980. Controlled delivery of an antitumor drug: localized action of liposome encapsulated cytosine arabinoside administered via the respiratory system. J. Pharmacol. Exp. Ther. 214:381.

179. Kataoka, T. and T. Kobayashi. 1978. Enhancement of chemotherapeutic effect by entrapping 1-beta-D-arabinofuranosylcytosine in lipid vesicles and its mode of action. Ann. N.Y. Acad. Sci. 308:387.

180. Kobayashi, T., T. Kataoka, S. Tsukagoshi, and Y. Sakurai. 1977. Enhancement of anti-tumor activity of 1-beta-D-arabinofuranosyl cytosine by encapsulation in liposomes. Int. J. Cancer 20:581.

181. Mayhew, E., D. Papahadjopoulos, Y. M. Rustum, and C. Dave. 1978. Use of liposomes for the enhancement of the cytotoxic effects of cytosine arabinoside. In *Annals of the New York Academy of Sciences,* Vol. 308, D. Papahadjopoulos (Ed.). New York Academy of Sciences, New York, pp. 371–386.

182. Mayhew, E., Y. Rustum, and F. Szoka. 1980. Efficacy of liposome entrapped cytosine arabinoside compared with infused free cytosine arabinoside against L-1210 tumor. Proc. Am. Assoc. Cancer Res. Am. Soc. Clin. Oncol. 21:293.

183. Mayhew, E., Y. M. Rustum, F. Szoka, D. Papahadjopoulos, and K. Paigen. 1979. Role of cholesterol in enhancing the antitumor activity of 1-beta-D-arabinofuranosylcytosine (ARA-C) entrapped in reverse phase evaporation vesicles (REV). Proc. Am. Assoc. Cancer Res. 20:186.

184. Stamp, D. and R. L. Juliano. 1979. Factors affecting the encapsulation of drugs within liposomes. Can. J. Physiol. Pharmacol. 57:535.

185. Tsukagoshi, S. and T. Kobayashi. 1977. Approaches to the chemotherapy of lymph node metastasis. In *Cancer Invasion and Metastasis: Biologic Mechanisms and Therapy,* S. B. Day, W. P. Myers, P. Stansly, S. Garattini, and H. G. Lewis (Eds.). Progress in Cancer Research and Therapy, Vol. 5. Raven Press, New York, pp. 465–474.

186. Tsukagoshi, S. and T. Kobayashi. 1977. Fundamental approaches to the chemotherapy of lymph node metastasis. Gann Monogr. Cancer Res. 20:183.

187. Dorsa, M. J. and W. Magee. 1978. Inhibition of cellular growth and DNA synthesis by 5'-fatty derivatives of 1-beta-arabinofuranosyl cytosine (ARA-C) incorporated into liposomes. Fed. Proc. 37:1309.

188. Kataoka, T., F. Ohhashi, and Y. Sakurai. 1979. Antitumor potency of acyl-1-beta-D-arabinofuranosyl cytosine trapped in liposome. *Proc. 38th Annu. Meet. Jpn. Cancer Assoc.*

189. Gregoriadis, G., P. J. Davisson, and S. Scott. 1977. Binding of drugs to liposome entrapped macromolecules prevents diffusion of drugs from liposomes in vitro and in vivo. Biochem. Soc. Trans. 5:1323.

190. Goldman, R., I. Facchinetti, D. Bach, A. Raz, and M. Shinitzky. 1978. A differential interaction of daunomycin, adriamycin and their derivatives with human erythrocytes and phospholipid bilayers. Biochim. Biophys. Acta 512:254.

191. Apple, M. A., H. Yanagisawa, snd C. A. Hunt. 1978. Effect of liposome encapsulation on the activity of potent new bis-daunorubicins. Proc. Am. Assoc. Cancer Res. 19:169.

192. Freiburg, J. A. 1978. Evaluation of liposomes as carriers of estracyt into tumor cells in vitro. Diss. Abstr. Int. 38:5267.

193. Simmons, S. P. and P. A. Kramer. 1977. Liposomal entrapment of floxuridine. J. Pharm. Sci. 66:984.

194. Shinozawa, S., K. Tsutsui, Y. Araki, and T. Oda. 1979. Antitumor effect of illudin S and neocarzinostatin entrapped in liposomes. *Proc. 38th Annu. Meet. Jpn. Cancer Assoc.*

195. Shinozawa, S., K. Tsutsui, and T. Oda. 1979. Enhancement of the antitumor effect of illudin S by including it into liposomes. Experentia 35:1102.

196. La Bonnardiere, C. 1978. Preliminary data on the protective effect of interferon bound to liposomes in the mouse murine hepatitis virus model. Ann. Microbiol. 129A:397.

197. Fidler, I. J. 1980. Therapy of spontaneous metastases by intravenous injection of liposomes containing lymphokines. Science 208:1469.

198. Fidler, I. J., A. Raz, W. E. Fogler, R. Kirsh, P. Bugelski, and G. Poste. 1980. Design of liposomes to improve delivery of macrophage-augmenting agents to alveolar macrophages. Cancer Res. 40:4460.

199. Fogler, W. E., A. Raz, and I. J. Fidler. 1980. In situ activation of murine macrophages by liposomes containing lymphokines. Cell Immunol. 53:214.

200. Poste, G. and R. Kirsh. 1979. Rapid decay of tumoricidal activity and loss of responsiveness to lymphokines in inflammatory macrophages. Cancer Res. 39:2582.

201. Poste, G., R. Kirsh, W. E. Fogler, and I. J. Fidler. 1979. Activation of tumoricidal properties in mouse macrophages by lymphokines encapsulated in liposomes. Cancer Res. 39:881.

202. Sone, S., G. Poste, and I. J. Fidler. 1980. Rat alveolar macrophages are susceptible to activation by free and liposome encapsulated lymphokines. J. Immunol. 124:2197.

203. Sone, S. and I. J. Fidler. 1980. Synergistic activation by lymphokines and muramyldipeptide of tumoricidal properties in rat alveolar macrophages. J. Immunol. 125:2454.

204. Chen, S. T. 1978. Interaction of anticancer agents with carriers facilitating cellular uptake. Diss. Abstr. Int. 39:668B.

205. Freise, J., P. Magerstedt, and F. W. Schmidt. 1979. The in vitro and in

vivo stability of the entrapment of methotrexate in negatively charged liposomes after sterile filtration. Z. Naturforsch. Sect. C Biosci. 34:114.

206. Freise, J., F. W. Schmidt, and P. Magerstedt. 1979. Effect of liposome-entrapped methotrexate on Ehrlich ascites tumor cells and uptake in primary liver cell tumor. J. Cancer Res. Clin. Oncol. 94:21.

207. Fry, D. W., J. C. White, and I. D. Goldman. 1979. Alterations of the carrier mediated transport of an anionic solute methotrexate by charged liposomes in Ehrlich ascites tumor cells. J. Membr. Biol. 50:123.

208. Kimelberg, H. K. and M. L. Atchison. 1978. Effects of entrapment in liposomes on the distribution degradation and effectiveness of methotrexate in vivo. In *Annals of the New York Academy of Sciences,* Vol. 308, D. Paphadjopoulos (Ed.). New York Academy of Sciences, New York, pp. 395–410.

209. Kimelberg, H. K., S. Biddlecome, T. F. Tracy, and R. S. Bourke. 1977. Distribution of liposome entrapped tritiated methotrexate after intravenous and cerebral intraventricular injection in cynomolgus monkeys. Fed. Proc. 36:862.

210. Patel, K. R., M. M. Jonah, and Y.-E. Rahman. 1980. Induction of methotrexate uptake into hepatoma H-129 cells by liposome encapsulation. Fed. Proc. 39:Abstr. 1370.

211. Woo, S. Y., D. P. Dilliplane, and L. F. Sinks. 1980. Enhancement of chemotherapeutic efficacy of methotrexate entrapped in liposomes against murine intracranial L1210 leukemia. Proc. Am. Assoc. Cancer Res. Am. Soc. Clin. Oncol. 21:307.

212. Todd, J. A., A. M. Levine, and Z. A. Tokes. 1978. Increased uptake of liposome entrapped methotrexate by chronic lymphocytic leukemia cells. Fed. Proc. 37:1543.

213. Todd, J. A., A. M. Levine, and Z. A. Tokes. 1980. Liposome encapsulated methotrexate interactions with human chronic lymphocytic leukemia cells. J. Natl. Cancer Inst. 64:715.

214. Rahman, Y. E., E. A. Cerny, M. M. Jonah, and J. L. Dainko. 1976. Liposome encapsulation of various anti-tumor drugs, and mouse tissue distribution of liposomal adriamycin. Fed. Proc. 35:786.

215. Shinozawa, S., Y. Araki, and T. Oda. 1980. Antitumor effect of neocarzinostatin entrapped in liposomes. Gann 71:107.

216. Shinozawa, S., Y. Araki, and T. Oda. 1980. Antitumor effect of neocarzinostatin entrapped in neutral, negatively and positively charged liposomes against Ehrlich solid carcinoma. Res. Commun. Chem. Pathol. Pharmacol. 30:181.

217. Gregoriadis, G. 1977. Liposome-entrapped proteins in therapeutic and preventive medicine. Hind. Antibiot. Bull. 20:14.

218. Rutman, R. J., N. G. Avadhani, and C. Ritter. 1977. Activation in vitro of nitrogen mustard by liposomal transport. Biochem. Pharmacol. 26:85.

219. Shinozawa, S., Y. Araki, and T. Oda. 1979. Distribution of tritiated prednisolone entrapped in lipid layer of liposome after intramuscular administration in rats. Res. Commun. Chem. Pathol. Pharmacol. 24:223.

220. Lasch, J., R. Koelsch, G. Brezesinski, V. Hermann, S. Riemann, and P. Bohley. 1977. Pepstatin- and leupeptin-loaded liposomes: a tool in protein breakdown studies. Acta Biol. Med. Ger. 36:11.

221. Todorov, D. and G. Deliconstantinos. 1980. Attempt to improve the cytotoxic effect of thalicarpine by entrapment in liposomal vesicles. Onkologiya 17:40.

222. Layton, D. and A. Trouet. 1980. A comparison of the therapeutic effects of free and liposomally encapsulated vincristine in leukemic mice. Eur. J. Cancer 16:945.

223. Layton, D., J. DeMeyere, M.-P. Collard, and A. Trouet. 1979. The accumulation by fibroblasts of liposomally encapsulated vinblastine. Eur. J. Cancer 15:1475.

224. Gregoriadis, G. 1979. Liposomes. In *Drug Carriers in Biology and Medicine*, G. Gregoriadis (Ed.). Academic Press, London, pp. 287–341.

225. Skipper, H. E., F. M. Schabel, L. B. Mellet, J. A. Montgomery, L. J. Wilkoff, H. H. Lloyd, and R. W. Brockman. 1970. Implications of biochemical, cytokinetic, pharmacologic and toxicologic relationships in the design of therapeutic schedules. Cancer Chemother. Rep. 54:631.

226. Kosloski, M. J., F. Rosen, R. J. Milholland, and D. Papahadjoupoulos. 1978. Effect of lipid vesicle liposome encapsulation of methotrexate on its chemotherapeutic efficacy in solid rodent tumors. Cancer Res. 38:2848.

227. Olson, F., E. Mayhew, D. Maslow, Y. Rustum, and F. Szoka. 1982. Characterization, toxicity and therapeutic efficacy of adriamycin encapsulated in liposomes. Eur. J. Canc. Clin. Oncol. 18:167.

228. Gregoriadis, G. and E. D. Neerunjun. 1975. Treatment of tumor bearing mice with liposome-entrapped actinomycin D prolongs their survival. Res. Commun. Chem. Pathol. Pharmacol. 10:351.

229. Kay, S. B., J. A. Boden, and B. E. Ryman. 1981. The effect of liposome (phospholipid vesicle) entrapment of actinomycin D and methotrexate on the in vivo treatment of sensitive and resistant solid murine tumours. Eur. J. Cancer 17:279.

230. Papahadjopoulos, D., G. Poste, W. J. Vail, and J. L. Fidler. 1976. Use of lipid vesicles as carriers to introduce actinomycin D into resistant tumor cells. Cancer Res. 36:2988.

231. Chabner, B. A., C. E. Myers, C. N. Coleman, and D. G. Johns. 1975. The chemical pharmacology of anti-neoplastic agent. N. Engl. J. Med. 292:1159.

232. Chabner, B. A., C. E. Myers, C. N. Coleman, and D. G. Johns. 1975. The Chemical pharmacology of antineoplastic agents. N. England J. Med. 292:1102.

233. DiMarco, A. 1975. Adriamycin (NSC-123127): mode and mechanism of action. Cancer Chemother. Rep. 6:91.

234. Donehower, R. C., C. E. Myers, and B. A. Chabner. 1979. New development on the mechanisms of action of antineoplastic drugs. Life Sci. 25:1.

235. Schwartz, H. S. 1974. Some determinants of the therapeutic efficacy of

actinomycin D, adriamycin and daunorubicin. Cancer Chemother. Rep. 58:55.

236a. Rustum, Y., E. Mayhew, J. Campbell, and F. Szoka. 1981. Inability of liposome encapsulated 1-β-D-arabinofuranosyl cytosine nucleotides to overcome drug resistance in L1210 cells. Eur. J. Canc. 17:809.

236b. Fidler, I. J., A. Raz, W. Fogler, L. C. Hoyes, and G. Poste. 1981. The role of plasma membrane receptors and the kinetics of macrophage activation by lymphokines encapsulated in liposomes. Cancer Res. 41:495.

236c. Sone, S. and I. J. Fidler. 1981. In vitro activation of tumoricidal properties in rat alveolar macrophages by synthetic muramyl dipeptide encapsulated in liposomes. Cell. Immunol. 57:42.

237. Gregoriadis, G. and R. A. Buckland. 1973. Enzyme-containing liposomes alleviate a model for storage disease. Nature 244:170.

238. Colley, C. M. and B. E. Ryman. 1976. The use of liposomally entrapped enzyme in the treatment of an artificial storage condition. Biochim. Biophys. Acta 451:417.

239. Butler, J. B., F. Tietze, F. Pellefigue, S. P. Spielberg, and J. D. Schulman. 1978. Depletion of cystine in cystinotic fibroblasts by drugs enclosed in liposomes. Pediatr. Res. 12:46.

240. Aminoff, D., W. F. V. Bruegge, W. C. Bell, K. Sarpolis, and R. Williams. 1977. Role of sialic acid in survival of erythrocytes with spleen and liver at the cellular level. Proc. Natl. Acad. Sci. USA 74:1521.

241. Rahman, Y. E. and B. J. Wright. 1975. Liposomes containing chelating agents. Cellular penetration and a possible mechanisms of metal removal. J. Cell Biol. 65:112.

242. Alving, C. R., E. A. Steck, W. L. Hanson, P. S. Loizeaus, W. J. Chapman, Jr., and V. B. Waits. 1978. Improved therapy of experimental leishmaniasis by use of a liposome-encapsulated antimonial drug. Life Sci. 22:1021.

243. Alving, C. R., E. A. Steck, W. L. Chapman, Jr., V. B. Waits, L. D. Hendricks, G. M. Swartz, Jr., and W. L. Hanson. 1978. Therapy of leishmaniasis. Superior efficacies of liposome encapsulated drugs. Proc. Natl. Acad. Sci. USA 75:2959.

244. New, R. R. C., M. L. Chance, S. C. Thomas, and W. Peters. 1978. Antileishmanial activity of antimonials entrapped in liposomes. Nature 272:55.

245. Black, C. D., G. J. Watson, and R. J. Ward. 1977. The use of Pentostam liposomes in the chemotherapy of experimental leishmaniasis. Trans. R. Soc. Trap. Med. Hyg. 71:550.

246. New, R. R. and M. L. Chance. 1980. Treatment of experimental cutaneous leishmaniasis by liposome-entrapped Pentostam. Acta Trop. 37:253.

247. Alving, C. R., E. A. Steck, W. L. Chapman, Jr., V. B. Waits, L. D. Hendricks, G. M. Swartz, Jr., and W. L. Hanson. 1980. Liposomes in leishmaniasis therapeutic effects of antimonial drugs 8-amino quinolines and tetracycline. Life Sci. 26:2231.

248. Gruenberg, J., D. Coral, A. L. Knupfer, and J. Deshusses. 1979. Inter-

actions of liposomes with *Trypanosoma brucei* plasma membrane. Biochem. Biophys. Res. Commun. 88:1173.

249. Alving, C. R., I. Schnieder, G. M. Swartz, Jr., and E. A. Steck. 1979. Sporozoite-induced malaria: therapeutic effects of glycolipids in liposomes. Science 205:1142.

250. Alving, C. R. 1982. Liposomes as carriers in leishmaniasis, malaria and vaccines. In *Drug Targeting,* G. Gregoriadis and A. Trouet (Eds.). Plenum Press, New York.

251. Dingle, J. T. 1977. Liposomes in the treatment of inflammatory joint disease. Rheumatoid Arthritis: Cell. Pathol. Pharmacol., Proc. Cambridge Rheum. Res. Group Symp., pp. 259-266.

252. Dingle, J. T. 1978. Articular damage in arthritis and its control. Ann. Intern. Med. 88:821.

253. Dingle, J. T., J. L. Gordon, B. L. Hazleman, C. G. Knight, D. P. Page Thomas, N. C. Phillips, I. H. Shaw, F. J. Fildes, J. E. Oliver, G. Jones, E. H. Turner, and J. S. Lowe. 1978. Novel treatment for joint inflammation. Nature 271:372.

254. Knight, C. G. and I. H. Shaw. 1979. Liposomes as carriers of anti-inflammatory steroids. Front Biol. 48:575.

255. Phillips, N. C., D. P. P. Thomas, C. G. Knight, and J. T. Dingle. 1979. Liposome incorporated cortico steroids. 2. Therapeutic activity in experimental arthritis. Ann. Rheum. Dis. 38:553.

256. Shaw, I. H., C. G. Knight, D. P. P. Thomas, N. C. Phillips, and J. T. Dingle. 1979. Liposome-incorporated corticosteroids. I. The interaction of liposomal cortisol palmitate with inflammatory synovial membrane. Br. J. Exp. Pathol. 60:142.

257. De Silva, M., B. L. Hazleman, D. P. P. Thomas, and P. Wraight. 1979. Liposomes in arthritis. A new approach. Lancet 1:1320.

258. Caprino, L., F. Antonetti, G. Bergesi, and F. Borrelli. 1980. Ex-ovo phospholipids as carriers of cortisol palmitate. Farm. Ed. Prat. 35:315.

259. Bird, H. A. 1979. Liposomes in arthritis. Lancet 2:247.

260. Rahman, Y. E. 1979. Potential of the liposomal approach to metal chelation therapy. In *Lysosomes in Applied Biology and Therapeutics,* Vol. 6, J. T. Dingle, P. I. J. Jacques (Eds.). North-Holland, Amsterdam, pp. 625-652.

261. Rahman, Y. E., M. W. Rosenthal, and E. A. Cerny. 1973. Intracellular plutonium: removal by liposome-encapsulated chelating agents. Science 180:300.

262. Guilmette, R. A., E. A. Cerny, and Y. E. Rahman. 1978. Pharmacokinetics of the iron chelator deferrioxamine as affected by liposome encapsulation potential in treatment of chronic hemosiderosis. Life Sci. 22:313.

263. Young, S. P., E. Baker, and E. R. Huehns. 1979. Liposome entrapped deferriozamine and iron transporting ionophores. A new approach to iron chelation therapy. Br. J. Haematol. 41:357.

264. Hemker, H. C., W. T. Hermens, A. D. Muller, and R. F. A. Muller. 1980. Oral treatment of hemophilia A by gastrointestinal absorption of factor-VIII entrapped in liposomes. Lancet 1:70.

265. Morgenthaler, J. J. 1980. Oral therapy for haemophilia. Lancet 1:546.

266. Caride, V. J. and B. L. Zaret. 1977. Liposome accumulation in regions of experimental myocardial infarction. Science 198:735.

267. Wilkman-Coffelt, J., J. Leung, and D. T. Mason. 1980. In vivo localization of liposomes in skeletal and cardiac muscle following experimental coronary ligation in guinea-pigs. Biochem. Med. 23:87.

268. Bonventre, P. F. and G. Gregoriadis. 1978. Killing of intraphagocytic *Staphylococcus aureus* by dihydrostreptomycin entrapped within liposomes. Antimicrob. Agents Chemother. 13:1049.

269. Hodges, N. A., R. Mounajed, C. J. Olliff, and J. M. Padfield. 1979. The enhancement of neomycin activity on *Escherichia coli* by entrapment in liposomes. J. Pharm. Pharmacol. 31:85.

270. Morgan, J. R. and K. E. Williams. 1980. Preparation and properties of liposome associated gentamicin. Antimicrob. Agents Chemother. 17:544.

271. Loeb, C., E. Benassi, P. Tanganelli, G. Besio, and M. Maffini. 1980. Antiepileptic effect of liposome-entrapped GABA on penicillin-induced epilepsy in rats. Adv. Epileptol, Epilepsy Int. Symp., 11th, p. 39.

272. Caride, V. J., W. Waylor, J. A. Cramer, and A. Gottschalk. 1976. Evaluation of liposome entrapped radioactive tracers as scanning agents. 1. Organ distribution of liposome technetium-99m diethylenetriamine penta acetic-acid in mice. J. Nucl. Med. 17:1067.

273. Espinola, L. G., J. Beaucaire, A. Gottschalk, and V. J. Caride. 1979. Radiolabeled liposomes as metabolic and scanning tracers in mice. 2. Indium-iii oxine compared with technetium-99m diethylenetriamine penta acetic-acid entrapped in multilamellar lipid vesicles. J. Nucl. Med. 20:434.

274. Hardy, J. G., I. W. Kellaway, J. Rogers, and C. G. Wilson. 1979. Distribution and fate in the rabbit of liposomes containing ^{131}I-sodium iodide. Br. J. Pharmacol. 67:459.

275. Hnatowich, D. J. and B. Clancy. 1980. Investigations of a new, highly negative liposome with improved biodistribution for imaging. J. Nucl. Med. 21:662.

276. Osborne, M. P., V. J. Richardson, K. Jeyasingh, and B. E. Ryman. 1979. Radionuclide-labeled liposomes—a new lymph node imaging agent. Int. J. Nucl. Med. Biol. 6:75.

277. Mackaness, G. B. and J. P. Hou. 1980. Contrast media containing liposomes as carriers. U.S. Patent No. 4192859, Mar. 11.

278. Belchetz, P. E., I. P. Braidman, J. C. W. Crawley, and G. Gregoriadis. 1977. Treatment of Gaucher's disease with liposome entrapped glucocerebroside beta glucosidase. Lancet 2:116.

279. Dhawan, V. M., D. B. Crawford, and R. P. Spencer. 1978. Technetium-99m-labeled liposomes preparation of radiopharmaceutical and its distribution in a hepatoma patient. Int. J. Nucl. Med. Biol. 5:118.

280. Richardson, V. J., B. E. Ryman, R. F. Jewkes, K. Jeyasingh, M. N. H. Tattersall, E. S. Newlands, and S. B. Kaye. 1979. Tissue distribution and tumor localization of technetium-99m labeled liposomes in cancer patients. Br. J. Cancer 40:35.

281. Richardson, V. J., B. E. Ryman, R. F. Jewkes, M. H. Tattersall, and E. S. Newlands. 1978. ^{99m}Tc-labelled liposomes preparation of radiopharmaceutical and its distribution in a hepatoma patient. Int. J. Nucl. Med. Biol. 5:118.

282. Gregoriadis, G., E. J. Wills, C. P. Swain, and A. S. Tavill. 1974. Drug carrier potential of liposomes in cancer chemotherapy. Lancet 1:1313.

283. Fraley, R., S. Subramani, P. Berg, and D. Papahadjopoulos. 1980. Introduction of liposome encapsulated DNA into cells. J. Biol. Chem. 255:10431.

284. Fraley, R. and D. Papahadjopoulos. 1981. New generation liposomes. The engineering of an efficient vehicle for intracellular delivery of nucleic acids. Trends Biochem. Sci. 63:77.

285. Fraley, R., S. Delaporta, and D. Papahadjopoulos. 1982. Liposome-mediated delivery of TMV RNA into tobacco protoplasts: a sensitive assay for monitoring liposome-protoplast interactions. Proc. Natl. Acad. Sci. USA 79:1859.

286. Mercer, W. E., D. J. Terefinko, and R. A. Schlegel. 1979. Red cell-mediated microinjection of macromolecules into monolayer culture of mammalian cells. Cells Biol. Int. Rep., p. 265.

287. Loyter, A., N. Zakai, and R. G. Kulka. 1975. Ultramicroinjection of macromolecules or small particles into animal cells. J. Cell Biol. 66:292.

288. Yamaizumi, M., M. Furusawa, T. Uchida, T. Nishimusa, and Y. Okada. 1978. Characterization of the ghost fusion method: a method for introducing exogenous substances into cultured cells. Cell Struct. Funct. 3:293.

Author Index

Numbers in parentheses are reference numbers and indicate that an author's work is referred to although his name may not be cited in the text. Italic numbers give the page on which the complete reference is listed.

A

Abra, R. M., 295(42), 303(42), *325*
Abraham, J., 99(142), *124*, 245 (371), *281*
Abrahams, P. J., 151(54), *199*
Abramoff, P., 148(22), *197*
Abramovitz, A. S., 213(285), 234 (263), 236(275,285), *273, 274*
Abramson, M. B., 33(13), *49*
Acuto, O., 215(135), 223(135), 238 (135), *264*
Adams, D. M., 298(66), 309(141), *327, 331*
Adamson, R. H., 148(24), *197*
Adler, F. L., 150(40,42), *198*
Adrian, G., 88(8), *115*, 163(133), 165(133), *204*
Adunyah, E. S., 58(35), *77*
Ager, M. E., 132(88), *141*
Aho, K., 231(234), *271*
Ainsworth, E. J., 70(127), *82*, 312 (158), *332*
Akiyama, Y., 213(284), 236(283, 284), *274*
Alderson, J. C. E., 99(76), 113(76), *120*, 245(368), *280*, 306(116), *330*
Alexander, A. M., 71(136), *83*

Alexander, H. E., 147(8,10), *196*
Allebach, E. S., 42(31), *50*, 167(151), 175(151), *205, 208*
Allen, T. M., 60(52), 65(94), *78, 80*, 100(105), *122*
Allerbach, E. I., 190(194), 191(193), *208*
Allfrey, V. G., 156(99), *202*
Allison, A. C., 19(95), *26*, 214(104, 106,401), 223(104,106), 249 (401,403,410), 254(401), 255 (105), *262, 283*
Altstiel, L. D., 74(183), *86*
Alving, C. R., 62(68,71,72), 72(144, 147), *79, 83, 84*, 210(13), 211 (13,17,18), 212(13,17,34,35, 37), 213(13,17,18,34,35,37,59, 65,70,72,75,77,81,82,84,85, 179,184,236,447), 214(81,82, 84,85,102), 215(17,35,37,59, 70,84,85,152), 216(37), 217 (17,65), 218(65,70,72), 219 (13,65,72,75,77), 220(81,82, 84-86,90-92), 221(13,82,85), 222(81,84,102), 223(110,111), 224(70,152), 226(77,84,85,179, 184), 227(13,17,37,65,75), 228 13,34,35), 229(17,18,75,82, 213b,214,215), 230(214,215,

N

Philips, F. W., 69(118), *82*
Phillips, N. C., 15(58), *24,* 298(63),
 317(255,256), *326, 339*
Phutrakul, S., 133(110), *142*
Pick, U., 35(34), 44(34), *51,* 128
 (33), *137*
Pierce, C. W., 214(430), 251(424),
 252(431), 253(424), *284*
Pillemer, L., 232(239), *271*
Pillion, D. J., 134(113), *142*
Pilwat, G., 159(122,123), *203*
Pink, D. A., 58(30), *76*
Piper, P. W., 158(117), *203*
Pirrotta, V., 155(87), *201*
Pitha, M. P., 154(80), *201*
Pitts, B. J. R., 132(93), *141*
Plakunova, V. G., 130(57), *139*
Podack, E. R., 234(258,259), 235
 (258,259,268,269,271), *272*
Porschen, W., 308(136), *331*
Porter, C. W., 93(36), *117,* 214
 (365), 245(365), *280*
Porter, M. T., 152(60), *199*
Portetelle, D., 154(82), *201*
Poste, G., 42(29), *50,* 73(167,168b,
 169), *85,* 87(3), 92(32), 93
 (36), 94(43,45,48), 95(45,48,
 57,58), 99(32,45), 114(45,
 140), *115, 117, 118, 124,*
 166(146), 173(146,148), 177
 (146), *205,* 214(365), 243
 (324-327,330,331,333,334),
 245(365), 249(333,413), *277,*
 278, 280, 283, 289(4), 292
 (17), 294(36), 302(93), 312
 (198,200-202), 314(230),
 315(198,200-202,236b), *323,*
 324, 325, 328, 335, 337, 338
Potempa, L. A., 231(231), *271*
Powell, M. E., 213(442), *285*
Preiser, H., 134(118-122), *143*
Prekumar, E., 154(80), *201*
Pressman, B. C., 5(13-15), *22*
Primus, F. J., 305(106), *329*
Prison, P., 72(148), *84*

Pritzl, P., 152(63), *200*
Pugliese, O., 215(135), 223(135), 238
 (135), *264*
Purchio, A. F., 146(1), *196*
Putman, D., 168(166), *206*

Q

Quinn, P. J., 54(10), 55(10), 56(10),
 57(10), 58(10,36), *75, 77,* 99
 (81), 112(81), *120,* 246(376),
 281

R

Race, R. R., 216(62), 219(62), *259*
Racela, A. S., 152(67), *200*
Racker, E., 31(5), *49,* 53(4), 58(33),
 75, 76, 126(4,5), 127(5,8,9),
 128(8,15,18,34), 129(40,44),
 130(9,51,54,63,66), 131(68,79),
 132(66,84,85), *135, 136, 137,*
 138, 139, 140, 141
Raftery, M. A., 58(38), *77*
Ragan, C. I., 129(43,45), *138*
Rahamimoff, R., 95(60), 114(60),
 119
Rahman, A., 68(113), *81,* 310(150),
 312(150,164), 313(150), *332,*
 333
Rahman, Y. E., 19(93), *26,* 70(127),
 82, 293(24), 306(24,121), 310
 (149), 312(157,158,210,214),
 313(149), 316(241), 317(260,
 261), 318(262), *324, 330, 332,*
 336, 338, 339
Ralston, E., 90(16), 94(16), 95(16),
 116
Ramm, L. E., 234(261), 237(261),
 273
Rampal, A. L., 133(107), *142*
Rando, R. R., 61(63a,63b), *78,* 306
 (117,128), *330*
Rapoport, S. I., 298(65), *327*
Rapport, M. M., 211(15), 212(15,

D

H

I